高等学校土木建筑
规划教材

前　言

　　《土木工程专业英语》是土木工程专业本科生的专业基础课程之一。最初开设此课程的目的，是使学生在完成大学基础英语学习后，通过对专业英语的后续学习，掌握一定量的专业词汇，培养能够顺利阅读并正确理解英文专业文献的能力；近年来，随着我国土木工程行业对外交流的增多，对本科毕业生专业英语的实际要求已不再停留在阅读层面，而是提高到了掌握听、说、读、写、译等综合技能上。然而，当前我国许多高校该课程的教学效果却并不十分理想，表现在不少具有较好基础英语功底的本科生，学习了专业英语之后，在毕业设计时仍无法顺利完成英文文献翻译和论文英文摘要撰写等基本环节。这表明，《土木工程专业英语》课程在教学思想和教学方法上亟待改进。

　　基于以上实际背景和教学现状，我们编写了此本《土木工程专业英语》本科教材。全书分为两部分：第一部分"专业英语学习篇"以土木工程为主线，系统介绍了土木工程及其下各专业方向的基本内容，包括土木工程、建筑工程、道路工程、桥梁工程、轨道交通工程、隧道工程、岩土与地下工程、机场工程、土木工程材料、土木工程设备、土木工程施工、土木工程管理、土木工程美学与环境工程等；该部分继承了传统教材的优点，强化学生专业词汇的积累、专业英语阅读能力的培养，体现了宽口径的"大土木"培养思想；并在每单元之后设有"技能训练"环节，引导学生变接受性被动学习为探索性主动学习。第二部分"专业英语实践篇"着眼于对学生听、说、读、写、译等专业英语综合技能的培养，系统介绍了用英语进行土木工程专业交流、翻译和写作时应掌握的技巧，并辅以实例介绍了英语词典、网络搜索引擎、计算机英语词典软件、在线词典、在线翻译、语料库等现代化工具在专业英语学习与实践中的应用。此外，本书还附有针对每单元课前和课后练习的"参考答案"，以及"专业英语词汇手册"。"专业英语词汇手册"部分相当于简明专业英语词典的作用，该部分亦可供土木工程专业技术人员遇到相关问题时查用。

　　本书由南京大学、哈尔滨工业大学、东南大学、中国药科大学、南京林业大学、河南大学、南通大学、南京工程学院、湖州师范学院、广东石油化工学院、河南城建学院、三江学院等高校的专业教师共同编写。在本书的编写过程中，参考了大量的英文专业书籍、论文和研究报告，在此特向这些资料的原作者们表示感谢。

　　本书的选材、译文、练习、实例等都是经过反复推敲、实践，但不当之处在所难免，敬请广大读者、同行专家不吝指正，以便改进。

郑家顺

2020 年 10 月

目　录

CONTENS

第一部分　专业英语学习篇

Part I *Special English Learning*

Unit 1　Civil Engineering and Civil Engineer

Lead-in

1. What are the qualities of civil engineers?
2. Do you think it's necessary to learn special English for civil engineering? Why?
3. Can you name these well-known engineering facilities?

_____ _____ _____ _____

_____ _____ _____ _____

Vocabulary warm-up: Matching

_____ 1. civil engineering	a. 结构工程
_____ 2. military engineering	b. 市政工程
_____ 3. environmental engineering	c. 土木工程
_____ 4. geotechnical engineering	d. 水利工程
_____ 5. structural engineering	e. 岩土工程
_____ 6. transportation engineering	f. 军事工程
_____ 7. municipal or urban engineering	g. 建筑工程
_____ 8. water resources engineering	h. 环境工程
_____ 9. materials engineering	i. 运输工程
_____ 10. coastal engineering	j. 测量工程
_____ 11. construction engineering	k. 海岸工程
_____ 12. surveying engineering	l. 材料工程

Text A　Introduction to Civil Engineering

Civil engineering is a professional engineering *discipline* that deals with the design, *construction* and maintenance of the physically and naturally built environment, including works such as bridges, roads, *canals*, *dams* and buildings. Civil engineering is the oldest engineering discipline after military engineering, and it is defined to distinguish non-military engineering from military engineering. It is traditionally broken into several sub-disciplines including environmental engineering, geotechnical engineering, structural engineering, transportation engineering, municipal or urban engineering, water resources engineering, materials engineering, coastal engineering, surveying and construction engineering. Civil engineering takes place at all levels: in the public sector from municipal to federal levels, and in the private sector from individual homeowners to international companies.

Engineering has been one part of life since the existence of human beings. The earliest practices of civil engineering may appear between 4000 BC and 2000 BC in Ancient Egypt and Mesopotamia when human beings started to abandon a *nomadic* existence, thus causing a need for the construction of the shelter. During that time, transportation became increasingly important and it led to the development of the wheels and sailing. The construction of Pyramids in Egypt (2700 ~ 2500 BC) might be considered as the first example of large-scale constructions. Other ancient historic civil engineering constructions include the *Parthenon* by Iktinos in Ancient Greece (447~438 BC), the *Appian Way* by Roman engineers (312 BC), the Great Wall of China by General Meng Tian under orders from Emperor Qin (220 BC), the *stupas* constructed in ancient *Sri Lanka* like the Jetavana ramaya, and the extensive *irrigation* works in Anuradhapura. The Romans developed civil structures throughout their empire, especially including *aqueducts*, *insulae*, harbors, bridges, dams and roads.

Until modern times there was no clear distinction between civil engineering and architecture, and the term engineer and architect were mainly geographical variations referring to the same person, often used interchangeably. In the 18th century, the term civil engineering began to be used to distinguish it from military engineering.

The first self-proclaimed civil engineer was John Smeaton who constructed the Eddystone Lighthouse. In 1771, Smeaton and some of his colleagues formed the Smeatonian Society of Civil Engineers, a group of leaders of the profession who met informally over dinner.

In 1818 the Institution of Civil Engineers was founded in London, and in 1820 the *eminent* engineer Thomas Telford became its first president. The institution received a Royal *Charter* in 1828, formally recognizing civil engineering as a profession. Its charter defined civil engineering as the art of directing the great sources of power in nature for the use and convenience of the man, as the means of production and of traffic in states, both for external and internal trade, as applied in the construction of roads, bridges, aqueducts, canals, river *navigation* and *docks* for internal intercourse and exchange, and in the construction of ports, harbors, moles, *breakwaters* and lighthouses, and in the art of navigation by artificial power for the purposes of commerce, and in the construction and application of machinery, and in the drainage of cities and towns.

In the mid-17th century, civil engineering was very original in this period, construction materials mainly used stones, grass reinforcement, *adobe* and other natural materials. Construction tools are primitive axes, hammers, knives, shovels and other hand tools, construction of facilities.

The history of modern civil engineering is generally believed from the mid-17th century to post-Second World War, a total of more than 300 years. During that period, civil engineering was becoming an independent discipline. In 1683, Italian scholar Galileo published *Dialogues Concerning Two New Sciences*, first expressing the formula of the design of the *beam* theory. In 1687, Newton summed up the three laws of *mechanics*. Mechanics of the Civil Engineering laid the basis of the analysis. Then in material mechanics, theory of elasticity and strength of materials based on the Navier in France in 1825 established a civil engineering structural design of the allowable stress method. Since then, civil engineering has become more systematic and theoretical guidance.

For materials, the invention of *Portland cement* in 1824 and *reinforced concrete* in 1867 became historic events in civil engineering. As the mass production of concrete and steel can be achieved, civil engineers can make use of these materials, the construction of large and complex engineering facilities.

After World War Ⅱ, the rapid development of science and technology made civil engineering relied on modern science and technology achieve further development. Steel and concrete, which are the most important building materials, had further development. Their strength improved several times, and reliability, *durability* and other properties were much better.

Civil engineering is the application of physical and scientific principles, and its history is intricately linked to advances in understanding of physics and mathematics throughout history. Because civil engineering is a wide ranging profession, including several separate specialized sub-disciplines, its history is linked to knowledge of structures, materials science, geography, geology, soils, hydrology, environment, mechanics and other fields.

Throughout ancient and medieval history, most architectural design and construction were carried out by artisans, such as stone *masons* and *carpenters*, who then became the role of master builder. Knowledge was retained in guilds and seldom supplanted by advances. Structures, roads and the *infrastructure* that existed were repetitive, and increases in scale were incremental.

One of the earliest examples of a scientific approach to physical and mathematical problems applicable to civil engineering is the work of Archimedes in the 3rd century BC, including Archimedes Principle, which underpins our understanding of buoyancy, and practical solutions such as Archimedes' screw. Brahmagupta, an Indian mathematician, used arithmetic in the 7th century AD, based on Hindu-Arabic numerals, for excavation (volume) computations.

Numerical analysis of the computer, which was difficult to compute a rough calculation in the past, can be simplified into a more precise calculation now. For example, by means of *finite element analysis* software, people can easily solve the problems to complete the complex human *statically indeterminate structure* calculation of internal forces and displacements. Finite element theory and the *structural dynamics* of the continuous development of the people can make a convenient and precise structure of the force and *deformation* calculation, making the design work greatly simplified. With the help of the computer-aided design (CAD application), the designers are free from manual graphics. More construction machineries allow a substantial increase of construction, increase of the degree of automation. With the use of more construction machineries, degree of automation makes the construction of a substantial increase.

Usually the construction of a project to build facilties has to go through three stages (investigation, design and construction) requiring the use of engineering *geological* investigation, hydro geological investigation, engineering survey, soil mechanics, engineering mechanics, engineering design, building materials, construction equipment, construction machinery, building the economy other disciplines and construction technology, construction organizations in the field of knowledge, as well as computer and mechanical testing techniques. Thus the scope of civil engineering is a broad comprehensive discipline. With the advance of science and technology development and engineering practice, the civil engineering disciplines have become broad categories, which are complex integrated systems.

Civil engineering is accompanied by the development of the human society. It reflects the construction of engineering facilities in various historical periods of society, economy, culture, science and technology development prospects. Therefore, civil society has become one of the witnesses of the historical developments.

In the ancient times, people began to build rudimentary houses, roads, bridges and communication channels to meet the need of simple life and production.

Many well-known engineering facilities showed human creativity in this historical period. For example, the Great Wall of China, Dujiangyan, *the Grand Canal*, Zhaozhou Bridge, Ying Xian Tower, the Pyramids, the Greek Parthenon, the Roman water supply, the Roman colosseum (Roman Jungle field), and many other famous churches and palaces.

After the Industrial Revolution, especially in the 20th century, the civil society put forward new demands and the social progress in all areas of civil engineering created good conditions. Thus civil engineering in this period got rapid development. In the worldwide, modern large-scale industrial plants, *skyscrapers*, nuclear power plants, highways and railways, bridges, long *tunnels* of large diameter pipelines, the Grand Canal, large dams, large airports, major seaports and marine engineering appeared. Modern civil society continued to create a new human physical environment, as the human society was an important part of modern civilization.

Civil engineering is a highly practical subject. In the early times, civil engineering practice was through summing up successful experiences, especially learning from failures. From the 17th century, Galileo and Newton, who were the guiders of modern civil engineering, practiced with the mechanical mechanics, structural mechanics, fluid mechanics, rock mechanics, and the theory as the basis for civil engineering disciplines. Civil engineering gradually developed into a science relying on experience.

In the development of civil engineering, engineering practice often precedes the theory; engineering accidents often show new unforeseen factors and trigger new theoretical research and development. Till now, a number of works dealing with the problem still rely on practical experience.

Why do the development of civil engineering technology rely on engineering practice, instead of scientific experiments and theoretical studies? There are two reasons. First, some of the objective situations are too complicated and difficult to faithfully carry out laboratory or field tests and theoretical analysis. Second, a new engineering practice can reveal new problems. For example, the construction of tall buildings, tall towers and long *span* bridges, masts, wind and *seismic* engineering issues highlightedly in order to develop this new theory and technology.

Features of modern civil engineering, with the rapid development of various construction requirements, are that people need to build large-scale, long span, tall, light, and sophisticated buildings with modern equipment. This raises new issues to the civil engineering and promotes the civil engineering discipline forward.

Glossary

Complete the glossary using words in the text.

1. _____	学科、科目	2. _____	建造,建设
3. _____	运河、管道	4. _____	水坝,堰;水库
5. _____	游牧的;流浪的	6. _____	帕台农神庙
7. _____	亚壁古道	8. _____	舍利塔
9. _____	斯里兰卡	10. _____	灌溉,水利
11. _____	围屋	12. _____	知名的
13. _____	许可证,执照	14. _____	航行
15. _____	码头	16. _____	防波堤
17. _____	灰质黏土	18. _____	梁,横梁
19. _____	力学,机械学		
20. _____	波特兰水泥,普通水泥,硅酸盐水泥		
21. _____	钢筋混凝土	22. _____	耐久性
23. _____	石匠,砖瓦匠	24. _____	木匠,工匠
25. _____	基础设施,公共建设		

26. _____	有限元分析	27. _____	超静定结构
28. _____	结构动力学	29. _____	损形,变形
30. _____	地质的,地质学的		
31. _____	大运河	32. _____	摩天大楼
33. _____	隧道;坑道;洞穴通道		
34. _____	跨度,跨径,墩距		
35. _____	地震的,由地震引起的;震撼世界的		

Exercises:

1. Fill in the Blanks.

1) Civil engineering is a professional engineering discipline that deals with the _____, _____ and _____ of the physically and naturally built environment, including works such as bridges, roads, _____, _____ and buildings.

2) Civil engineering takes place at all levels: in the public sector from _____ to _____ levels, and in the private sector from _____ homeowners to _____ companies.

3) Civil engineering is the application of _____ principles, and its history is intricately linked to advances in understanding of _____ _____ throughout history. Because civil engineering is a wide ranging profession, including several separate specialized sub-disciplines, its history is linked to knowledge of structures, _____ science, geography, geology, soils, hydrology, environment, _____ and other fields.

2. Write down the words with the following prefix or suffix.

geo-   _____

sub-   _____

non-   _____

-port _____

-ology _____

3. Translate the following paragraphs into Chinese.

1) Until modern times there was no clear distinction between civil engineering and architecture, and the term engineer and architect were mainly geographical variations referring to the same person, often used interchangeably. In the 18th century, the term civil engineering began to be used to distinguish it from military engineering.

2) Civil engineering is a highly practical subject. In the early times, civil engineering practice was through summing up successful experiences, especially learning from failures. From the 17th century, Galileo and Newton, who were the guiders of modern civil engineering, practiced with the combined mechanical, gradually mechanics, structural mechanics, fluid mechanics, rock mechanics, and the theory as the basis for civil engineering disciplines. Civil engineering gradually developed into a

science relying on experience.

Text B   The Civil Engineer

Civil engineering might be generalized as the basis of all engineering. In the time of the Romans, engineers were building roads, arch bridges, aqueducts, and cemented structures. Much later as the inventive genius of man developed the use of electricity and steam machinery, other forms of engineering came about. At the present time, the civil engineer is concerned with the building of things that do not move. For example, he may specialize in the construction of one or more of the followings: bridges, railroad beds, highways, tunnels, office buildings, monuments, airports, railroad stations, dams, canals, water-supply *conduits* (水道, 沟渠), water-treatment plants, plants for the disposal of refuse and *sewerage* (污物处理), hotels, industrial plants, rocket-launching pads, parking garages, transmission-line towers, and *radar* (雷达) towers.

1. Planning of Structures

The principal job of the civil engineer is the physical design of structures. For a given structure, such as a bridge, an exhaustive study must first be made of the forces acting on the structure. These are called loads, which may consist of wind load, weight of the parts of the structure itself, moving loads such as trains or trucks, snow or rain, and conditions of exposure temperature which may be encountered during the useful life of the structure. A system of structural units such as beams, *girders* (大梁), and columns must be devised to resist these forces.

The civil engineer must investigate the materials available to resist these forces. Materials widely used are steel, concrete, stone, bricks and combinations of materials, such as reinforced concrete. Furthermore, new materials, such as plastics, are constantly being developed. The final appearance of the structure is important and is determined by the use of an ornamental material such as marble. The finished structure must also be resistant to *corrosion* (腐蚀, 侵蚀), so various paint and surface treatment are used by the civil engineer. In this process, he is helped by the architect, who is concerned with the building appearance and functions such as enabling crowds of people to move through a railroad station or in an office building.

2. Traffic and Highway Construction

The civil engineer is concerned with providing optimum conditions for the efficient flow of traffic in a metropolitan area. Well in advance of a design and construction, a master plan of the city must be prepared, laying out the throughways, parkways, and parking facilities. Accesses must be provided for business concerns and stores. This represents the work of the traffic and highway engineer. He must establish traffic volume forecast by counting the probable number of automobiles going and coming to urban and city areas. The number that can be stored in a given space must be determined before a parking garage can be designed. In addition, bridges and tunnels must be designed to carry the traffic over and under both natural and man-made obstacles.

3. Sanitation

The civil engineer is concerned with furnishing us with a supply of drinking wastes and removing sanitary wastes. Sources for surface water must be sought in the highlands, where dams are designed and reservoirs are set up to store the rain water. The civil engineer must make sure that the dams are watertight and strong enough to hold back the pressure of the water. The rain is stored in special buckets and kept track of by weight and time. These records are analyzed to determine how much water will be available for a given period of time. Also the frequency of occurrence of intense storms is analyzed to determine the maximum floods that might occur and wipe out the dam.

One growing problem for the civil engineer is the disposal of waste products. In the early days, sewage and industrial waste products were readily dumped into the nearest river or buried. Now the large increase of population, both in the cities and along the rivers, has forced the engineer to determine effective means of alleviating the pollution of the waters.

The engineer must lay out a system of pipes, usually made of *vitrified*（玻璃状的）clay, to carry the waste products by liquid flow to a central plant. The system usually flows by gravity with occasional pumping when necessary. Provision must be made for gases that may collect in the top of the pipes. In the plant, the impurities are settled out and the liquid wastes are treated by various methods to render them harmless for disposal in the receiving waters. The solids removed must be disposed of by burying on land or at sea. In recent years, the increasing quantity of radioactive wastes is giving the civil engineer some difficult problems to solve.

In the case of many rivers and lakes, fresh water of various degrees of pollution must be drawn out and purified for use as drinking water. The engineer must provide processes whereby the solid particles are filtered out and the water is sterilized, usually by means of *chlorine*（氯）to reach top stories of buildings. Quantities of water must be stored and be available to put out fires and to provide a supply during emergencies.

4. Flood Control and Electric Power Generation

For this aspect of his work, which is not unrelated to the foregoing, the civil engineer must plan a series of dams to develop water power along a river as well as hydroelectric and steam-generating power plants. Also, certain rivers are made navigable by water stored in reservoirs and fed into canal locks. The famous Erie Canal system is a local illustration. When storms come up, the reservoir's elevation must be drawn down to provide room for the flood and reduce its damage. This is called flood control and is one phase of a four-way river development plan — water supply, navigation, flood control, and hydroelectric power. In planning the projects, the civil engineer must consider other factors, such as disturbance of fish or of wild fowl and hunting areas.

In all of these projects carried out by the civil engineer, the associated financial problems must first be considered. The community's need for the project must be demonstrated, and the most economical way of carrying it out must be determined. Then money for the project can be raised by government or financiers.

In addition to planning and directing the construction of projects, civil engineers are frequently consulted for public works and power plants. Many civil engineers are engaged with the federal, state and local governments in connection with public works projects.

Text C On Being Your Own Engineer

Here at this university and in this department that has trained so many outstanding civil engineers, you have achieved a standard of excellence that results in your recognition at this Honors Day ceremony. It gives me the greatest pleasure to congratulate you on these achievements. Here in your undergraduate career, you have become leaders in the pursuit of engineering knowledge, the first essential step in becoming a civil engineer. Excellence in undergraduate studies correlates highly with a successful engineering career in later years. I sincerely hope that the satisfaction of a successful career continues to be yours and that these honors and recognitions that you so rightfully receive today will be only the first of many satisfactions that will come to you in your practice of civil engineering.

Yet a successful undergraduate career is not always or inevitably followed by leadership in your profession. In a changing world, in a dynamic profession such as civil engineering, how can you be sure today that you will be among the leaders of your profession 20 or 30 years from now? How can you even be sure to pick the most aptitude for? Do you dare leave these matters to chance, do you dare let nature simply take its course? Nobody can predict the future and nobody can guarantee success in the future. But there are, nevertheless, many positive things you can do to shape you own career. I should like to think about some of these with you today.

I believe every engineer, perhaps even while an undergraduate but certainly upon graduation, needs to form and follow his own plan for the development of his professional career. Perhaps it is an unpleasant thought, but I believe it is only realistic that nobody else is quite as interested in your career as you yourself should be. If you don't plan it yourself, it is quite possible that nobody will. On the other hand, there are too many factors; there are too many changes in a dynamic profession to permit laying out a fixed plan. The plan that you follow must be flexible and it must continually be evaluated.

To be sure, every career depends to some extent on chances, on the breaks, good or bad. But if you have followed a sound plan, you will be ready for the good breaks when they come. Those who feel they have never had favorable opportunities usually have not been ready and have not even recognized opportunities when they come.

Civil engineering projects don't exist in the classroom or in the office or in the laboratory. They exist out in the field, in the society. They are the highways, the train systems, the landslides to be corrected, the waste disposal plants to be constructed, the bridges, the airports; they have to be built by men and machines. In my view, nobody can be a good designer, a good researcher, a leader in the civil engineering profession unless he understands the methods and the problems of the builders. This understanding ought to be firsthand, and if you are going to get it, you have to plan for it. Without this experience in the field, your designs may be impractical, your research may be irrelevant, or your teaching may not prepare your students properly for their profession. There are several ways in which you can get contractor. Or on the other hand, you might be an inspector for a resident engineer for the designer or owner. It doesn't matter in what capacity you work, and it doesn't take a very long time to get worthwhile experience in the field, but sometime early in your career, you should plan to get it. Since the real projects are out there in the field, you will have to go where they are to get the construction experience, and you may have to get the construction experience, and you may have to put up with a little inconvenience in order to get it.

Real problems of civil engineering design include both concepts and details. In fact, details often make or break a project. A beautifully designed cantilever bridge in Vancouver Harbor collapsed during construction because a few stiffeners were omitted on the webs of some temporary supporting beams. Spectacular failures such as this don't always follow from neglected details, but poor design, poor engineering often do. I believe every civil engineer needs a personal knowledge of the details in his branch of civil engineering. If he's going to be a *geotechnical* （岩土工程技术的）engineer, for example, he needs to know among other things exactly how borings are made and samples taken under a variety of circumstances. If he's going to be a structural engineer, he needs to know how steel structures are actually fabricated and erected. He needs to know, in other words, the state of the commercial art that plays such a large part in his profession. He needs to know how things are customarily done so that he can tell whether, for example, a commercially available sampling tool will do the job at a modest competitive price or whether some unusual tool must be developed for the particular requirements of the job. So it seems to me that you should plan to get this sort of experience also: to spend some time on a drilling rig if you plan to be a geotechnical engineer; to work for a steel fabricator or in a design office if you intend to be a structural engineer.

How can you get this varied experience, these various components of civil engineering that are so dissimilar? I think, for the most part, you have to do it by choosing your jobs carefully and changing your jobs if and when it seems necessary. You may be lucky in your very first job and go to work for an organization that designs, that supervises construction, that makes its own laboratory tests, that supervises borings, and so on. If this should be sure, you would be fortunate, but this is not usually the case. Even such an organization may tend to let you get stuck in one phase of their work, and you may have to persuade them from time to time to let you work in other parts of their activities. More likely you will have to change organizations, possibly even to move to another part of the country or of the world. Unfortunately you can't order the jobs that you want, when you want them, and where you want them. But you can look at every opportunity to see if it fits in your plan and to judge if the time is right to make a change. The breadth of experience is so important in a civil engineer's background that it can't be obtained in any other way than by a variety of activities within a given job. You owe it to yourself and to your career to see that you get this varied background; you ought to avoid being a job-hopper. Each of your employers will have an investment on you. At least for a while, when you start to work for him, he will not be getting his money's worth from you. You owe him a return on his investment, you owe him good work, and you owe staying with a reasonable minimum time while you're getting that experience.

On my first real job, I had the good fortune to be working under Karl Terzaghi. He had a good many requirements, but one of the most important was that I should keep a notebook in which I should record not just what I had done that day, but what I had seen, what I had observed. When I went down into a tunnel heading, I should come back and sketch how the heading was being executed and how it was being braced. I soon discovered that very often, when I came back, I couldn't remember exactly what had gone on in the heading. I couldn't remember exactly how the bracing fit together. In other words, my eyes had seen what was going on, but my brain didn't really register. My powers of observation were poor. But as I continued to keep this notebook, I discovered that more and more I could remember what I had seen, and more and more my powers of observation developed, I recommend this to you as one way to make your experiment more meaningful.

An investment of ten years or so after your degree, including perhaps graduate studies as well, in accordance with a carefully planned but flexible program, will go a long way toward assuring success in your engineering career. But there is another important aspect to be considered. And worthwhile career is demanding. It makes demands on your time and effort, and also on your family. And there are other demands on your life besides your career. Your wife or your husband will have her or his own goals and even may also have a career in mind. The demands of others in your life and the fulfillment of their goals and careers will require cooperation, adjustment, give and take. Moves from one place to another will require leaving friends, and require that your children change schools. Tensions and conflicts are inevitable while compromises and reasons are necessary. You and your partner will need the best possible understanding. Many a marriage has foundered on the career ambitions of one or both partners and, conversely, many a career has foundered on unreasonable or misunderstanding social or financial demands of the partner. There is seldom a perfect solution to this problem, but there are many good solutions. The important thing is to face up to the problems early and to keep working on them. The best engineers, I think, have achieved a reasonable balance among their goals in life. Often they can truly say that their partner in life has also been their partner in their career.

Your generation has a most exciting prospect. Don't believe for a minute the prophecies that technology has outlived its usefulness. You will have, fortunately, much more to consider than technology. You will need to be true conservationists, true ecologists in the positive sense. You will need to be involved in the social cost-benefit assessments of civil engineering work above and beyond the dollar cost-benefits. Progress in these directions will be the challenge and the great achievement of your generation, and it is an exciting prospect. But to succeed, you must be fully prepared, not poorer, but better grounded technically than your predecessors. In the next ten years, the choices you make and the experiences you get will be crucial. As honor students, you have taken the first necessary step with skill and distinction. All of us, your teachers, your husbands, wives, and friends wish you even greater success in the future. Indeed you must succeed, or this world will be a poorer place rather than a richer place in which to live.

Integrated Skills: Reading a Speech

Text C is the script of a speech. Now imagine you are the person who will deliver it. When you read the script, you should:

1. adjust the volume of the microphone, and ask the audience "Can you hear me in the back of the room?";

2. read the script loudly, slowly and clearly;

3. face all the audience, and make eye contact with them as much as possible;

4. use bigger size and double space to help locate the script after your eye contact with the audience;

5. avoid using too formal vocabulary or expressions which are difficult to understand;

6. never forget to smile.

Unit 2 Building Types, Components and Design

Lead-in

1. What are the components of a building?
2. What factors should be considered to build a skyscraper?
3. Can you name these famous tall buildings?

_____ _____ _____ _____

_____ _____ _____ _____

Vocabulary warm-up: Matching

_____ 1. mechanical system	a. 卫生设备系统
_____ 2. electrical system	b. 灌溉系统
_____ 3. heating and cooling system	c. 喷泉系统
_____ 4. lighting system	d. 机械系统
_____ 5. plumbing system	e. 供水系统
_____ 6. water supply system	f. 电气系统
_____ 7. condensate collection system	g. 凝水回收系统
_____ 8. irrigation system	h. 加热与冷却系统
_____ 9. fountain system	i. 系统调查
_____ 10. systematic investigation	j. 照明系统

Text A Components of a Building and Tall Buildings

Materials and structural forms are combined to make up the various parts of a building, including the load-carrying frames, skin, floors, and *partitions*. The building also has mechanical and electrical systems, such as elevators, heating and cooling systems, and lighting systems. The superstructure is the part of a building above ground, and the substructure and foundation is the part of a building below ground.

The skyscraper owes its existence to two developments of the 19th century: the steel skeleton construction and the passenger elevator. Steel as a construction material dates from the introduction of the Bessemer *converter* in 1855. Gustave Eiffel (1832—1923) introduced steel construction in France. His designs for the Galerie des Machines and the Tower for the Paris Exposition of 1889 expressed the lightness of the steel *framework*. The Eiffel Tower, 984 feet (300 meters) high, was the tallest structure built by man and was not *surpass*ed until 40 years later by a series of American skyscrapers.

The first elevator was installed by Elisha Otis in a department store in New York in 1857. In 1889, Eiffel installed the first elevators on a grand scale in the Eiffel Tower, whose hydraulic elevators could transport 2,350 passengers to the summit every hour.

Load-carrying frame. Until the late 19th century, the *exterior* walls of a building were used as *bearing walls* to support the floors. This construction is essentially a post and *lintel* type, and it is still used in frame construction for houses. Bearing-wall construction limited the height of buildings because of the enormous wall thickness required; for instance, the 16-story Monadnock Building built in the 1880's in Chicago had walls 5 feet (1.5 meters) thick at the lower floors. In 1883, William Le Baron Jenney (1832—1907) supported floors on cast-iron columns to form a cage-like construction. The skeleton construction, consisting of steel beams and columns, was first used in 1889. As a consequence of skeleton construction, the enclosing walls become a *"curtain wall"* rather than serving a supporting function. Masonry was the curtain wall material until the 1930's, when light metal and glass curtain walls were used.

After the introduction of the steel skeleton, the height of buildings continued to increase rapidly. All tall buildings were built with a skeleton of steel until World War Ⅱ. After the war, the shortage of steel and the improved quality of concrete led to tall buildings being built of reinforced concrete. Marina Towers (1962) in Chicago is the tallest concrete building in the United States; its height—588 feet (179 meters) — is exceeded by the 650-foot (198-meter) Post Office Tower in London and by other towers.

A change in attitude about skyscraper construction has brought a return to the use of the bearing wall. In New York city, the Columbia Broadcasting System Building, designed by Eero Saarinen in 1962, has a perimeter wall consisting of 5-foot (1.5-meter) wide concrete columns spaced 10 feet (3 meters) from column center to center. This perimeter wall, in effect, constitutes a bearing wall. One reason for this trend is that stiffness against the action of wind can be economically obtained by using the walls of the building as a tube; the World Trade Center buildings are another examples of this tube approach. In contrast, rigid frames or vertical trusses are usually provided to give lateral stability.

Skin. The skin of a building consists of both transparent elements (windows) and *opaque* elements (walls). Windows are traditionally glass, although plastics are being used, especially in schools where breakage creates a maintenance problem. The wall elements, which are used to cover the structure and are supported by it, are built of a variety of materials: bricks, precast concrete, stone, opaque glass, plastics, steel, and aluminum. Wood is used mainly in house construction; it is not generally used for commercial, industrial, or public buildings because of the fire hazard.

Floors. The construction of the floors in a building depends on the basic structural frame that is used. In steel skeleton construction, floors are either slabs of concrete resting on steel beams or a *deck* consisting of *corrugated* steel with a concrete topping. In concrete construction, the floors are either slabs of concrete on concrete beams or a series of closely spaced concrete beams (ribs) in two directions topped with a thin concrete slab, giving the appearance of a waffle on its underside. The kind of floor that is used depends on the span between supporting columns or walls and function of the space. In an apartment building, for instance, where walls and columns are spaced at 12 to 18 feet (3.7 to 5.5 meters), the most popular construction is a solid concrete slab with no beams. The underside of the slab serves as the ceiling for the space below it. Corrugated steel decks are often used in office buildings because the corrugations, when enclosed by another sheet of metal, form *duct*s for telephone and electrical lines.

Mechanical and Electrical Systems. A modern building not only contains the space for which it is intended (offices, classrooms, apartments) but also contains *ancillary* space for mechanical and electrical systems that help to provide a comfortable environment. These ancillary spaces in a skyscraper office building may constitute 25% of the total building area. The importance of heating, *ventilating*, electrical, and *plumbing systems* in an office building is shown by the fact that 40% of the construction budget is allocated to them. Because of the increased use of sealed buildings with windows that cannot be opened, elaborate mechanical systems are provided for ventilation and *air conditioning*. Ducts and pipes carry fresh air from central fan rooms and air conditioning machinery. The ceiling, which is suspended below the upper floor construction, conceals the ductwork and contains the lighting units. Electrical wiring for power and for telephone communication may also be located in this ceiling space or may be buried in the floor construction in pipes or *conduits*.

There have been attempts to incorporate the mechanical and electrical systems into the architecture of buildings by frankly expressing them; for example, the American Republic Insurance Company Building (1965) in Des Moines, Iowa, exposes both the ducts and the floor structure in an organized and elegant pattern and dispenses with the suspended ceiling. This type of approach makes it possible to reduce the cost of the building and permits innovations, such as in the span of the structure.

Soil and Foundations. All buildings are supported on the ground, and therefore the nature of the soil becomes an extremely important consideration in the design of any building. The design of a foundation depends on many soil factors, such as type of soil,

soil *stratification*, thickness of soil layers and their compaction, and groundwater conditions. Soils rarely have a single composition; they generally are mixtures in layers of varying thickness. For evaluation, soils are graded according to particle size, which increases from silt to clay to sand to gravel to rock. In general, the larger particle soils will support heavier loads than the smaller ones. The hardest rock can support loads up to 100 tons per square foot (976.5 metric tons/sq meter), while the softest silt can support a load of only 0.25 ton per square foot (2.44 metric tons/sq meter). All soils beneath the surface are in a state of compaction; that is, they are under a pressure that is equal to the weight of the soil column above it. Many soils (except for most sands and gavels) exhibit elastic properties — they deform when they are compressed under the load and rebound when the load is removed. The elasticity of soils is of ten time-dependent, that is, deformations of the soil occur over a length of time, which may vary from minutes to years after a load is imposed. Over a period of time, a building may settle if it imposes a load on the soil greater than the natural compaction weight of the soil. Conversely, a building may heave if it imposes loads on the soil smaller than the natural compaction weight. The soil may also flow under the weight of a building; that is, it tends to be squeezed out.

Due to both the compaction and flow effects, buildings tend to settle. Uneven settlements, exemplified by the leaning towers in Pisa and Bologna, can have damaging effects — the building may lean, walls and partitions may crack, windows and doors may become inoperative, and in the extreme, a building may collapse. Uniform settlements are not so serious, although extreme conditions, such as those in Mexico City, can have serious consequences. Over the past 100 years, a change in the groundwater level there has caused some buildings to settle more than 10 feet (3 meters). Because such movements can occur during and after construction, careful analysis of the soils under a building is vital.

The great variability of soils has led to a variety of solutions to the foundation problem. Where firm soil exists close to the surface, the simplest solution is to rest columns on a small slab of concrete (spread footing). Where the soil is softer, it is necessary to spread the column load over a greater area; in this case, a continuous slab of concrete (rafts or mats) under the whole building is used. In cases where the soil near the surface is unable to support the weight of the building, piles of wood, steel or concrete are driven down to firm soil.

The construction of a building proceeds naturally from the foundation up to the superstructure. The design process, however, proceeds from the roof down to the foundation (in the direction of gravity). In the past, the foundation was not subjected to systematic investigation. A scientific approach to the design of foundations has been developed in the 20th century. Karl Terzaghi of the United States pioneered studies that made it possible to make accurate predictions of the behavior of foundations, using the science of soil mechanics coupled with exploration and testing procedures. Foundation failures of the past, such as the classical example of the leaning tower in Pisa, have become almost nonexistent. The foundation still is a hidden but costly part of many buildings.

Glossary

Complete the glossary using words in the text.

1. _____　隔墙,隔板
2. _____　炼钢炉,吹风转炉
3. _____　构架,框架,结构
4. _____　超过,胜过
5. _____　外部的,外面的
6. _____　承重墙
7. _____　楣,门窗的过梁
8. _____　悬墙,幕墙
8. _____　透明的,不透光的
10. _____　桥面,层面
11. _____　起皱的,起波纹的
12. _____　管道,通道,预应力筋孔道
13. _____　辅助的,附属的
14. _____　使通风,使通气,给……装置通风设备
15. _____　卫生设备系统
16. _____　空气调节
17. _____　管道,导管,水道,水管
18. _____　分层,层理

Exercises

1. Fill in the blanks.

1) Materials and structural forms are combined to make up the various parts of a building, including the _____ frames, skin, floors, and partitions. The building also has mechanical and electrical systems, such as _____, heating and cooling systems, and lighting systems. The _____ is the part of a building above ground, and the _____ and foundation is the part of a building below ground.

2) Due to both the _____ and flow effects, buildings tend to settle. _____ settlements, ___ exemplified by the leaning towers in Pisa and Bologna, can have damaging effects — the building may _____, walls and partitions may _____, windows and doors may become _____, and in the extreme,

a building may _____ .

2. Write down the words with the following prefix or suffix.

super-　_____

multi-　_____

trans-　_____

sur-　_____

-bility　_____

3. Translate the following paragraphs into Chinese.

1) There have been attempts to incorporate the mechanical and electrical systems into the architecture of buildings by frankly expressing them; for example, the American Republic Insurance Company Building exposes both the ducts and the floor structure in an organized and elegant pattern and dispenses with the suspended ceiling.

2) The wall elements, which are used to cover the structure and are supported by it, are built of a variety of materials: bricks, precast concrete, stone, opaque glass, plastics, steel, and aluminum. Wood is used mainly in house construction; it is not generally used for commercial, industrial, or public buildings because of the fire hazard.

4. Translate the following into English.

1) 精心设计的建筑能为人们的工作和生活提供良好的空间环境。

2) 由于框架结构的使用，封闭墙体更多地成为围护墙而不是用作支撑构件。

Text B　Building Types and Design

A building is closely bound up with people, for it provides people with the necessary space to work and live in.

As classified by their uses, buildings are mainly of two types: *industrial buildings* (工业建筑) and *civil buildings* (民用建筑). Industrial buildings are used by various factories or industrial production while civil buildings are those that are used by people for dwelling, employment, education and other social activities.

Industrial buildings are factory buildings that are available for processing and manufacturing of various kinds, in such fields as the mining industry, the metallurgical industry, machine building, the chemical industry and the textile industry. Factory buildings can be classified into two types: single-story ones and multi-story ones. The construction of industrial buildings is the same as that of civil buildings. However, industrial and civil buildings differ in the materials used and in the way they are used.

Civil buildings are divided into two broad categories: residential buildings and public buildings. Residential buildings should suit family life. Each flat should consist of at least three necessary rooms: a living room, a kitchen and a toilet. Public buildings can be used in politics, cultural activities, administration work and other services, such as schools, office buildings, child-care centers, parks, hospitals, shops, stations, theatres, *gymnasiums* (体育馆，健身房), hotels, exhibition halls, bath pools, and so on. All of them have different functions, which in turn require different design types as well.

Housing is the living *quarters* (住处) for human beings. The basic function of housing is to provide shelter from the elements, but people today require much more than this of their housing. A family moving into a new neighborhood will want to know, if the available housing meets its standards of safety, health, and comfort. A family will also ask how near the housing is to grain shops, food markets, schools, stores, the library, a movie theater, and the community center.

In the mid – 1960's, a most important value in housing was sufficient space both inside and out. A majority of families preferred single-family homes on about half an *acre* (英亩) of land, which would provide space for spare-time activities. In highly industrialized countries, many families preferred to live as far out as possible from the center of a *metropolitan* (大城市的) area, even if the wage earners had to travel some distance to their work places. Quite a large number of families preferred country housing to suburban housing, because their chief aim was to get far away from noise, crowding, and confusion. The *accessibility* (可达性, 可接近性) of pubic transportation had ceased to be a decisive factor in housing because most workers drove their cars to work. People were chiefly interested in the arrangement and size of rooms and the number of bedrooms.

Before any of the buildings can begin, plans have to be drawn to show what the building will be like, the exact place in which it is to go and how everything is to be done.

An important point in building design is the *layout* (计划, 方案, 布局) of room, which should provide the greatest possible convenience in relation to the purposes for which they are intended. In a dwelling house, the layout may be considered under three categories: "day", "night", and "services". Attention must be paid to the provision of easy communication between these areas. The "day" rooms generally include a dining-room, a sitting-room and a kitchen, but other rooms, such as a *study* (书房), may be added, and there may be a hall. The living-room, which is generally the largest, often serves as a dining-room, too, or the kitchen may have a dining *alcove* (凹室, 壁龛). The "night" rooms consist of the bedrooms. The "services" comprise the kitchen, bathrooms, *larders (食品室, 储藏室)*, and water-closets. The kitchen and larder connect the services with the day rooms.

It is also essential to consider the question of outlook from the various rooms, and those most in use should preferably face south as much as possible. It is, however, often very difficult to meet the optimum requirements, both on account of the surroundings and the location of the roads. In resolving these complex problems, it is also necessary to follow the local town-planning regulations which are concerned with public *amenities* (舒适, 适宜, 愉快), density of population, height of buildings, proportion of green space to dwellings, *building lines* (建筑红线), the general appearance of new properties in relation to the neighborhood, and so on.

There is little standardization in industrial buildings although such buildings still need to comply with local town-planning regulations. The modern trend is towards light, airy factory buildings with the offices, reception rooms, telephone exchange, etc., the house in one low building overlooking the access road, the workshop, also light and airy, being less accessible to public view.

Generally of reinforced concrete or metal construction, a factory can be given a "shed (小棚，小屋)" type *ridge*（脊，岭）roof, incorporating windows facing north so as to give evenly distributed natural lighting（自然采光）without sun glare.

Text C   Introduction of Burj Khalifa

Burj Khalifa, known as Burj Dubai prior to its inauguration, is a skyscraper in Dubai, United Arab Emirates; and it is the tallest man-made structure ever built, at 828 m. The construction of Burj Khalifa began on 21 September 2004, with the exterior of the structure completed on 1 October 2009. The building officially opened on 4 January 2010, and is part of the new 2 km² flagship development called Downtown Dubai at the "First Interchange" along Sheikh Zayed Road, near Dubai's main business district.

1. Conception

Burj Khalifa has been designed to be the centrepiece of a large-scale, mixed-use development that will include 30,000 homes, nine hotels such as the Address Downtown Dubai, 3 hectares of parkland, at least 19 residential towers, the Dubai Mall, and the 12-hectare man-made Burj Khalifa Lake.

The building has returned the location of Earth's tallest free-standing structure to the Middle East — where the Great Pyramid of Giza claimed this achievement for almost four millennia before being surpassed in 1311 by Lincoln Cathedral in England.

The decision to build Burj Khalifa is reportedly based on the government's decision to diversify from an oil-based economy to one that is service-and-tourism-oriented. According to officials, it is necessary for projects like Burj Khalifa to be built in the city to garner more international recognition, and hence investment. "He (Sheikh Mohammed bin Rashid Al Maktoum) wanted to put Dubai on the map with something really sensational," said Jacqui Josephson, a tourism and VIP delegations executive at Nakheel Properties.

2. Architecture and Design

The tower is designed by Skidmore, Owings and Merrill, which also designed the Willis Tower (formerly the Sears Tower) in Chicago, Illinois and the new One World Trade Center in New York City, among numerous other famous high-rises. The building resembles the bundled tube, form of the Willis Tower, but is not a bundled tube structure. Its design is reminiscent of Frank Lloyd Wright's vision for The Illinois, a mile-high skyscraper designed for Chicago. According to Marshall Strabala, an SOM architect who worked in the building's design team, Burj Khalifa was designed based on the 73-floor Tower Palace Three, an all-residential building in Seoul, South Korea. In its early planning, Burj Khalifa was intended to be entirely residential.

Subsequent to the original design by Skidmore, Owings and Merrill, Emaar Properties chose Hyder Consulting to be the supervising engineer. Hyder was selected for its expertise in structural and MEP (mechanical, electrical and plumbing) engineering. Hyder Consulting's role was to supervise construction, certify SOM's design, and be the engineer and architect of record to the UAE authorities. Emaar Properties also engaged GHD, an international multidisciplinary consulting firm, to act as an independent verification and testing authority for concrete and steelwork.

The design of Burj Khalifa is derived from patterning systems embodied in Islamic architecture. According to the structural engineer, Bill Baker of SOM, the building's design incorporates cultural and historical elements particular to the region. The Y-shaped plan is ideal for residential and hotel usage, with the wings allowing maximum outward views and inward natural light. The design architect, Adrian Smith, said the triple-lobed footprint of the building was inspired by the flower Hymenocallis. The tower is composed of three elements arranged around a central core. As the tower rises from the flat desert base, setbacks occur at each element in an upward spiraling pattern, decreasing the cross section of the tower as it reaches the sky. There are 27 terraces in Burj Khalifa. At the top, the central core emerges and is sculpted to form a finishing spire. A Y-shaped floor plan maximizes views of the Persian Gulf. Viewed from above or from the base, the form also evokes the onion domes of Islamic architecture. During the design process, engineers rotated the building 120 degrees from its original layout to reduce stress from prevailing winds. At its tallest point, the tower sways a total of 1.5 m.

To support the unprecedented height of the building, the engineers developed a new structural system called the buttressed core, which consists of a hexagonal core reinforced by three buttresses that form the "Y" shape. This structural system enables the building to support itself laterally and keeps it from twisting.

The spire of Burj Khalifa is composed of more than 4,000 tonnes of structural steel. The central pinnacle pipe weighing 350 tonnes was constructed from inside the building and jacked to its full height of over 200 m using a strand jack system. The spire also houses communications equipment.

More than 1,000 pieces of art will adorn the interiors of Burj Khalifa, while the residential lobby of Burj Khalifa will display the work of Jaume Plensa, featuring 196 bronze and brass alloy cymbals representing the 196 countries of the world. The visitors in this lobby will be able to hear a distinct timbre as the cymbals, plated with 18-carat gold, are struck by dripping water, intended to mimic the sound of water falling on leaves.

The exterior cladding of Burj Khalifa consists of 142,000 m² of reflective glazing, and aluminium and textured stainless steel spandrel panels with vertical tubular fins. The cladding system was designed to withstand Dubai's extreme summer temperatures. Additionally, the exterior temperature at the top of the building is thought to be 6℃ cooler than at its base. Over 26,000 glass panels were used in the exterior cladding of Burj Khalifa. Over 300 cladding specialists from China were brought in for the cladding work on the tower.

Burj Khalifa is expected to hold up to 35,000 people at any one time. A total of 57 elevators and 8 escalators are installed. Engineers had considered installing the world's first triple-deck elevators, but the final design calls for double-deck elevators. The double-deck elevators are equipped with entertainment features such as LCD displays to serve visitors during their travel to the observation deck. The building has 2,909 stairs from the ground floor to the 160th floor.

3. Water Supply System

The Burj Khalifa's water system supplies an average of 946,000 L of water per day.

At the peak cooling times, the tower requires cooling equivalent to that provided by 10,000 t of melting ice in one day. The building has a condensate collection system, which uses the hot and humid outside air, combined with the cooling requirements of the building and results in a significant amount of condensation of moisture from the air. The condensed water is collected and drained into a holding tank located in the basement car park; this water is then pumped into the site irrigation system for use on the Burj Khalifa Park.

4. Maintenance

To wash the 24,348 windows, a horizontal track has been installed on the exterior of Burj Khalifa at level 40, 73 and 109. Each track holds a 1,500 kg bucket machine which moves horizontally and then vertically using heavy cables. Above level 109, up to tier 27 traditional cradles from davits are used. The top of the spire, however, is reserved for specialist window cleaners, who brave the heights and high winds dangling by ropes to clean and inspect the top of the pinnacle. Under normal conditions, when all building maintenance units will be operational, it will take 36 workers three to four months to clean the entire exterior facade.

Unmanned machines will clean the top 27 additional tiers and the glass spire. The cleaning system was developed in Australia at a cost of $ 8 million.

5. The Dubai Fountain

Outside, and at a cost of Dh 800 million (US $ 217 million), a record-setting fountain system was designed by WET Design, the California-based company responsible for the fountains at the Bellagio Hotel Lake in Las Vegas. Illuminated by 6,600 lights and 50 colored projectors, it is 275 m long and shoots water 150 m into the air, accompanied by a range of classical to contemporary Arabic and world music. On 26

October 2008 Emaar announced that based on results of a naming contest the fountain would be called the Dubai Fountain.

6. Construction

The tower was constructed by South Korean company, Samsung Engineering & Construction, which also did work on the Petronas Twin Towers and Taipei 101. Samsung Engineering & Construction built the tower in a joint venture with Besix from Belgium and Arabtec from UAE. Turner is the Project Manager on the main construction contract.

Under UAE law, the Contractor and the Engineer of Record, Hyder Consulting, are jointly and severally liable for the performance of Burj Khalifa.

The primary structural system of Burj Khalifa is reinforced concrete. Over 45,000 m^3 of concrete, weighing more than 110,000 tonnes were used to construct the concrete and steel foundation, which features 192 piles, with each pile 1.5 metre diameter and 43 metre long buried more than 50 m deep. Burj Khalifa's construction used 330,000 m^3 of concrete and 55,000 tonnes of steel rebar, and construction took 22 million man-hours. A high density, low permeability concrete was used in the foundations of Burj Khalifa. A cathodic protection system under the mat was used to minimize any detrimental effects from corrosive chemicals in local ground water.

The previous record for pumping concrete on any project was set during the extension of the Riva del Garda Hydroelectric Power Plant in Italy in 1994, when concrete was pumped to a height of 532 m. Burj Khalifa exceeded this height on 19 August 2007, and as of May 2008 concrete was pumped to a delivery height of 606 m, the 156th floor. The remaining structure above is built of lighter steel.

Burj Khalifa is highly compartmentalized. Pressurized, air-conditioned refuge floors are located approximately every 35 floors where people can shelter on their long walk down to safety in case of an emergency or fire.

Special mixes of concrete are made to withstand the extreme pressures of the massive building weight; as is typical with reinforced concrete construction, each batch of concrete used was tested to ensure it could withstand certain pressure.

The consistency of the concrete used in the project was essential. It was difficult to create concrete that could withstand both the thousands of tonnes bearing down on it and the Persian Gulf temperatures that can reach 50℃. To combat this problem, the concrete was not poured during the day. Instead, during the summer months ice was added to the mixture and it was poured at night when the air is cooler and the humidity is higher. A cooler concrete mixture cures evenly throughout and is therefore less likely to set too quickly and crack. Any significant cracks could have put the entire project in jeopardy.

The unique design and engineering challenges of building Burj Khalifa have been featured in a number of television documentaries, including the Big, Bigger, Biggest series on the National Geographic and Five channels, and the Mega Builders series on the Discovery Channel.

Integrated Skills: Free Debate

With the development of economy, many companies, cities and even countries begin to consider tall buildings as symbols of economic powers, thus leads to a worldwide competition in building height. The competition began in America represented by Chicago, New York and so on in the late 19th century; and the competitive center has shifted currently to Asia. Some people believe that the development of tall buildings is positive for promoting the engineering technology and creating many miracles of human civilization. Others hold the view that the construction of tall buildings is negative because it will waste a lot of money and resources. What is your opinion about that? Please hold a debate in your class based on the above subject.

During the free debate, you should:

1. accumulate the related information before the debate;

2. beconfident, loud and clear during the debate;

3. insist on your opinion, though there might be no definite right or wrong in the debate;

4. predict the opponents' argument, and get ready for the rebuttal;

5. listen to the opponents carefully, and find their loopholes.

Unit 3 Building Structures and Seismic Resistance

Lead-in

1. What do you think is the most destructive natural disaster for the buildings?
2. What structure do you think can best resist a strong earthquake?

Vocabulary warm-up: Matching

_____ 1. rigid frame structures	a. 框架—剪刀墙结构
_____ 2. infilled frame structures	b. 框筒结构
_____ 3. wall-frame structures	c. 桁架筒结构
_____ 4. framed-tube structures	d. 刚架结构
_____ 5. tube-in-tube or hull-core structures	e. 束筒结构
_____ 6. braced-tube structures	f. 填充框架结构
_____ 7. bundled-tube structures	g. 筒中筒/核心筒结构

Text A   Earthquake Resistant Structural Systems

1. *Rigid Frame* Structures

Rigid frame structures typically comprise floor diaphragms supported on beams which link to continuous columns (Figure 3-1). The joints between beams and columns are usually considered to be "rigid". The frames are expected to carry the gravity loads through the flexural action of the beams and the *propping* action of the columns. Negative moments are induced in the beam adjacent to the columns causing the mid-span positive moment to be significantly less than those in a simply supported span. In structures in which gravity loads dictate the design, economies in member size that arise from this effect tend to be offset by the higher cost of the rigid joints.

Lateral loads, imposed within the plane of the frame, are resisted through the development of bending moments in the beams and columns. Framed buildings often employ moment resistant frames in two orthogonal directions, in which case the column elements are common to both

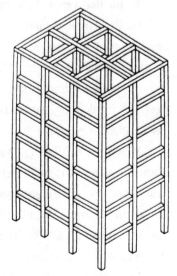

Figure 3-1   Rigid frame structure

frames.

Rigid frame structures are well suited to accommodate high levels of inelastic deformation. When a capacity design approach is employed, it is usual to assign the end zones of the flexural beams to accept the post-elastic deformation expected, and to design the column members such that their dependable strength is in excess of the over-strength capacity of the beam hinges, thereby ensuring they remain within their elastic response range regardless of the *intensity* of ground shaking. Rigid frame structures are, however, often quite flexible. When they are designed to be fully ductile, special provisions are often needed to prevent the premature onset of damage to non-structural components.

Rigid frame construction is ideally suited for *reinforced* concrete buildings because of the inherent *rigidity* of reinforced concrete frame joints. The rigid frame form is also used for steel frame buildings. But moment resistant connections in steel tend to be costly. The sizes of the columns and girders at any level of a rigid frame are directly influenced by the *magnitude* of the external shear at that level, and they therefore increase toward the base. Consequently, the design of the floor framing cannot be repetitive as it is in some *braced frames*. A further result is that sometimes it is not possible in the lowest storeys to accommodate the required depth of girder within the normal *ceiling* space.

While rigid frames of a typical scale that serve alone to resist lateral loading have an economic height limit of about 25 storeys, smaller-scale rigid frames in the form of a perimeter tube, or typically scaled rigid frames in combination with shear walls or braced bents, can be economic up to much greater heights.

2. Infilled Frame Structures

Infilled frames (Figure 3-2) are the most usual form of construction for tall buildings of up to 30 storeys in height. Column and girder framing of reinforced concrete, or sometimes steel, is infilled by panels of brickwork, or cast-in-place concrete.

When an infilled frame is subjected to lateral loading, the infill behaves effectively as a strut along its compression diagonal to brace the frame. Because the infills serve also as external walls or internal partitions,

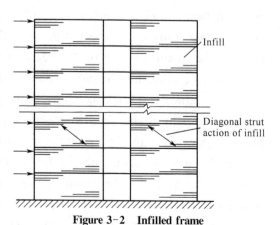

Figure 3-2 Infilled frame

the system is an economical way of stiffening and strengthening the structure.

The complex interactive behavior of the infill in the frame, and the rather random quality of masonry, make it difficult to predicate with accuracy the stiffness and strength of an infilled frame. For these reasons, the use of the infills for bracing buildings has mainly been supplementary to the rigid frame action of concrete frames.

3. Shear Walls

A shear wall is a vertical structural element that resists lateral forces in the plane of the wall through shear and bending. The high in-plane stiffness and strength of concrete and masonry walls make them ideally suitable for bracing buildings as shear walls.

A shear wall acts as a beam cantilevered out of the ground or foundation and, just as with a beam, part of its strength derives from its depth. Figure 3-3 shows two examples of a shear wall, one in a simple one-storey building and another in a multi-storey building. In Figure 3-3a, the shear walls are oriented in one direction, so only lateral forces in this direction can be resisted. The roof serves as the horizontal diaphragm and must also be designed to resist the lateral loads and transfer them to the shear walls.

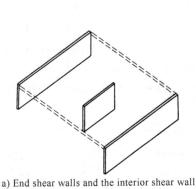
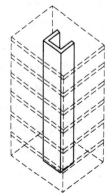

a) End shear walls and the interior shear wall　　　b) Interior shear walls for bracing in two directions

Figure 3-3　Shear wall

Figure 3-3a also shows an important aspect of shear walls in particular and vertical elements in general. This is the aspect of symmetry that has a bearing on whether torsional effects will be produced. The shear walls in Figure 3-3a show the shear walls symmetrical in the plane of loading.

Figure 3-3b illustrates a common use of shear walls at the interior of a multi-storey building. Because walls enclosing *stairways*, elevator shafts, and mechanical chases are mostly solid and run the entire height of the building, they are often used for shear walls. Although not as efficient from a strictly structural point of view, interior shear walls do leave the exterior of the building open for windows.

Notice that in Figure 3-3b there are shear walls in both directions, which is a more realistic situation because both wind and earthquake forces need to be resisted in both directions. In this diagram, the two shear walls are symmetrical in one direction, but the single shear wall produces a non-symmetric condition in the other since it is off center. Shear walls do not need to be symmetrical in a building, but symmetry is preferred to avoid torsional effects.

If, in a low-to-medium-rise building, shear walls are combined with frames, it is reasonable to assume that the shear walls attract all the lateral loading so that the frame may be designed for only gravity loading. It is essentially important in shear wall structures to try to plan the wall layout so that the lateral load tensile stresses are suppressed by the gravity load stresses. This allows them to be designed to have only the minimum reinforcement.

Since shear walls are generally both stiff and can be inherently robust, it is practical to design them to remain nominally elastic under design intensity loading, particularly in regions of low or moderate seismicity. Under increased loading intensities, post-elastic deformations will develop within the lower portion of the wall (generally considered to extend over a height of twice the wall length above the foundation support system). Good post-elastic response can be readily achieved within this region of reinforced concrete or masonry shear walls through the provision of adequate confinement of the principal reinforcing steel and the prohibition of lap splices of reinforcing bars.

Shear wall structures are generally quite stiff and, as such *inter-storey drift* problems are rare and generally easily contained. The shear wall tends to act as a rigid body rotating about a *plastic* hinge which forms at the base of the wall. Overall structural deformation is thus a function of the wall rotation. Inter-storey drift problems which do occur are limited to the lower few floors.

A major shortcoming with shear walls within buildings is that the size provides internal (or external) access barriers which may contravene the architectural requirements. This problem can be alleviated by coupling adjacent more slender shear walls so a *coupled shear wall* structure is formed. The *coupling beams* then become shear links between the two walls and with careful detailing can provide a very effective, ductile control mechanism (Figure 3-4).

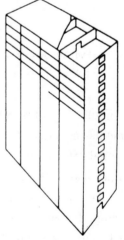

Figure 3-4　Coupled shear wall structure

4. Braced Frames

A braced frame is a truss system of the concentric or eccentric type in which the lateral forces are resisted through axial stresses in the members. Just as with a truss, the braced frame depends on diagonal members to provide a load path for lateral forces from each building element to the foundation. Figure 3-5 shows a simple one-storey braced frame. At one end of the building two bays are braced and at the other end only one bay is braced. This building is only braced in one direction and the diagonal member may be either in tension or compression, depending on which way the force is applied.

Figure 3-5b shows two methods of bracing a multistorey building. A single diagonal compression member in one bay can be used to brace against lateral loads coming from either direction. Alternately, tension diagonals can be used to accomplish the same result, but they must be run both ways to account for the load coming from either direction.

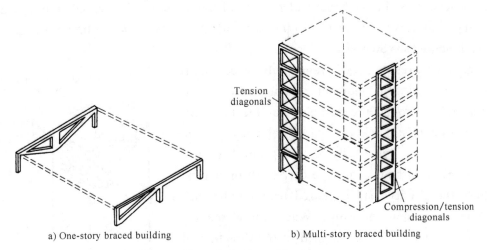

a) One-story braced building   b) Multi-story braced building

Figure 3-5   Braced frame

Braced framing can be placed on the exterior or interior of a building, and may be placed in one structural bay or several. Obviously, a braced frame can present design problems for windows and doorways, but it is a very efficient and rigid lateral force resisting system.

Two major shortcomings of braced systems are that the inclined diagonal orientation often conflicts with conventional occupancy use patterns; and secondly they often require careful detailing to avoid large local torsional eccentricities being introduced at the connections with the diagonal brace being offset from the frame node.

5. *Wall-Frame* Structures

When shear walls are combined with rigid frames (Figure 3-6), the walls, which tend to deflect in a flexural configuration, and the frames, which tend to deflect in a shear mode, are constrained to adopt a common shape by the horizontal rigidity of the girders and slabs. As a consequence, the walls and frames interact horizontally, especially at the top, to produce a stiffer and stronger structure. The interacting wall-frame combination is appropriate for buildings in the 40-to-60-storey range, well beyond of rigid frame or shear wall alone.

In addition, less well-known feature of the wall-frame structure is that, in a carefully "tuned" structure, the shear in the frame can be made approximately uniform over the height,

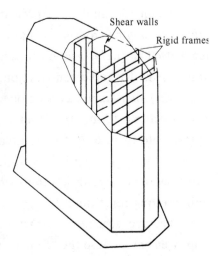

Figure 3-6   Wall-frame structure

allowing the floor framing to be repetitive.

Although the wall-frame structure is usually perceived as a concrete structural form, with shear walls and concrete frames, a steel counterpart using braced frames and steel rigid frames offers similar benefit of horizontal interaction. The braced frames behave with an overall flexural tendency to interact with the shear mode of the rigid frames.

6. Framed-Tube Structures

The lateral resistance of framed-tube structures is provided by very stiff moment resisting frames that form a "tube" around the perimeter of the building. The frames consist of closely spaced columns, 2~4 m between centers, joined by deep *spandrel* girders (Figure 3-7). Although the tube carries all the lateral loading, the gravity load is shared between the tube and interior columns or walls. When lateral loading acts, the perimeter frames aligned in the direction of loading act as the "web" of the massive tube cantilever, and those normal to the direction of the loading act as the "flanges".

The close spacing of the columns throughout the height of the structure is usually unacceptable at the entrance level. The columns are therefore merged, or

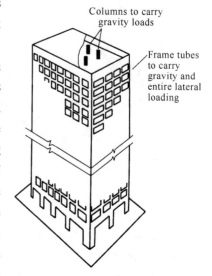

Columns to carry gravity loads

Frame tubes to carry gravity and entire lateral loading

Figure 3-7　Framed-tube structure

terminated on a *transfer beam*, a few storeys above the base so that only a few, larger, more widely spaced columns continue to the base. The tube form was developed originally for buildings of rectangular plane; however, for other plane shapes, and has occasionally been used in circular and triangular configurations.

The tube is suitable for both steel and reinforced construction and has been used for buildings ranging from 40 to more storeys. The highly repetitive pattern of the frames lends itself to prefabrication in steel, and to the use of rapidly gang forms in concrete, which make for rapid construction.

The *framed tube* has been one of the most significant modern developments in high-rise structural form. It offers a relatively efficiently, easily constructed structure, appropriate for use up to the greatest of heights. Aesthetically, the tube's externally evident form is regarded with mixed enthusiasm: some praise the logical clearly expressed structure while others criticize the girder-like facade as small-windowed and uninteresting repetitious.

The tube structure's structural efficiency, although high, still leaves scope for improvement because the "flange" frames tend to suffer from "*shear lag*"; this results in mid-face "flange" columns being less stressful than the corner columns and, therefore, not contributing as fully as they could to the flange action.

7. *Tube-in-Tube or Hull-Core* Structures

This variation of the framed tube consists of an outer framed tube, the "hull" together with an internal elevator and service core (Figure 3-8). The hull and the inner core act jointly in resisting both gravity and lateral loading. In a steel structure the core may consist of braced frames, whereas in a concrete structure it would consist of an assembly of shear walls.

To some extent, the outer framed tube and the inner core interact horizontally as the shear and flexural components of a wall-frame structure, with the benefit of increasing lateral stiffness. However, the structural tube usually adopts a highly dominant role because of its much greater structural depth.

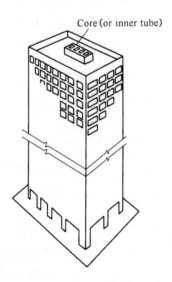
Core (or inner tube)
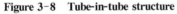

Figure 3-8   Tube-in-tube structure   **Figure 3-9   Braced-tube structure**

8. *Braced-Tube* Structures

Another way of improving the efficiency of the framed tube, thereby increasing its potential for greater height as well as allowing greater spacing between the columns, is to add diagonal bracing to the faces of the tube. This arrangement was first used in a steel structure in 1969, in Chicago's John Hancock Building (Figure 3-9). Because the diagonals of a braced tube are connected to the columns at each intersection, they virtually eliminate the effects of shear lag in both the flange and web frames. As a result, the structure behaves under lateral loading more like a braced frame, with greatly diminished bending in the members of the frames. Consequently, the spacing of the columns can be larger and the depth of the spandrels less, thereby allowing larger size windows than that in the conventional tube structure.

In the braced-tube structure the bracing contributes also to the improved performance of the tube in carrying gravity loading: differences between gravity load stresses in the columns are evened out by the braces transferring loading from the more highly to the less highly stressed columns.

9. *Bundled-Tube* Structures

This structural form has been used for the Sears Tower in Chicago. The Sears Tower consists of four parallel rigid steel frames in each orthogonal direction, interconnected to form nine bundled tubes. As in the single-tube structure, the frames in the direction of lateral loading serve as "webs" of the vertical cantilever, with the normal frame acting as "flanges".

The introduction of internal webs greatly reduces the shear lag in the flanges; consequently their columns are more evenly stressed than in the single-tube structure, and their contribution to the lateral stiffness is great. This allows columns of the frames to be spaced further apart and to be less obtrusive. In the Sears Tower, the advantage was taken of the bundled form to discontinue some of the tubes, and so reduce the plan of the building at stages up to the height.

Glossary

Complete the glossary using words in the text.

1. _____	框架	2. _____	支撑
3. _____	（地震）烈度	4. _____	加强，配筋于
5. _____	刚度	6. _____	震级
7. _____	支撑框架	8. _____	天花板
9. _____	楼梯	10. _____	塑性的
11. _____	层间位移	12. _____	联肢剪力墙
13. _____	连梁	14. _____	框架-剪力墙结构
15. _____	上下层窗间墙	16. _____	转换梁
17. _____	框筒	18. _____	剪力滞后
19. _____	筒中筒	20. _____	桁架筒
21. _____	束筒		

Exercises

1. Write down the words with the following prefix or suffix.

post- _____

mid- _____

seis- _____

-sphere _____

-tude _____

2. Translate the following paragraphs into Chinese.

1) If, in low-to-medium-rise building, shear walls are combined with frames, it is reasonable to assume that the shear walls attract all the lateral loading so that the frame

may be designed for only gravity loading. It is essentially important in shear wall structures to try to plan the wall layout so that the lateral load tensile stresses are suppressed by the gravity load stresses. This allows them to be designed to have only the minimum reinforcement.

2）Two major shortcomings of braced systems are that the inclined diagonal orientation often conflicts with conventional occupancy use patterns; and secondly they often require careful detailing to avoid large local torsional eccentricities being introduced at the connections with the diagonal brace being offset from the frame node.

3. Answer the questions.

1）Please name the types of earthquake resistant structural systems.

2）How does a rigid frame structure resist the gravity load and lateral load?

3）Why are shear walls in both directions preferred?

4）How are the loads shared between frame and tube in a framed-tube structure?

Text B Reinforced Concrete Structures

Concrete and reinforced concrete are used as building materials in every country. In many countries, including the United States and Canada, reinforced concrete is a dominant structural material in engineered construction. The universal nature of reinforced concrete construction stems from the wide availability of reinforcing bars and the *constituents*（组成，成分）of concrete, gravel, sand, and cement, the relatively simple skills required in concrete construction, and the economy of reinforced concrete compared to other forms of construction. Concrete and reinforced concrete are used in bridges, buildings of all sorts, underground structures, water tanks, television towers, offshore oil exploration and production structures, dams, and even in ships.

1. Mechanics of Reinforced Concrete

Concrete is strong in compression but weak in tension. As a result, *cracks*（裂缝）develop whenever loads, or restrained *shrinkage*（收缩）or temperature changes, give rise to tensile stresses in excess of the tensile strength of the concrete. In the plain concrete beam, the moments due to applied loads are resisted by an internal tension-compression *couple*（力偶）involving tension in the concrete. Such a beam fails very suddenly and completely when the first crack forms. In a reinforced concrete beam, steel bars are embedded in the concrete in such a way that the tension forces needed for moment equilibrium after the concrete cracks can be developed in the bars.

The construction of a reinforced concrete member involves building a form or mold in the shape of the member being built. The form must be strong enough to support the weight and *hydrostatic*（静水的）pressure of the wet concrete, and any forces applied to it by workers, concrete buggies, wind, and so on. The reinforcement is placed in this form and held in place during the concreting operation. After the concrete has hardened, the forms are removed.

2. Factors Affecting the Choice of Concrete for a Structure

The choice of whether a structure should be built of concrete, steel, masonry, or *timber*（木材，木料）depends on the availability of materials and on a number of value decisions.

1）Economy

Frequently, the foremost consideration is the overall cost of the structure. This is, of course, a function of the costs of the materials and the labor necessary to erect them. Frequently, however, the overall cost is affected as much or more by the overall construction time since the *contractor*（承包商）and owner must allocate money to carry out the construction and will not receive a return on this investment until the building is ready for occupancy. As a result, financial savings due to rapid construction may more than offset increased material costs. Any measures the designer can take to standardize the design and forming will generally pay off in reduced overall costs.

In many cases the long-term economy of the structure may be more important than the first cost. As a result, *maintenance*（维护，维修）and *durability*（耐久性）are important considerations.

2）Suitability of Material for Architectural and Structural Function

A reinforced concrete system frequently allows the designer to combine the architectural and structural functions. Concrete has the advantage that it is placed in a plastic condition and is given the desired shape and texture by means of the forms and the finishing techniques. This allows such elements as flat plates or other types of slabs to serve as load-bearing elements while providing the finished floor and ceiling surfaces. Similarly, reinforced concrete walls can provide architecturally attractive surfaces in addition to having the ability to resist gravity, wind, or seismic loads. Finally, the choice of the size or shape is governed by the designer and not by the availability of standard manufactured members.

3）Fire Resistance

The structure in a building must withstand the effects of a fire and remain standing while the building is evacuated and the fire is extinguished. A concrete building inherently has a 1-to-3-hour *fire rating*（耐火等级）without special fireproofing or other details. Structural steel or timber buildings must be fireproofed to attain similar fire ratings.

4）Rigidity

The occupants of a building may be disturbed if their building oscillates in the wind or the floors vibrate as people walk by. Due to the greater stiffness and mass of a concrete structure, vibrations are seldom a problem.

5）Low Maintenance

Concrete members inherently require less maintenance than structural steel or timber members do. This is particularly true if dense, *air-entrained concrete*（加气混凝

土) has been used for surfaces exposed to the atmosphere, and if care has been taken in the design to provide adequate drainage off and away from the structure.

6) Availability of Materials

Sand, gravel, cement, and concrete mixing facilities are very widely available, and reinforcing steel can be transported to most job sites more easily than structural steel can. As a result, reinforced concrete is frequently used in remote areas.

On the other hand, there are a number of factors that may cause one to select a material other than reinforced concrete. These include:

1) Low Tensile Strength

As stated earlier, the tensile strength of concrete is much lower than its compressive strength (about 1/10), hence concrete is subject to cracking. In structural uses this is overcome by using reinforcement to carry tensile forces and limit crack widths to acceptable values. Unless care is taken in design and construction, however, these cracks may be unsightly or may allow penetration of water.

2) Forms and *Shoring*（支撑）

The construction of a cast-in-place structure involves three steps not encountered in the construction of steel or timber structures. These are the construction of the forms, the removal of these forms, and propping or shoring the new concrete to support its weight until its strength is adequate. Each of these steps involves labor and/or materials which are not necessary with other forms of construction.

3) Relatively Low Strength per Unit of Weight or Volume

The compressive strength of concrete is roughly 5% to 10% that of steel, while its unit density is roughly 30% that of steel. As a result, a concrete structure requires a larger volume and a greater weight of material than a comparable steel structure does. As a result, long-span structures are often built from steel.

4) Time-dependent Volume Changes

Both concrete and steel undergo approximately the same amount of thermal expansion and contraction. Because there is less mass of steel to be heated or cooled, and because steel is a better conductor than concrete, a steel structure is generally affected by temperature changes to a greater extent than is a concrete structure. On the other hand, concrete undergoes drying shrinkage, which, if restrained, may cause deflections or cracking. Furthermore, deflections will tend to increase with time, possibly doubling, due to creep of the concrete under sustained loads.

3. Building *Codes*（规范）

The first set of building regulations for reinforced concrete were drafted under the leadership of Professor Morsch of the University of Stuttgart and were issued in Prussia in 1904. Design regulations were issued in Britain, France, Austria, and Switzerland between 1907 and 1909.

The American Railway Engineering Association appointed a Committee on Masonry in 1890. In 1903 this committee presented *specifications*（说明书，规程）for Portland cement concrete. Between 1908 and 1910 a series of committee reports led to the Standard Building Regulations for the Use of Reinforced Concrete published in 1910 by the National Association of Cement Users which subsequently became the American Concrete Institute.

A Joint Committee on Concrete and Reinforced Concrete was established in 1904 by the American Society of Civil Engineers, American Society for Testing and Materials, the American Railway Engineering Association, and the Association of American Portland Cement Manufactures. This group was later joined by the American Concrete Institute. Between 1904 and 1910 the Joint Committee carried out research. A preliminary report issued in 1913 listed the more important papers and books on reinforced concrete published between 1898 and 1911. The final report of this committee was published in 1916. The history of reinforced concrete building codes in the United States was reviewed in 1954 by Kerekes and Reid.

The design and construction of buildings is regulated by municipal bylaws called building codes. These exist to protect the public health and safety. Each city and town is free to write or adopt its own building code, and in that city or town, only that particular code has legal status. Because of the complexity of building code writing, cities in the United States generally base their building codes on one of three model codes: the Uniform Building Code, the Standard Building Code, or the Basic Building Code. These codes cover such things as use and occupancy requirements, fire requirements, heating and ventilating requirements, and structural design.

The definitive design specification for reinforced concrete buildings in North America is the Building Code Requirements for Reinforced Concrete（ACI-318-95）, which is explained in a Commentary.

This code, generally referred to as the ACI Code, has been incorporated in most building codes in the United States and serves as the basis for comparable codes in Canada, New Zealand, Australia, and parts of Latin America. The ACI Code has legal status only if adopted in a local building code.

Each nation or group of nations in Europe has its own building code for reinforced concrete. The CEB-FIP Model Code for Concrete Structures is intended to serve as the basis for future attempts to unify European codes. This code and the ACI Code are similar in many ways.

Text C   Earthquake Induced Vibration of Structures

1. Seismicity and Ground Motions

The most common cause of earthquakes is thought to be the violent slipping of rock masses along major geological *fault*（断层）lines in the Earth's crust, or *lithosphere*（岩石圈）. These fault lines divide the global crust into about 12 major *tectonic*（地壳构造的）

plates, which are rigid, relatively cool slabs about 100 km thick. Tectonic plates float on the molten *mantle* (地幔) of the Earth and move relatively to one another at the rate of 10 to 100 mm/year.

The basic mechanism causing earthquakes in the plate boundary regions appears to be that the continuing deformation of the crustal structure eventually leads to stresses which exceed the material strength. A rupture will then initiate at some critical points along the fault line and will propagate rapidly through the highly stressed material at the plate boundary. In some cases, the plate margins are moving away from one another. In those cases, molten rock appears from deep in the Earth to fill the gap, often manifesting itself as *volcanoes* (火山). If the plates are pushing together, one plate tends to dive under the other and, depending on the density of the material, it may resurface in the form of mountains and valleys. In both these scenarios, there may be volcanoes and earthquakes at the plate boundaries, both being caused by the same mechanism of movements in the Earth's crust. Another possibility is that the plate boundaries will slide sideways past each other, essentially retaining the local surface area of the plate. It is believed that about three quarters of the world's earthquakes are accounted for by this rubbing-striking-slipping mechanism, with ruptures occurring on faults on boundaries between tectonic plates. *Earthquake occurrence maps* (地震发生图) tend to outline the plate boundaries. Such earthquakes are referred to as *interplate earthquakes* (边缘地震).

Earthquakes also occur at locations away from the plate boundaries. Such events are known as *intraplate earthquakes* (板内地震) and they are much less frequent than interplate earthquakes. Because tectonic plates are not homogeneous or **isotropic** (同向性的), areas of local high stress are developed as the plate attempts to move as a rigid body. Accordingly, rupture within the plate and the consequent release of energy, are believed to give rise to these intraplate events.

The point in the Earth's crustal system where an earthquake is initiated (the point of rupture) is called the *hypocenter* or *focus* (震源) of the earthquake. The point on the Earth's surface directly above the focus is called the *epicenter* (震中) and the depth of the focus is the focal depth. Earthquake occurrence maps usually indicate the location of various epicenters of past earthquakes and these epicenters are located by *seismological* (地震学上的) analysis of the effect of earthquake waves on strategically located receiving instruments called seismometers.

When an earthquake occurs, several types of seismic wave are radiated from the rupture. The most important of these are the *body waves* (体波) (primary P and secondary S waves). P waves are essentially sound waves traveling through the Earth, causing particles to move in the direction of wave propagation with alternate expansions and compressions. They tend to travel through the Earth with velocities of up to 8000 m/sec (up to 30 times faster than sound waves through air). S waves are shear waves with particle motion transverse to the direction of propagation. S waves tend to travel at about 60% of the velocity of P waves, so they always arrive at seismometers after the P waves. The time lag between arrivals often provides *seismologists* (地震学家) with useful information about the distance of the epicenter from the recorder.

2. Measurement of Earthquakes

Earthquakes are complex multi-dimensional phenomena, the scientific analysis of which requires measurements. Prior to the invention of modern scientific instruments, earthquakes were qualitatively measured by intensity, which differed from point-to-point. With the deployment of seismometers, an instrumental quantification of the entire earthquake event — the unique magnitude of the event — became possible.

1) Magnitude

The total strain energy released during an earthquake is known as the magnitude of the earthquake and it is measured on the Richter scale. It is defined quite simply as the amplitude of the recorded vibrations on a particular kind of seismometer located at a particular distance from the epicenter

$$M_L = \log A - \log A_o \qquad (3-1)$$

where M_L is local magnitude, A is the maximum trace amplitude in microns recorded on a seismometer at a site 100 km from the epicenter, A_o is a standard value as a function of distance.

2) Intensity

The magnitude of an earthquake by itself, which reflects the size of an earthquake at its source, is not sufficient to indicate whether the structural damage can be expected at a particular site. The distance of the structure from the source has an equally important effect on the response of a structure, as do the local ground conditions.

Seismic intensity is a measure of the effect, or the strength, of an earthquake hazard at a specific location based on the observed human behavior and the structural damage.

Numerous intensity scales were developed in pre-instrumental times. The most common in use today is the Modified Mercalli Intensity (MMI). MMI is a subjective scale defining the level of shaking at specific sites on a scale of I (barely felt) to XII (total destruction). (MMI is expressed in Roman numerals to connote its approximate nature.) For example, moderate shaking that causes few instances of fallen plaster or cracks in chimneys constitutes MMI Ⅵ. It is difficult to find a reliable relationship between magnitude, which is a description of the earthquake's total energy level, and intensity, which is a subjective description of the level of shaking of the earthquake at specific sites, because shaking severity can vary with building types, design and construction practices, soil types, and distance from the event.

Modern *seismometers* (地震仪) (or seismographs) are sophisticated instruments utilizing, in part, electromagnetic principles. These instruments can provide digitized or graphical records of earthquake-induced accelerations in both the horizontal and vertical directions at a particular site.

Accelerometers provide records of earthquake accelerations and the records may be appropriately integrated to provide velocity records and displacement records. Peak accelerations, velocities, and displacements are all in turn significant for structures of differing stiffness.

3. Influence of Local Site Conditions

Local geological and soil conditions may have a significant influence on the amplitude and frequency content of ground motions in the following ways:

1) Interaction between the bedrock earthquake motion and the soil will modify the actual ground accelerations' input to the structure. This manifests itself by an increase in the amplitude of the ground motion over and above that at the bedrock, and a *filtering* (过滤) of the motion so that the range of frequencies at present becomes narrow with the high-frequency components being eliminated. This condition particularly arises in areas where soft *sediments* (沉淀物,沉积) and *alluvial* (冲积的,淤积的) soil overly form bedrock. The degree of *amplification* (放大) is dependent on the strength of shaking at the bedrock. Because of nonlinear effects in the soil, the amplification ratio is less in strong shaking than under base motions of lower amplitude.

2) The soil properties in the proximity of the structure contribute significantly to the effective stiffness of the structural foundation. This may be a significant parameter in determining the overall structural response, especially for structures that would be characterized as stiff under other environmental loading.

3) The strength (and response) of the local soil under earthquake shaking may be critical to the overall stability of the structure.

It is also important that the information on relevant geological features, such as faulting, should be assessed. Geological information on suspected active faults near the site can assist in providing a basis for evaluating the intensity of a likely earthquake. It is usual to use this information, together with the regional seismicity data, to determine the likely level of seismic activity.

4. Response of Structures to Ground Motions

The effect of ground motions on the various categories of structures is dictated almost entirely by the distribution of mass and stiffness in the structure. It is important to appreciate that, in an earthquake, loads are not applied to the structure. Rather, earthquake loading arises because of accelerations generated by the foundation level(s) of the structure being influenced by transient ground motions. Specifically, the product of the structural mass and the total acceleration produces the *inertia loading* (惯性力) experienced by the structure. This is an expression of Newton's Second Law. It is important to appreciate that the total acceleration is the absolute acceleration of the structure, namely, the sum of the ground acceleration and that of the structure relative to the ground.

If the structure is stiff, there is little, if any, additional acceleration relative to the ground motion and, therefore, the earthquake loading experienced is essentially proportional to the building mass.

For structures that are flexible, for example, those in the high-rise or long-span category, the absolute acceleration is low. This occurs because the ground acceleration and the acceleration of the building relative to the ground tend to oppose one another. In this case, the earthquake loading is approximately proportional to the square root of the mass.

For structures in the cantilever category, which are essentially vertical, it is the horizontal accelerations that are significant; whereas for structures that are largely horizontal in extent, the effect of the vertical accelerations is dominant. Moreover, if the plane distributions of mass and stiffness are dissimilar in vertical structures, significant twisting motions may arise.

The *peak ground acceleration* (地面运动峰值加速) is of importance in the response of stiff structures and *peak ground displacements* (地面运动峰值位移) are of importance in the response of flexible structures, with peak ground velocity being of importance for structures of intermediate stiffness. Stiff structures tend to move in *unison* (调和;和谐, 一致) with the ground while flexible structures, such as high-rise buildings, experience the ground moving beneath them, their upper floors tending to remain motionless.

Integrated Skills: Summary

Please write a summary of the above Text B or Text C with no more than 500 words.

Unit 4　Road Design

Lead-in

1. How many types of roads can you list? What are they?
2. What factors should be considered while designing the roads?

Vocabulary Warm-up: Matching

_____ 1. belt highway	a. 高速公路
_____ 2. bypass highway	b. 车行道
_____ 3. freeway	c. 主干道
_____ 4. frontage road	d. 环路,环线
_____ 5. collector-distributor road	e. 直行车道
_____ 6. carriageway	f. 人行道
_____ 7. major artery	g. 支路,旁线
_____ 8. pedestrian walkway	h. 放射式道路
_____ 9. radial road	i. 集散道路
_____ 10. through lane	j. 临街道路

Text A　Horizontal and Vertical Alignment

1. Background

One definition of a visually attractive and unobtrusive highway is the degree to which the horizontal and *vertical alignments* of the *route* have been integrated into its surrounding natural and human environments. This takes careful planning and design, as noted in the *AASHTO* Green Book: Coordination of *the horizontal alignment* and *the profile* should not be left to chance but should begin with *preliminary design*, during which adjustments can readily be made... The designer should study long, continuous stretches of highway in both plan and profile and visualize the whole in three dimensions.

This application of a holistic approach to highway design, where the road is integrated into its surroundings, separates the outstanding project from one that merely satisfies basic engineering design criteria. An excellent description of this holistic design process is contained in the publication *Aesthetics in Transportation*, from which the following is excerpted: A general rule for designers is to achieve a "flowing" line, with a smooth and natural appearance in the land, and a sensuous, rhythmic continuity for the driver. This effect results from following the natural *contours* of the land, using graceful and gradual horizontal and vertical *transitions*, and relating the *alignment* to permanent features such as rivers or mountains.

2. Horizontal and Vertical Alignment Considerations

The greatest opportunities for influencing the horizontal and vertical alignments of a highway occur during the planning and preliminary engineering phases associated with a new-location facility. The designs of such facilities have the most dramatic effects on the natural and human environments through which they pass.

A more typical design problem faced by today's highway engineers is the improvement of an existing highway or street. In many instances, the basic alignments may have been established well over 100 years. Regardless, the same basic design principles with respect to horizontal and vertical alignments can, however, be applied to both new and existing facilities.

Important points to consider regarding horizontal and vertical alignments are that they should be consistent with the *topography*, preserve developed properties along the road, and incorporate community values. The superior alignments are ones that follow the natural contours of the land and do not affect aesthetic, scenic, historic, and cultural resources along the way. Construction costs may be reduced in many instances when less *earthwork* is needed, and resources and development are preserved. It is not always possible, however, to avoid having an impact on both the natural and human environments. That is why the superior alignments incorporate input received by the community through a participatory design process.

When possible, the alignment should be designed to enhance attractive scenic views, such as rivers, rock formations, parks, historic sites and outstanding buildings. The designation of certain highways as scenic byways recognizes the importance of preserving such features along our nation's roadways.

Equally important as the consideration of the horizontal alignment is that of the facility's vertical alignment. A number of factors influence the vertical alignment of a highway, including the followings: ① Natural *terrain*; ② Minimum *stopping sight distance* for the selected design speed; ③The number of trucks and other heavy vehicles in the traffic stream; ④The basic roadway *cross-section*; i. e. , *two lanes* versus *multiple lanes*; ⑤ Natural environmental factors, such as wetlands and historic, cultural, and community resources.

3. Combination of the Horizontal and Vertical Alignment

The interrelationship of horizontal and vertical alignment is best addressed in the route location and preliminary design phases of a project. At this stage, appropriate *tradeoffs* and balances between design speed and the character of the road — traffic volume, topography, and existing development — can be made. A mistake often made by inexperienced engineers is designing the horizontal alignment first and then trying to *superimpose* the design onto a vertical profile. Because they must be complementary, horizontal and vertical geometries must be designed concurrently. Uncoordinated horizontal and vertical geometries can ruin the best parts and accentuate the weak points of each element. Excellence in the combination of their designs increases efficiency and safety, encourages uniform speed, and improves appearance — almost always without additional cost.

One tool to assist in coordinating horizontal and vertical geometries is the use of computer-aided design (CAD). *CAD* enables highway designers to quickly assess the interrelationships between the horizontal and vertical alignment, particularly in areas of difficult terrain.

Proper consideration of these basic design considerations will help to ensure that both new-location facilities and improvements to existing highways fit harmoniously into their surroundings.

1) Issues

There are numerous examples around this country of excellence in integration of the horizontal and vertical alignments of highways into their surroundings. Unfortunately, here are also examples of new or *widened highways* that have scarred a rural *landscape* or disrupted an established community. While these past actions cannot easily or inexpensively be rectified, future problems can be avoided by applying the principles outlined above and the creative approaches detailed below.

2) Avoiding the Impact on Adjacent Natural and Human Environments

Particularly during the era of Interstate Construction from the 1950's to the 1980's, a number of instances of new highway construction had a devastating impact on communities and areas of environmental sensitivity. It is readily acknowledged that there will be some degree of physical impact on the surroundings associated with the construction of any new location highway or major reconstruction or widening of an existing highway facility. However, from the perspective of horizontal and vertical alignment, much of this impact can and should be alleviated.

3) Solutions

The impact on the surrounding environments can be minimized by careful attention to details during the route location and preliminary design phases and a willingness of all concerned parties to work together toward a common goal. For example, minor adjustments to the originally proposed horizontal and vertical alignments (combined with the use of short sections of the *retaining wall*) along the Lincoln Beach Parkway (U. S. Route 101) in Oregon eliminated the need to acquire any of the adjacent homes and businesses.

Similarly, a minor horizontal alignment shift at the beginning of the project allowed for the Hollister Bypass (SR 156) in San Benito County, CA, to avoid affecting a number of historic properties.

The use of a *"cut-and-cover"* design, whereby the roadway is placed below the existing ground level and covered over with a park, buildings, or other public space can help to avoid the negative impact. Lake Place Park in Duluth, MN and other public parks were the results of cut-and-cover tunnels that not only saved historic properties but also gave pedestrians improved access to Lake Superior.

In many cases, there is a potential for designing a divided highway with independent horizontal and vertical alignments for each direction of traffic.

4. Coordination Between Horizontal and Vertical Alignments

When horizontal and vertical alignments are designed separately from one another, unnecessarily large cuts and fills may be required, resulting in very dramatic and often visually undesirable changes to the natural landscape.

Solutions：

One of the ways to ensure the most effective coordination of horizontal and vertical alignments is through the use of a multidisciplinary design team during the planning and engineering phases of a project. On such projects as 1-66 in Fairfax and Arlington Counties in Virginia, the combined expertise of *landscape architects*, *urban designers*, *structural engineers*, and historic preservationists, in addition to civil engineers and highway designers, has resulted in superior highway improvement projects.

The concept of using a multidisciplinary design team is not new; it was pioneered in the early 1900's during the planning and design of the Bronx River Parkway in Westchester County, NY. After a period of using primarily on large-scale or controversial projects, this approach has come back into more general application as a way to achieve community consensus.

Glossary

Complete the glossary using words in the text.

1. _____ 纵断面线型　　2. _____ 路线;航线;通道
3. _____ 美国国家公路与运输协会
4. _____ 平面线型
5. _____ 纵断面;侧面;轮廓
6. _____ 初步设计　　7. _____ 等高线;轮廓
8. _____ 过渡;转变;转换
9. _____ 线型,定线　　10. _____ 地势;地形
11. _____ 土方工程,挖土
12. _____ 地形,地势;领域;地带
13. _____ 停车视距　　14. _____ 横断面;横截面
15. _____ 双车道　　16. _____ 多车道
17. _____ 折中方案;权衡
18. _____ 添加,附加;重叠;安装
19. _____ 计算机辅助设计
20. _____ 扩宽的公路　　21. _____ 景观,风景

22. _____   挡土墙,护壁;护岸

23. _____   下挖加盖板   24. _____   景观设计师

25. _____   城市规划设计师

26. _____   结构工程师

Exercises

1. Write down the words with the following prefix or suffix.

uni-   _____

out-   _____

ex-   _____

vis-   _____

-ography   _____

2. Translate the following into Chinese.

1) Coordination of the horizontal alignment and the profile should not be left to chance but should begin with preliminary design, during which adjustments can readily be made ... The designer should study long, continuous stretches of highway in both plan and profile and visualize the whole in three dimensions.

2) A general rule for designers is to achieve a "flowing" line, with a smooth and natural appearance in the land, and a sensuous, rhythmic continuity for the driver. This effect results from following the natural contours of the land, using graceful and gradual horizontal and vertical transitions, and relating the alignment to permanent features such as rivers or mountains.

3) The use of a "cut-and-cover" design, whereby the roadway is placed below the existing ground level and covered over with a park, buildings, or other public space can help to avoid negative impact. Lake Place Park in Duluth, MN and other public parks were the result of cut-and-cover tunnels that not only saved historic properties but also gave pedestrians improved access to Lake Superior.

3. Translate the following into English.

在选线和初步设计阶段,最好就能确定路线平纵线形间的关系。在这个阶段设计车速与道路交通流量、地形、已经存在的开发之间需要均衡。由于平纵线形是相互补充的,设计时必须保持同步,而不是割裂开来,有些没有经验的设计师常常是错误地先设计平面线形,然后再添加纵面线形。平纵线形的不协调将影响公路的部分路段整体性,突出路线弱点。良好的平纵线形设计将有利于提高道路行车效率及安全性,保证均一的车速,并且不需要增加额外费用,就可以改善道路外观。

Text B   Cross-section Elements

1. Background

The cross section of a road includes some or all of the following elements: ①Traveled way (the portion of the roadway provided for the movement of vehicles, exclusive of shoulders); ②Roadway (the portion of a highway, including shoulders, provided for vehicular use); ③Median area (the physical or painted separation provided on divided highways between two adjacent roadways); ④Bicycle and pedestrian facilities; ⑤Utility and landscape areas; ⑥Drainage channels and side slopes; ⑦Clear zone width (i. e., the distance from the edge of the traveled way to either a fixed obstacle or nontraversable slope).

Considered as a single unit, all these cross-section elements define the highway right-of-way. The right-of-way can be described generally as the publicly owned parcel of land that encompasses all the various cross-section elements (see Figure 4−1 and Figure 4−2).

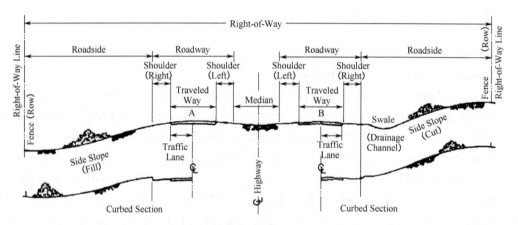

Figure 4−1　Two-lane rural highway cross-section design features and terms

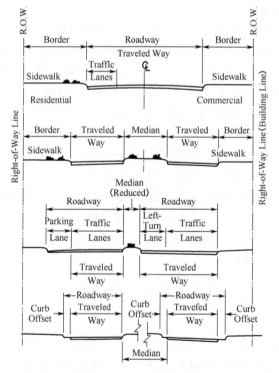

Figure 4−2　Urban highway cross-section design features and terms

Some decisions about the cross section are made during project development, such as the capacity and number of lanes for the facility. Other decisions, such as functional classification, are made earlier in the process. Within these parameters, the Green Book guidelines recommend a range of values for the dimensions to use for cross-sectional elements. Deciding which of the elements to include and selecting the appropriate dimensions within these ranges are the duties of the designer.

In selecting the appropriate cross-section elements and dimensions, designers need to consider a number of factors, including the followings: ①Volume and composition (percent of trucks, buses, and recreational vehicles) of the vehicular traffic expected to use the facility; ②The likelihood that bicyclists and pedestrians will use the route; ③Climatic conditions (e. g., the need to provide storage space for plowed snow); ④The presence of natural or human-made obstructions adjacent to the roadway (e. g., rock cliffs, large trees, wetlands, buildings, power lines); ⑤ Type and intensity of development along the section of the highway facility that is being designed; ⑥Safety of the users.

The most appropriate design for a highway improvement is the one that balances the mobility needs of the people using the facility (motorists, pedestrians, or bicyclists) with the physical constraints of the corridor within which the facility is located.

2. Cross-section Elements

1) Travel Lanes

The number of lanes needed for a facility is usually determined during the concept stage of project development. It is usually the number of lanes necessary to accommodate the expected traffic volumes at a level of service determined to be appropriate for the facility. The number of lanes can only be added in integer units, i. e., a two-lane highway can be widened to three or four lanes. Each additional lane represents an increase in the traffic-carrying capability of the facility.

Knowing future projected travel demands, the designer, using the analysis procedures in the Highway Capacity Manual, can provide input into the decision-making process during project development to determine the appropriate number of travel lanes for the level of service desired. Community input also plays a part in this decision. A community may decide through public involvements that a lower level of service is acceptable for the situation than the level of service normally provided for new construction projects.

In urban and suburban areas, signalized intersections are usually the predominant factors controlling the capacity of the highway or street. There may be more latitude in determining the number of lanes for these types of facilities. For example, a two-lane facility approaching an intersection can be expanded to four lanes (one left-turn lane, two through lanes, one right-turn lane) at the intersection itself and then returned to two lanes beyond the intersection. The need to distribute traffic safely will determine the need for any expansion of the approach roadway. The added lanes at the intersection can be in a variety of configurations to serve the travel desires of the traffic.

2) Lane Width

The width of travel lanes is limited by the physical dimensions of automobiles and trucks to a range between 2.7 and 3.6 m. Generally, as the design speed of a highway increases, so must the lane width allow for the lateral movement of vehicles within the lane. However, constricted right-of-way and other design restrictions can have an impact on this decision. Chapter IV of the AASHTO Green Book recognizes the need for flexibility in these cases: Although lane widths of 3.6 m are desirable on both rural and urban facilities, there are circumstances that necessitate the use of lanes less than 3.6 m wide. In urban areas where right-of-way and existing development become stringent controls, the use of 3.3 m lanes is acceptable. Lanes of 3.0 m width are acceptable on low-speed facilities. Lanes of 2.7 m width are appropriate on low-volume roads in rural and residential areas.

3) Medians

An important consideration in the design of any multilane highway is whether to provide a median and, if one is provided, what the dimensions should be. The primary functions of highway medians are to: separate opposing traffic flows; provide a recovery area for out-of-control vehicles; allow space for speed changes and left-turning and U-turning vehicles; minimize headlight glare; provide width for future lanes (particularly in suburban areas); provide a space for landscape planting that is in keeping with safety needs and improves the aesthetics of the facility; provide a space for barriers.

Depending on agency practice and specific location requirements, medians may be depressed, raised, or flush with the surface of the traveled way. Medians should have a dimension that is in balance with the other elements of the total highway cross section. The general range of median widths is from 1.2 m, usually in urban areas, to 24 m or more, in rural areas. An offset of at least a 500 mm should be provided between any vertical element located within the median, such as a curb or barrier, and the edge of the adjacent traveled lane.

The design and width of medians again require tradeoffs for designers. In locations where the total available right-of-way is restricted, a wide median may not be desirable if it requires narrowing the areas adjacent to the outside edge of the traveled way. A reasonable border width is required to serve as a buffer between private development along the road and the edge of the traveled way, and space may be needed for sidewalks, highway signs, utilities, parking, drainage channels and structures, proper slopes and clear zones, and any retained native plant material. On the other hand, wider medians provide more space for the plant material, offer a refuge for pedestrians at intersections, and help soften the look of the roadway. Including and designing medians requires public input to find the design that meets the needs of the community.

The use of two-way left-turn lanes on urban streets in densely developed suburban commercial areas has increased as an alternative to raised medians with left-turn or U-turn bays. Although not as aesthetically pleasing as raised, planted medians, continuous left-turn lanes can improve capacity. Two-way left-turn lanes generally are not recommended in residential areas because they do not afford a safe refuge for pedestrians. Also, the number of driveways can create unsafe vehicle maneuvers.

4) Shoulders

Although the physical dimensions of automobiles and trucks limit the basic width of travel lanes, the treatment of that portion of the highway to the right of the actual traveled way, that is, the "roadway edge", provides the designer with a greater degree of flexibility. This is true in both urban and rural areas, although different design elements are more appropriate in each location.

Shoulder widths typically vary from as little as 0.6 m on minor rural roads, where there is no surfacing, to about 3.6 m on major highways, where the entire shoulder may be stabilized or paved.

The treatment of shoulders is important from a number of perspectives, including safety, the capacity of the highway section, impact on the surrounding environment, and both the initial capital outlay and ongoing maintenance-operating costs. The shoulder design should balance these factors. For example, a designer must consider the impact of the shoulder width and other roadside elements in the surrounding environment and, at the same time, how these dimensions will affect capacity. Even with a maximum lane width of 3.6 m, the absence of a shoulder or the presence of an obstruction at the edge of the travel lane can result in a reduction in capacity of as much as 30 percent, compared to an area where shoulder or clear zone exists that is a minimum 1.8 m wide. On the other hand, significant environmental, scenic, or historic resources may be adversely affected by a widened shoulder.

Another consideration is the accommodation of pedestrians and nonmotorized vehicles. In many parts of the country, highway shoulders provide a separate traveled way for pedestrians, bicyclists, and others (when no sidewalks are provided).

In addition to the dimensions of shoulders, designers have choices to make of the materials used. Shoulders may be surfaced for either their full or partial widths. Some of the commonly used materials include gravel, shell, crushed rock, mineral or chemical additives, bituminous surface treatments, and various forms of asphaltic or concrete pavements.

In a number of states, particularly in the southern part of the country where snow removal is not an issue, grass or turf surfaces have been provided on top of compacted earth embankments. The advantages of grass shoulders are that they provide both a natural storm water detention system and are aesthetically pleasing. The disadvantages can be that they are often less safe than paved shoulders and force pedestrians and bicyclists to share the road with motorists, if no off-street facility is provided.

Shoulders represent an important element in roadway drainage systems by carrying surface runoff away from the travel lanes into either open or closed drainage systems. A variety of design treatments have been used to accommodate roadway drainage across shoulder areas. In rural and suburban areas, the most common technique allows surface runoff to cross over the shoulder and go directly into drainage ditches parallel to the roadway edge.

In rural areas where significant physical and/or environmental constraints exist, more "urban" style solutions have been used. For example, along an older section of Maryland State Route 51, passing through the Green Ridge State Forest in Allegany County, steep, narrow cuts along the existing alignment severely limited the total roadway width. Asphalt curbing and a closed drainage system were constructed in conjunction with a recent pavement rehabilitation project. This allowed for a modest widening of the travelway and elimination of an area of steep and narrow ditches, without the need to engage in major earthwork.

5) Clear Zones

An important consideration in defining the appropriate cross section for a particular highway facility is the width of the clear zone. As defined in Chapter IV of the *AASHTO Green Book*, the clear zone is "... the unobstructed, relatively flat area provided beyond the edge of the traveled way for the recovery of errant vehicles."

The width of the clear zone is influenced by several factors, the most important of which are traffic volume, design speed of the highway, and slope of the embankments. The *AASHTO Roadside Design Guide* is a primary reference for determining clear zone widths for freeways, rural arterials, and high-speed rural collectors based on these factors. For low-speed rural collectors and rural local roads, the *AASHTO Green Book* suggests providing a minimum clear zone width of 3.0 m. For urban arterials, collectors, and local streets with curbs, the space available for clear zones is typically restricted.

6) Curbs

Used primarily in urban and suburban environments, curbs can serve some or all of the following functions: Drainage control; Roadway edge delineation; Right-of-way reduction; Aesthetics; Delineation of pedestrian walkways; Reduction of maintenance operations; Assistance in roadside development.

There are basically two types of curbs: barrier and mountable. Flexibility in the use of either type is a handy tool for a highway designer when defining the cross section of an improvement project. Barrier-type curbs are not, however, recommended for projects with design speeds above 65 km/h.

Curbs can be constructed from a variety of materials, including concrete, asphalt, and cut stone. Figure 4-3 illustrates a variety of commonly used barrier and mountable curbs.

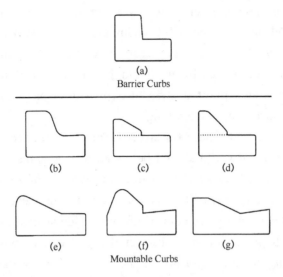

(a)
Barrier Curbs

(b)　　　　(c)　　　　(d)

(e)　　　　(f)　　　　(g)
Mountable Curbs

Figure 4-3　Examples of barrier and mountable curbs

7) Sidewalks and Pedestrian Paths

The safe and efficient accommodation of pedestrians along the traveled way is equally important as the provisions for vehicles. Too often, pedestrians are a secondary consideration in the design of roadways, particularly in suburban areas. Although sidewalks are an integral part of city streets, they are much more rare in rural areas and provided only sporadically in suburban areas, despite data suggesting that providing sidewalks along highways in rural and suburban areas results in a reduction in pedestrian accidents.

When considering the placement of sidewalks, designers have several options. The sidewalk can either be placed flush with the roadside edge (if a curb is provided) or next to a buffer area, such as a planted strip (usually of grass or plant material), located between the sidewalk and roadside. The pros and cons of each option should be weighed and considered by the designer, using input from the community. For example, a planted strip has these advantages:

① Pedestrians are kept at a greater distance from moving vehicles and thus are safer (in urban areas with on-street parking, parked cars help to act as a shield for pedestrians from moving traffic, so a buffer space may not be necessary to address that concern.).

② Planted strips tend to add to the aesthetics of the facility and help reduce the apparent width of hard surface space.

③ Planted strips provide a space for snow storage.

Buffers, or planted strips, may have the disadvantage of requiring additional right-of-way that may negatively affect restricted right-of-way corridors.

Another important consideration, and one in which the designer is given some flexibility, is in the width of the sidewalk and planted strip. Typically, sidewalks in residential or low-density commercial areas vary in width from 1. 2 to 2. 4 m. The Americans with Disabilities Act Accessibility Guidelines of August 1992 set the minimum passing width on a sidewalk at 1.525 m at least every 61 m. If a planted strip is provided between the sidewalk and the curb, it should be at least 0. 6 m wide to allow for maintenance activities. This planted strip also provides space for street lights, fire hydrants, street hardware, and landscaping.

Sidewalks can also provide space for street furniture and necessary traffic poles and signals; however, additional width should be added to sidewalks to accommodate these fixtures. The wider the sidewalk, the greater the number of pedestrians that can be accommodated and the less difficult it is for them to maneuver around these fixed objects. When considering the placement of objects inside sidewalks, it is important not to overlook the need to maintain as unobstructed a pathway as possible. For instance, locating utility poles to the sides and not in the center of sidewalks is important. This detail facilitates the movements of people with disabilities as well.

Adding sidewalks to a facility where none previously exists can be beneficial to a community. When the Lincoln Beach Parkway section of the Pacific Coast Highway (U. S. Route 101) was reconstructed in the early 1990's, sidewalks were added along both sides of the facility. Not only did this result in a more aesthetically pleasing alternative to the shoulder section for the two travel lanes that previously existed, but the sidewalks made it safer for residents to walk between their homes and local commercial facilities. Residents can now interact with each other much more easily, which has fostered a higher level of community spirit.

8) Accommodating Bicycles

Bicycles are recognized by many as a viable mode of transportation in the United States, both for commuting and recreation. Transportation designers should consider the needs of these users in the design of facilities. Basically, there are five types of bicycle facilities:

① Shared lane — a "standard-width" travel lane that both bicycles and motor vehicles share.

② Wide outside lane — an outside travel lane with a width of at least 4. 2m to accommodate both bicyclists and motorized vehicles.

③ Bicycle lane — a portion of the roadway designated by striping, signing, and/or pavement markings for preferential or exclusive use by bicycles and/or other nonmotorized vehicles.

④ Shoulder — a paved portion of the roadway to the right of the traveled way designed to serve bicyclists, pedestrians and others.

⑤ Multiuse path — a facility that is physically separated from the roadway and intended for use by bicyclists, pedestrians, and others.

There are three primary factors to consider when designing facilities to accommodate bicycles and other nonmotorized vehicles:

① What type of bicyclist is the route most likely to serve, i. e. , advanced bicyclists, basic bicyclists, or children?

Advanced bicyclists are the experienced riders who make up the majority of the current users of collector and arterial streets, who wish to operate at maximum speed with minimum delays, and require sufficient space on the roadway shoulder to be treated as vehicles. Designated bicycle lanes along a roadway give riders an even greater degree of comfort along arterial and collector streets. Basic bicyclists and children generally prefer the most comfortable, although sometimes circuitous, access to destinations, using low-speed, low-traffic-volume streets or a separate, multiuse path.

② What type of roadway project is involved, i. e. , new construction, major reconstruction, or minor rehabilitation?

Recommended design treatments are most easily implemented when new construction or major reconstruction is planned. Although retrofit and/or enhancement projects may be relatively limited in scope, opportunities to make at least minor improvements to better accommodate the needs of pedestrians and bicycles should be investigated. Marginal roadway improvements undertaken as part of 3R projects, such as widening the pavement area from 0.3 to 0.6 m will enhance the roadway for bicycle use.

③ What are the current and future traffic operations and design characteristics of the route that will affect the choice of bicycle design treatments?

Six factors are recognized by transportation planners and engineers **as** having the greatest effect on bicycle use:

① Traffic volume — higher traffic volumes represent greater potential risk for bicycles.

② Average motor vehicle operating speed — operating speed is more important than the posted speed limit; motor vehicle operating speed can negatively affect the bicyclist's comfort unless mitigated by special design treatments.

③ Traffic mix — the presence of trucks, buses, and other large vehicles can increase risk and have a negative impact on the comfort of bicyclists.

④ On-street parking — additional width is needed for bicycle lanes on roads that have on-street parking.

⑤ Sight distance — this must be sufficient to allow a motor vehicle operator to either change lane position or slow to the bicyclist's speed when overtaking the bicycle, primarily on rural highways.

⑥ Number of intersections — the number and frequency of intersections should be considered when assessing the use of bike lanes. Intersections pose special challenges to bicycle and motor vehicle operators and require special treatments.

9) Landscape Design and Selection of Plant Material

Landscape design is an important element in the design of all highway facilities and should be considered early in the process, so that it is in keeping with the character or theme of the highway and its environment. *The AASHTO Green Book* mentions three objectives of landscape design: To provide vegetation that will be an aid to aesthetics and safety; To provide vegetation that will aid in lowering construction and maintenance costs; To provide vegetation that creates interest, usefulness, and beauty for the pleasure and satisfaction of the traveling public.

Landscape design for urban highways and streets plays an additional role in mitigating the many nuisances associated with urban traffic and can help a roadway achieve a better "fit" with its surroundings.

10) Trees

An important aspect of roadside landscape design is the treatment of trees. Single-vehicle collisions with trees account for nearly 25 percent of all fixed object fatal accidents annually and result in the deaths of approximately 3,000 people each year. This problem is most apparent on roads that have existing trees, where designers do not have direct control over placement. For landscape projects, where the type and location of trees and other vegetation can be carefully chosen, the potential risks can be minimized.

Integrating trees into the design of a facility has many advantages. Trees provide a visual "edge" to the roadway that helps guide motorists. Trees also add to the aesthetic quality of a highway. In urban and suburban areas, trees soften the edges of arterial and collector streets. If sight distance is a concern, taller trees with lower branches that are trimmed or low-growing (shorter than 1 m) herbaceous and woody plants can be another option along both the roadway edge and in raised medians.

It is important to select the appropriate species of tree for the highway environment. In particular, trees need to be chosen that can survive poor air quality, infertile and compacted soils, and extreme temperature fluctuations. Remember that maintenance, particularly during the first year after installation, is essential to the long-term health and viability of trees and other plants. Utilize the skills and knowledge of the city or town urban forester or arborist, the local agricultural extension service, or a landscape architect to identify the plant material that will be best suited for the location.

In addition to selecting a type of tree for its hardiness, the size and placement of trees is another important consideration. Generally, a tree with a trunk diameter greater than 100 mm measured 100 mm above the ground line is considered a "fixed object" along the roadway. Because most trees grow larger than this, their placement along the roadway needs to be carefully considered. Factors that affect this decision include the design speed, traffic volume, roadway cross section and placement of guardrail. Trees should not be placed in the clear zone for any new construction or major reconstruction, nor should they be considered safe because they are placed just outside the clear zone. The safe placement of trees to prevent errant drivers from hitting them should be made in conjunction with a highway designer who is knowledgeable about safety. However, the decision to create a clear zone that requires the removal of existing trees is an issue that should be presented to the public and addressed by the multidisciplinary team early on.

Trees are an important aspect of community identity and carry a great deal of emotional ties with the residents. If communities consider existing trees a valuable resource, alternatives to complete eradication should be pursued. These include installation of traffic barriers, lowering of the design speed, or even complete redesign of the facility to incorporate the trees. It is not unusual for a community to value one specific tree and desire to preserve it. In general, transportation designers must balance safety with other community values when considering facility design and tree preservation.

11) Utilities

One element of cross-section design that is often overlooked is the accommodation of public utilities. Overhead utilities typically include electric, telephone, and cable television. For new construction in urban areas, electric, telephone, and other telecommunication lines are now often placed underground.

Motor vehicle collisions with utility poles result in approximately 10 percent of all fixed-object fatal crashes in the United States annually. Utility poles also have a negative affect on the aesthetics of a roadway. It is important, therefore, whether designing in rural or urban locations, to consider accommodating utilities early in the design process.

The most desirable design solution, in terms of safety for overhead utilities, is to locate the utility poles where they are least likely to be struck by a vehicle. (The same is true for sign and luminaire supports.) The 1996 *AASHTO Roadside Design Guide* notes the following options for the location and design of utilities: Bury power and telephone lines underground; Increase lateral pole offset; Increase pole spacing; Combine pole usage with multiple utilities; Use a breakaway pole design; Use traffic barriers to shield poles.

Burying power and telephone lines, although the safest and most aesthetically pleasing option, is also the most expensive. For example, during the reconstruction of 1. 66 km of Carson Street in the city of Torrance, CA, all the existing overhead utilities were placed underground at a cost of about $ 2.3 million, or approximately 37 percent of the total project cost. Because of these tradeoffs, the design and location of utilities requires public input and should be considered early in the design of each project.

12) Traffic Barriers

The options available to designers for traffic barriers include deciding whether or not to include them in the design and, if they are included, deciding which type to choose. The purpose of the barrier, as stated in the *AASHTO Green Book*, is to "minimize the severity of potential accidents involving vehicles leaving the traveled way where the consequences of errant vehicles striking a barrier are less than leaving the roadway." In addition to preventing collisions with fixed objects along the roadside, traffic barriers are themselves obstacles and have some degree of accident potential. The use of traffic barriers should consider these tradeoffs.

A wide variety of traffic barriers are available for installation along highways and streets, including both longitudinal barriers and crash cushions. Longitudinal barriers (such as guardrails and median barriers) are designed primarily to redirect errant vehicles and keep them from going beyond the edge of the roadway. Crash cushions primarily serve to decelerate errant vehicles to a complete stop (such as impact attenuators at freeway exit gore areas).

The design of the traffic barrier is an important detail that contributes to the overall look or theme of roadway design; therefore, in addition to safety, the selection of an appropriate barrier design should include aesthetic considerations. In addition, all traffic barriers should meet crash-testing guidelines for the type of roadway being designed. Crash-testing guidelines have different levels, depending on the facility and the type of vehicles that will use the facility. For example, on parkways with restricted truck traffic, many aesthetic barriers have been designed and crash tested. The criteria used for these types of barriers are less stringent than the criteria for facilities with truck traffic. Because aesthetic considerations are usually a factor on parkways, many of these barriers are designed to add to the visual quality of the road. Even for roads that are not parkways, however, there are still many barrier designs that meet the criteria for facilities with truck traffic. Given these options, designers must balance their decisions based on safety, cost, and aesthetics.

A sample of available traffic barrier designs includes: ①A three-strand cable barrier system allowing deflections on impact of up to 4.6 m; ②Various steel beam barriers allowing deflections on impact of up to 1.2 m; ③Steel-backed timber barriers that allow deflections on impact of up to 2.4 m; ④New Jersey shaped concrete barriers; ⑤Stone masonry walls consisting of a reinforced concrete core faced with stone masonry.

An in-depth discussion of the factors associated with the decision to install traffic barriers and guidance on the selection of a particular barrier design is presented in the AASHTO Roadside Design Guide.

A concern among some states when selecting a barrier design is cost. Aesthetic barriers might have a higher upfront cost than standard steel barriers and may be more expensive to maintain. One solution to this concern is to be consistent in the type of aesthetic barrier used throughout a state. For instance, a state might want to limit the type of barriers used to only two, an inexpensive barrier for highways where aesthetics are not a major concern and an aesthetic type for highways where visual quality is important. In this way, states can cut back on the cost to maintain multiple barrier designs.

Weathering steel guardrails are an example of an inexpensive barrier that may be considered acceptable in certain surroundings. For many states, weathering steel has been a good solution, because its rustic color helps the guardrail blend into the environment. Weathering steel has, however, had durability problems in a few areas.

13) Accommodating Transit

Highways operate as truly multimodal transportation facilities, particularly in large urban areas. Accommodating public transit and other high-occupancy vehicles (HOVs) is an important consideration. On one end of the scale, this may involve including sidewalks to allow local residents to walk to and from bus stops. As higher levels of vehicle traffic and transit usage are expected, bus turnouts may need to be considered. At the higher end of the scale, such as on major urban freeways, dedicated bus lanes and/or HOV lanes may need to be incorporated into the design. The management of the local public transit operator should be consulted during the planning stage, if possible, so that these facilities can be incorporated into the design from the beginning.

3. Issues

Some of the challenging aspects of highway design have to do with cross-section elements. Decisions that designers need to make may include the number of lanes proposed for the improvement, the width of travel lanes and shoulder areas, the type of drainage proposed, or the desirability of including sidewalks or bicycle/pedestrian paths as part of the project.

1) Restricted Right-of-Way

Many roads currently exist that were not built to today's standards. These roads may be located in restricted right-of-way corridors that have scenic or historic resources adjacent to the roadway. It is necessary to try to avoid impacting these resources when considering highway improvements.

Solution:

One option, as has been discussed previously, is to reconsider the functional classification and design speed of a particular section of highway, because these decisions go a long way toward defining the basic design parameters that can be used in connection with an improvement of the facility. Lowering the design speed or changing the functional classification results in a lowering of the minimum width dimensions for the cross-sectional elements.

Another option is to maintain the road as a 3R project. Design criteria established by states are generally lower for 3R projects than for reconstruction projects. A third option is to seek design exceptions. Whichever alternative is chosen, the designer should try to maintain consistency in the roadway cross section. If only a small stretch of highway is located within restricted right-of-way, it would be unsafe to narrow that stretch while maintaining a much higher roadway width before and after it.

A successful resolution of the design of a highway cross section was found during the planning and design for the State Route 9A project along the Hudson River in Manhattan. The existing at-grade "interim" facility had two 3.6 m lanes in each direction, separated by a 4.6 m flush median with a Jersey barrier.

The preferred alternative, which is now under construction, replaces a rather unattractive urban street, with a six-to-eight-lane divided urban boulevard that has a landscaped median. The new design incorporates extensive landscaping and separate bikeways and pedestrian walkways. The width of the travel lanes was reduced from 3.6 m on the existing surface street to 3.4 m on the new urban boulevard. This cross section accommodates traffic demands and dramatically enhances the physical environment of the project area. More information about this project is in the case study section of this Guide.

2) The Design of Cross-Section Details

Some highway facilities may be designed with the greatest concern to fit into their surrounding environments, but if the details are not carefully thought out, they can still leave the impression of an unappealing roadway.

Solution:

The design of all elements of the highway cross section adds greatly to its appearance. Design details include the design and width of the median and traffic barriers and the selection of plant material. All these elements contribute to the theme of the roadway and should be considered as a unit. The best method for achieving a unified look is to work with a multidisciplinary design team from the beginning of the project development process through the last detail of the design.

Details are some of the first elements users of a facility will notice. For example, designers may go through a lot of trouble to preserve vegetation along the roadway because of its importance to the community and its scenic qualities, but if designers use concrete barriers as shields in front of this vegetation, that one element may catch the users' attention.

Another option that aids designers in the details of cross-section elements is the use of computer-imaging technology. The series of figures on the following page illustrates the application of various combinations of basic design elements to define a number of widening options for a portion of State Highway 23 in Rockville, MN. These options include the use of different median types and widths and incorporate different levels of right-of-way acquisition.

The Minnesota DOT has found the use of such computer-imaging techniques to be particularly useful in illustrating the impact of alternative design concepts on existing facilities for project area residents and businesses. Minnesota DOT has made this approach a standard element in all major project planning and preliminary engineering assignments.

With the increasing need to ensure meaningful and continuous public involvement on all such projects, the use of computer imaging to illustrate design alternatives to communities will help to alleviate potential conflicts and misunderstandings and lead to the best design decisions.

Text C   Introduction of the New Nano Interchange Design as a Directional Freeway-to-Freeway Interchange

1. Introduction

The supply of freeways in urban areas across developing countries has generally not kept pace with traffic demand. Interchanges are essential components in freeway operations because they control freeway access and handle the movement of traffic between freeways. Since, in many cases, interchanges are inferior in design quality compared to the associated freeways and there are the likelihood of increased speed differentials between freeway mainlines and ramp junctions, the interchanges have contributed to operation and safety problems and to high fuel consumption. In addition, system interchanges also consume large amounts of right of way (ROW), especially in dense urban areas of developing countries, and attempts to limit the amount of ROW have contributed to system interchanges that are poorly designed and have the potential capacity problems. Freeway designers need to provide drivers with large capacities while saving space in dense urban areas.

There are many alternative forms for service (freeway to surface street) interchanges with four approaches. However, engineers have fewer choices where a four-approach system (freeway to freeway) interchange is needed. A directional four-level interchange is the most widely accepted in the United States as a system interchange without loops, shown in Figure 4-4.

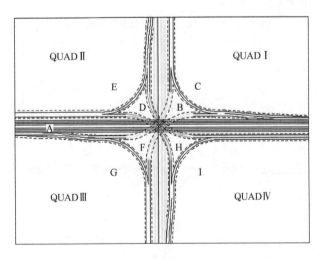

Figure 4-4   Four-level interchange

A few years ago, Dr. Joseph E. Hummer began thinking about the challenge of addressing deficiencies in existing conventional system interchange design in dense urban areas, and he conceived a new system interchange design, the nano interchange. Specifically created with intentions of minimizing the amount of right of way (or the "footprint") needed for an urban interchange, the nano interchange may be an alternative design for densely populated and developed urban areas. Distinguishing design features of the nano interchange include direct connections for all movements, combination of left- and right-hand entrances and exits, and four levels of freeway and ramp structures.

A study is needed to evaluate if the innovative nano interchange will function well enough to compensate for its disadvantages, specifically in compact urban areas where real estate is precious and expensive. In this research effort, the capabilities and applicability of the new interchange will be estimated based on analyses of traffic operations, safety, construction costs, and right of way.

2. The Nano Interchange Concept and Design

A nano interchange is designed with the expectation that it would provide shorter travel distances, higher speeds, lower amounts of ROW, and higher levels of service compared to some other designs. The nano interchange design allows direct connections to be made because each freeway is in a double-deck configuration where one direction is at a higher elevation than the other. The most noticeable geometric characteristic of the nano interchange is that it uses direct connections for all the left turn and the right turn movements, made possible by the double-deck configuration. However, the nano interchange has some problems, such as: ①confounding driver expectations by having left exits and entrances, because most drivers expect to have right exits and entrances; ②incurring high construction costs.

Figure 4-5 shows a schematic of the nano interchange types. Mainline 4 is on the

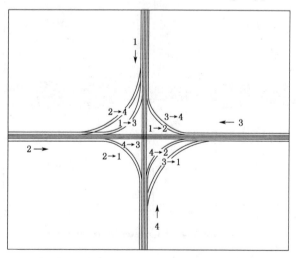

Figure 4-5 Plan view of a nano interchange

highest level, with mainlines 3, 2, and 1 positioned accordingly from high to low, respectively. Figure 4-5 also shows the ramp configuration of the nano interchange; such that ramp 1→2 connects mainline 1 to mainline 2, etc. Figure 4-6 shows a three-dimensional geometric configuration of the nano interchange created in VISSIM. The reversed curve segment of a mainline freeway heading into the interchange is shown in Figure 4-7.

Figure 4-6　Three-dimensional view of a nano interchange

Figure 4-7　Reversed curve segment heading into a nano interchange

3. Results of Evaluation in Operation, Safety, Construction Cost and ROW

The nano interchanges had the highest construction costs but required the least total acres of right of way, compared to the directional four-level interchanges of the same ramp design speed, as shown in Table 4-1. ROW savings for the nano over the conventional were from 15 to 33 acres. While the nano interchange with a ramp design speed of 35 mph had the smallest right of way, it had the most expensive construction cost. However, in urban areas where population densities and property prices are very high, the construction plus right-of-way costs may not be much higher for the nano than the four-level.

Table 4-1   Construction Costs and Right of Way Estimates

Design Speed, mph	Construction Costs, Million $, 2006		Right of Way, acres	
	Conventional	Nano	Conventional	Nano
35	83	289	54	39
45	120	266	69	48
55	150	272	101	68

There are differences in the interchange performance-related MOEs (Measure of effectiveness) between the nano and conventional interchanges by ramp design speed and percentage of heavy vehicles. All the MOEs were averaged over all the volume scenarios for each interchange type. Generally, the entire interchange performances of the nano interchange are worse than those of the conventional interchange for each ramp design speed. For 10% heavy vehicles, differences in travel times and average speeds between the nano and conventional interchange diminish as ramp design speeds increase. Meanwhile, differences in delays increase as ramp design speeds increase. For 20% heavy vehicles, differences in the three MOEs between the nano and conventional interchanges are greater with an increase in ramp design speeds. The nano interchanges appear to have smaller ramp flow-related travel times than conventional interchanges.

The individual performance estimations indicate that the relatively poor interchange performance of the nano interchanges over the conditions tested are primarily due to turbulence in diverging influence areas, which are located in the rising grade segments, and in merging influence areas.

The operational results for the entire interchange as well as for individual segments show that conventional interchanges perform better than nano interchanges for all the volume scenarios tested. The primary operational problems of the nano interchanges for the tested volume conditions are as follows. First, the diverging influence areas of the nano interchange, located mostly on rising grades, generally cause the worst operational performance, as the compound effects of the geometric features and the poor climbing performance of heavy vehicles make the diverging maneuver more difficult. Second, the *geometry* (几何,几何学) of the ramp itself appears to affect the approach speeds of the ramp flows in some merging influence areas of the nano interchange. The operational results show that merging influence areas which are connected to longer lengths and steeper grades of ramp appear to be inferior for average speeds, as compared to other merge influence areas. Third, as one of the geometric features of the nano interchange, the direct left-turn and right-turn ramps provide shorter travel distances for ramp flows. However, the short travel distances could not compensate for the inferior operational performances of the rest of the network as compared to the conventional interchanges.

It was expected that the lower the percentage of heavy vehicles, the better the operational efficiency of the nano interchange. Also, it should be noted that the traffic situations were simulated under the assumption that all exiting or entering vehicles were guided safely and made reasonable diverging or merging maneuvers. In the real world, left exits and entrances could cause some drivers to make erratic maneuvers. Specifically, because the nano interchange has left-hand exits and entrances, the nano interchange is more likely to have a higher potential for driver error than conventional interchanges. This factor thus could result in a poorer operational performance than the simulation results show.

Table 4 - 2 shows the number of each conflict type for each interchange type recorded using SSAM with a time to collision (TTC) equal or less than 1.5 seconds and maximum post-encroachment time (PET) equal or less than 5.0 seconds. These results indicate that while for the ramp volume of 1,600 vph, the number of conflict for the nano interchange is about 1.5 times higher than for the conventional interchange, there were not large differences in the number of the conflict between the two designs for ramp volumes of 1,000 vph. Further, the interchanges with the high ramp design speed appeared to have the lower probability of the conflicts than the low ramp design speed. Therefore, the surrogate safety measures provide evidence that as ramp and mainline freeway volumes increase, the nano interchange has high potential conflict points which can attribute to have an increase in the probability of collisions compared to the conventional interchange.

Table 4-2   Numbers of conflict estimated using SSAM

Ramp Volume, vph		Ramp Design Speed, mph	55		45		35	
		Interchange Type	Nano	Con.	Nano	Con.	Nano	Con.
1,600 1,600	Conflict Type	Total	609	465	692	463	701	509
		Rear-end	330	220	324	189	326	218
		Lane-change	279	245	368	274	375	291
1,000 1,000	Conflict Type	Total	392	339	479	316	474	367
		Rear-end	173	132	200	108	206	135
		Lane-change	219	207	280	208	268	232

Integrated Skills: Presentation

The combination of road and landscape design (see Figure 4-8) and the relationship between road alignments and traffic accidents are two of the hottest research topics recently. You should collect information about one of the above topics which you are most interested in, and then make a presentation in your class. Visual tools such as a

PowerPoint file should be prepared in order to achieve an effective presentation.

Figure 4-8　The combination of road and landscape design in Ning-Chang Expressway

Unit 5　Road Subgrade and Pavement Engineering

Lead-in

1. How do you categorize the pavements?

2. Can you explain different reasons for choosing one type of pavement or the other, practical, economical, and political?

Vocabulary Warm-up: Matching

_____ 1. PCC	a. 升级配沥青磨耗层
_____ 2. HMA / SMA	b. 沥青玛蹄脂碎石
_____ 3. OGFC	c. 有接缝的普通混凝土路面
_____ 4. ATPM / ATPB	d. 聚合物水泥混凝土
_____ 5. PEM	e. 连续配筋混凝土路面
_____ 6. JPCP / JRCP / CRCP	f. 热拌沥青混合料
_____ 7. JRCP	g. 有接缝的配筋混凝土路面
_____ 8. CRCP	h. 沥青处治排水性材料

Text A　Pavement Types

1. Introduction

Hard surfaced *pavements*, which make up about 60 percent of U. S. roads and 70 percent of Washington State roads are typically categorized into *flexible* and *rigid* pavements:

Flexible pavements are those which are surfaced with *bituminous* (or *asphalt*) materials (see Figure 5-1). These types of pavements are called "flexible" since the total pavement structure "*bends*" or "*deflects*" due to *traffic loads*. A flexible pavement structure is generally composed of several *layers* of materials which can accommodate this "flexing".

Rigid pavements are those which are surfaced with *portland cement concrete* (PCC) (see Figure 5-2). These types of pavements are called "rigid" because they are

Figure 5-1　Flexible pavement

*substantially **stiff**er* than flexible pavements due to PCC's high ***stiffness***.

Figure 5-2　Rigid pavement

Each of these pavement types ***distributes*** load over the ***subgrade*** in a different fashion. Rigid pavement, because of PCC's high stiffness, tends to distribute the load over a relatively wide area of subgrade (see Figure 5-3). The ***concrete slab*** itself supplies most of a rigid pavement's structural capacity. Flexible pavement uses more flexible ***surface course*** and distributes loads over a smaller area. It relies on a combination of layers for ***transmitting*** load to the subgrade (see Figure 5-4).

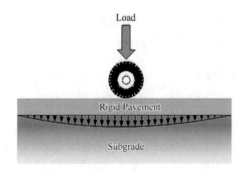

Figure 5-3　Rigid pavement load distribution　　　Figure 5-4　Flexible pavement load distribution

In general, both flexible and rigid pavements can be designed for long life (e. g., in excess of 30 years) with only minimal ***maintenance***. Both types have been used for just about every classification of road. Certainly there are many different reasons for choosing one type of pavement or the other, some practical, some economical, and some political. As a point of fact, 93 percent of U. S. paved roads and about 87 percent of Washington State paved roads are surfaced with bituminous (asphalt) materials.

2. Flexible Pavement

Flexible pavements are so named because the total pavement structure deflects, or flexes, under loading. A flexible pavement structure is typically composed of several layers of materials. Each layer receives the loads from the above layer, spreads them out, and then passes on these loads to the next layer below. Thus, the further down in the pavement structure a particular layer is, the less load (in terms of force per area) it must carry.

In order to take maximum advantage of this property, material layers are usually arranged in order of descending load bearing capacity with the highest load bearing capacity material (and most expensive) on the top and the lowest load bearing capacity material (and least expensive) on the bottom. The typical flexible pavement structure consists of:

Surface course. This is the top layer and the layer that comes in contact with traffic. It may be composed of one or several different HMA sublayers.

Base course. This is the layer directly below the HMA layer and generally consists of *aggregate* (either *stabilize*d or unstabilized).

Subbase course. This is the layer (or layers) under the base layer. A subbase is not always needed.

There are many different types of flexible pavements. This section covers three of the more common types of HMA mix types used in the U.S. HMA mix types differ from each other mainly in *maximum aggregate size*, *aggregate gradation* and *asphalt binder* content/type. This Guide focuses on *dense-grade*d HMA in most flexible pavement sections because it is the most common HMA pavement material in the U.S.. This section provides a brief exposure to:

Dense-graded HMA. Flexible pavement information in this Guide is generally concerned with dense-graded HMA. Dense-graded HMA is a versatile, all-around mix making it the most common and well-understood mix type in the U.S.

Stone matrix asphalt (SMA). SMA, although relatively new in the U.S., has been used in Europe as a surface course for years to support heavy traffic loads and resist studded *tire wear*.

Open-graded HMA. This includes both *open-graded* friction course (OGFC) and asphalt *treated permeable* materials (ATPM). Open-graded mixes are typically used as *wearing course*s (OGFC) or underlying *drainage layer*s (ATPM) because of the special advantages offered by their *porosity*.

1) Dense-Graded Mixes

A dense-graded mix (see Figure 5-5) is a well-graded HMA mixture intended for general use. When properly designed and constructed, a dense-graded mix is relatively *impermeable*. Dense-graded mixes are generally referred to by their nominal maximum aggregate size. They can further be classified as either fine-graded or coarse-graded. *Fine-graded mixes* have more fine and sand sized particles than *coarse-*

Figure 5-5   Pavement adopting dense-graded mix

graded mixes (see Table 5-1 for definitions of fine- and coarse-graded mixes).

Table 5-1　Fine-and coarse-graded definitions for dense-graded HMA

Mixture Nominal Maximum Aggregate Size	Coarse-Graded Mix	Fine-Graded Mix
37.5 mm	<35% passing the 4.75 mm (No. 4 Sieve)	>35% passing the 4.75 mm (No. 4 Sieve)
25.0 mm	<40% passing the 4.75 mm (No.4 Sieve)	>40% passing the 4.75 mm (No.4 Sieve)
19.0 mm	<35% passing the 2.36 mm (No. 8 Sieve)	>35% passing the 2.36 mm (No. 8 Sieve)
12.5 mm	<40% passing the 2.36 mm (No. 8 Sieve)	>40% passing the 2.36 mm (No. 8 Sieve)
9.5 mm	<45% passing the 2.36 mm (No. 8 Sieve)	>45% passing the 2.36 mm (No. 8 Sieve)

Purpose: Dense-graded mixes are suitable for all pavement layers and for all traffic conditions. They work well for structural, friction, leveling and *patching* needs.

Materials: Well-graded aggregate, asphalt binder (with or without modifiers), RAP.

Mix Design: Superpave, Marshall or Hveem procedures.

Other Info: Particulars about dense-graded HMA are covered by flexible pavement sections in the rest of this Guide.

2) Stone Matrix Asphalt (SMA) Mixes

Stone matrix asphalt (SMA) is a gap-graded HMA that is designed to maximize *deformation* (*rutting*) resistance and *durability* by using a structural basis of stone-on-stone contact (see Figure 5-6). Because the aggregates are all in contact, rut resistance relies on aggregate properties rather than asphalt binder properties.

Since aggregates do not deform as much as asphalt binder underload, this stone-on-stone contact greatly reduces rutting. SMA is generally more expensive than a typical dense-

Figure 5-6　Pavement adopting SMA

graded HMA (about 20%~25%) because it requires more durable aggregates, higher asphalt content and, typically, a modified asphalt binder and fibers. In the right situations it should be cost-effective because of its increased rut resistance and improved durability. SMA, originally developed in Europe to resist rutting and studded tire wear, has been used in the U.S. since about 1990.

Purpose: Improved rut resistance and durability. Therefore, SMA is almost exclusively used for surface courses on high volume interstates and U.S. roads.

Materials: Gap-graded aggregate (usually from *coarse aggregate*, *manufactured* sands and *mineral filler*, all combined into a final *gradation*, asphalt binder (typically with a modifier).

Mix Design: Superpave or Marshall procedures with modifications.

Other Info: Because SMA mixes have a high asphalt binder content (on the order of 6 percent), as the mix sits in the HMA storage silos, transport trucks, and after it is placed, the asphalt binder has a tendency to drain off the aggregate and down to the bottom — a phenomenon known as *"mix draindown"*. Mix draindown is usually combated by adding *cellulose* or *mineral fiber*s to keep the asphalt binder in place. *Cellulose fiber*s are typically shredded newspapers and magazines, while mineral fibers are spun from molten rock. A laboratory test is run during mix design to ensure the mix is not subject to excessive draindown. In mix design a test for *void*s in the coarse aggregate is used to ensure there is stone-on-stone contact. Other reported SMA benefits include wet weather friction (due to a coarser *surface texture*, lower tire noise (due to a coarser surface texture) and less severe *reflective cracking*. Mineral fillers and *additive*s are usually added to minimize asphalt binder drain-down during construction, increase the amount of asphalt binder used in the mix and to improve mix durability.

3) Open-Graded Mixes

An open-graded HMA mix is designed to be water permeable (dense-graded and SMA mixes usually are not permeable). Open-graded mixes use only *crushed stone* (or *gravel*) and a small percentage of manufactured sands. There are three types of open-graded mixes typically used in the U.S. :

① **Open-graded friction course** (**OGFC**). Typically 15 percent air voids, no minimum air voids specified, lower aggregate standards than PEM. See Figure 5-7.

② Porous European mix (PEM). Typically $18 \sim 22$ percent air voids, specified minimum air voids, higher aggregate standards than OGFC and requires the use of asphalt binder modifiers.

③ Asphalt treated permeable bases (ATPB). Less stringent specifications than OGFC or PEM since it is used only under dense-graded HMA, SMA or PCC for drainage.

Figure 5-7　Pavement adopting OGFC

Purpose: OGFC and PEM-Used as for surface courses only. They reduce tire splash/ spray in wet weather and typically result in smoother surfaces than dense-graded HMA. Their high air voids trap road noise and thus reduce tire-road noise by up to 50-percent (10 dB). ATPB-Used as a drainage layer below dense-graded HMA, SMA or PCC.

Materials: Aggregate (crushed stone or gravel and manufactured sands), asphalt binder (with modifiers).

Mix Design: Less structured than for dense-graded or SMA mixes. Open-graded mix design generally consists of ① material selection, ② gradation, ③ *compaction* and void determination ④ asphalt binder drain-down evaluation.

Other Info: Both OGFC and PEM are more expensive per ton than dense-graded HMA, but the unit weight of the mix when in-place is lower, which partially offsets the higher per-ton cost. The open gradation creates pores in the mix, which are essential to the mix's proper function. Therefore anything that tends to clog these pores, such as low-speed traffic, excessive dirt on the roadway or deicing sand, should be avoided.

3. Rigid Pavement

Rigid pavements are so named because the pavement structure deflects very little under loading due to the high *modulus of elasticity* of their surface course. A rigid pavement structure is typically composed of a PCC surface course built on top of either the subgrade or an underlying base course. Because of its relative rigidity, the pavement structure distributes loads over a wide area with only one, or at most two, structural layers.

The typical rigid pavement structure consists of:

Surface course. This is the top layer, which consists of the PCC *slab*.

Base course. This is the layer directly below the PCC layer and generally consists of aggregate or stabilized subgrade.

Subbase course. This is the layer (or layers) under the base layer. A subbase is not always needed and therefore may often be omitted.

Almost all rigid pavement is made with PCC, thus this Guide only discusses PCC pavement. Rigid pavements are differentiated into three major categories by their means of crack control:

Jointed *plain concrete* pavement (JPCP). This is the most common type of rigid pavement. JPCP controls cracks by dividing the pavement up into individual slabs separated by *contraction joint*s. Slabs are typically one lane wide and between 3.7 m and 6.1 m long. JPCP does not use any reinforcing steel but does use *dowel bar*s and *tie bar*s.

Jointed *reinforced concrete* pavement (JRCP). As with JPCP, JRCP controls cracks by dividing the pavement up into individual slabs separated by contraction joints. However, these slabs are much longer (as long as 15 m) than JPCP slabs, so JRCP uses reinforcing steel within each slab to control within-slab cracking. This pavement type is no longer constructed in the U.S. due to some *long-term performance* problems.

Continuously reinforced concrete pavement (CRCP). This type of rigid pavement uses reinforcing steel rather than contraction joints for crack control. Cracks typically appear ever 1.1~2.4 m, and are held tightly together by the underlying reinforcing steel.

1) Jointed Plain Concrete Pavement (JPCP)

Jointed plain concrete pavement (JPCP) uses contraction joints to control cracking and does not use any reinforcing steel. *Transverse joint* spacing is selected so that temperature and moisture stresses do not produce intermediate cracking between joints. This typically results in a spacing no longer than about 6.1 m. Dowel bars are typically used at transverse joints to assist in load transfer. Tie bars are typically used at *longitudinal joint*s.

Crack Control: Contraction joints, both transverse and longitudinal.

Joint Spacing: Typically between 3.7 m and 6.1 m. Due to the nature of concrete, slabs longer than about 6.1 m will usually crack in the middle. Depending upon environment and materials slabs shorter than this may also crack in the middle.

Reinforcing Steel: None.

Load Transfer: Aggregate *interlock* and dowel bars. For low-volume roads aggregate interlock is often adequate. However, high-volume roads generally require dowel bars in each transverse joint to prevent excessive *faulting*.

Other Info: A majority of U.S. State DOTs build JPCP because of its simplicity and proven performance.

2) Jointed Reinforced Concrete Pavement (JRCP)

Jointed reinforced concrete pavement (JRCP) uses contraction joints and reinforcing steel to control cracking. Transverse joint spacing is longer than that for JPCP and typically ranges from about 7.6 m to 15.2 m. Temperature and moisture stresses are expected to cause cracking between joints, hence reinforcing steel or a steel mesh is used to hold these cracks tightly together. Dowel bars are typically used at transverse joints to assist in load transfer while the reinforcing steel/wire mesh assists in load transfer across cracks.

Crack Control: Contraction joints as well as reinforcing steel.

Joint Spacing: Longer than JPCP and up to a maximum of about 15 m. Due to the nature of concrete, the longer slabs associated with JRCP will crack.

Reinforcing Steel: A minimal amount is included mid-slab to hold cracks tightly together. This can be in the form of deformed reinforcing bars or a thick wire mesh.

Load Transfer: Dowel bars and reinforcing steel. Dowel bars assist in load transfer across transverse joints while reinforcing steel assists in load transfer across mid-panel cracks.

Other Info: During construction of the interstate system, most agencies in the Eastern and Midwestern U.S. built JRCP. Today only a handful of agencies employ this design. In general, JRCP has fallen out of favor because of inferior performance when compared to JPCP and CRCP.

3) Continuously Reinforced Concrete Pavement (CRCP)

Continuously reinforced concrete pavement (CRCP, see Figure 5-8) does not require any contraction joints. Transverse cracks are allowed to form but are held tightly together with continuous reinforcing steel. Research has shown that the maximum allowable design crack width is about 0.5 mm to protect against *spalling* and water penetration.

Cracks typically form at intervals of $1.1\sim2.4$ m. Reinforcing steel usually constitutes about $0.6\sim0.7$ percent of the cross-sectional pavement area and is located near mid-depth in the slab. Typically, No. 5 and No. 6 deformed reinforcing bars are used.

Figure 5-8 CRCP construction

During the 1970's and early 1980's, CRCP design thickness was typically about 80 percent of the thickness of JPCP. However, a substantial number of these thinner pavements developed *distress* sooner than anticipated and as a consequence, the current trend is to make CRCP the same thickness as JPCP. The reinforcing steel is assumed to only handle nonload-related stresses and any structural contribution to resisting loads is ignored.

Crack Control: Reinforcing steel.

Joint Spacing: Not applicable. No transverse contraction joints are used.

Reinforcing Steel: Typically about $0.6\sim0.7$ percent by cross-sectional area.

Load Transfer: Reinforcing steel, typically No. 5 or 6 bars, grade 60.

Other Info: CRCP generally costs more than JPCP or JRCP initially due to increased quantities of steel. Further, it is generally less forgiving of construction errors and provides fewer and more difficult *rehabilitation* options. However, CRCP may demonstrate superior long-term performance and cost-effectiveness. Some agencies choose to use CRCP designs in their heavy urban traffic corridors.

Glossary

Complete the glossary using words in the text.

1. _____	路面	2. _____	柔性
3. _____	刚性	4. _____	沥青的
5. _____	沥青(n.);以沥青铺(vt.)		
6. _____	弯曲(n./v.)	7. _____	弯曲;弯沉(v.)
8. _____	交通荷载	9. _____	层,层次
10. _____	波特兰水泥	11. _____	混凝土;凝结物
12. _____	坚硬的;刚性的	13. _____	刚度

14. _____ 分布;扩散　　15. _____ 路基;地基

16. _____ 水泥混凝土路面板

17. _____ 路面面层　　　18. _____ 传递

19. _____ 养护,维修　　20. _____ 集料,粒料

21. _____ 稳定　　　　　22. _____ 集料最大粒径

23. _____ 集料级配　　　24. _____ 沥青胶结料

25. _____ 密级配　　　　26. _____ 轮胎

27. _____ 磨耗　　　　　28. _____ 开级配

29. _____ 处理;处治

30. _____ 有渗透性的,透水的

31. _____ 磨耗层　　　　32. _____ 排水层

33. _____ 孔隙率

34. _____ 无渗透性的;不透水的

35. _____ 细粒式混合料　36. _____ 粗粒式混合料

37. _____ 修补;补丁　　　38. _____ 变形

39. _____ 车辙　　　　　40. _____ 耐久性

41. _____ 粗集料　　　　42. _____ 机制砂

43. _____ 矿物填料,矿粉　44. _____ 级配

45. _____ 混合料析漏　　46. _____ 木质素;纤维素

47. _____ 矿棉纤维　　　48. _____ 木质素纤维

49. _____ 空隙　　　　　50. _____ 路表纹理

51. _____ 反射裂缝,反射开裂

52. _____ 外加剂

53. _____ 轧制石料,轧制碎石

54. _____ 碎石;砾石;砂砾

55. _____ 压实　　　　　56. _____ 弹性模量

57. _____ 厚板,平板

58. _____ 素混凝土,无配筋混凝土

59. _____ 缩缝　　　　　60. _____ 传力杆

61. _____ 系杆

62. _____ 配筋混凝土,钢筋混凝土

63. _____ 长期性能　　　64. _____ 横向接缝

65. _____ 纵向接缝　　　66. _____ 嵌锁

67. _____ 错台　　　　　68. _____ 散裂

69. _____ 病害　　　　　70. _____ 修复;恢复

Exercises

1. Write down the words with the following prefix or suffix.

de-　　_____

con-　_____

max-　_____

cross-　_____

-icity　_____

2. Answer the following questions.

1) What are flexible pavements?

2) What are rigid pavements?

3) What does pavement structure typically consist of?

4) What types do flexible pavements include?

5) What are main differences between JRCP and CRCP?

3. Translate the following into Chinese.

1) Because SMA mixes have a high asphalt binder content (on the order of 6 percent), as the mix sits in the HMA storage silos, transport trucks, and after it is placed, the asphalt binder has a tendency to drain off the aggregate and down to the bottom — a phenomenon known as "mix draindown". Mix draindown is usually combated by adding cellulose or mineral fibers to keep the asphalt binder in place. Cellulose fibers are typically shredded newspapers and magazines, while mineral fibers are spun from molten rock.

2) CRCP generally costs more than JPCP or JRCP initially due to increased quantities of steel. Further, it is generally less forgiving of construction errors and provides fewer and more difficult rehabilitation options. However, CRCP may demonstrate superior long-term performance and cost-effectiveness. Some agencies choose to use CRCP designs in their heavy urban traffic corridors.

Text B　Perpetual Asphalt Pavements

1. Background

The concept of Perpetual Pavements was introduced in 2000 by the Asphalt Pavement Alliance (APA). They defined a Perpetual Pavement as "an asphalt pavement designed and built to last longer than 50 years without requiring major structural rehabilitation or reconstruction, and needing only periodic surface renewal in response to distresses confined to the top of the pavement" (APA, 2002). At that time, it was recognized that many well-built, thick asphalt pavements that were categorized as either full-depth or deep-strength pavements had been in service for decades with only minor periodic surface rehabilitation to remove defects and improve ride quality. The advantages of such pavements include:

1) Low life-cycle cost by avoiding deep pavement repairs or reconstruction;

2) Low user-delay costs since minor surface rehabilitation of asphalt pavements only requires short work windows that can avoid peak traffic hours;

3) Low environmental impact by reducing the amount of material resources over the

pavement's life and recycling any materials removed from the pavement surface.

A somewhat unified approach to designing Perpetual Pavements was adopted by a number of experts (Thompson and Carpenter, 2004; Timm and Newcomb, 2006) based on mechanistic-empirical concepts originally proposed by Monismith (1992) in the design of the I-710 freeway in California. The premise to this approach was that pavement distresses with deep structural origins could be avoided if pavement responses such as stresses, strains, and deflections could be kept below thresholds where the distresses begin to occur. Thus, an asphalt pavement could be designed for an indefinite structural life by designing for the heaviest vehicles without being overly conservative.

This contrasts to empirical methods that predated the Perpetual Pavement design approach. In those design procedures, greater volumes of heavy vehicles resulted in greater pavement thickness. This was due largely to the way these empirical methods were developed. For instance, the 1993 American Association of State Highway and Transportation Officials (AASHTO) Guide for the Design of Pavement Structures was based on the results of a road test conducted from 1958 to 1961. In this study, pavements were subjected to 1 million axle load applications, and failures were monitored over time. The heaviest single axle load used at the AASHO Road Test (30,000 lb) applied about 8 million equivalent single axle loads (ESAL) (18,000 lb equivalents) to the thickest asphalt section. Since that time, pavement structures have been designed for heavy traffic volumes that exceed the 8 million ESAL level by 25 times, thus forcing pavement designers to extrapolate the road test results far beyond the conditions for which they were developed. The result of this extrapolation was ever-increasing thickness with traffic volume, instead of recognizing the pavement thickness at which the heaviest loads could be sustained without additional structure. Thus, the idea of Perpetual Pavements came into existence as much to prevent over-design as to provide a long-life structure.

2. Overview of Design

Pavement engineers have been producing long-lasting asphalt pavements since the 1960s. Research has shown that well-constructed and well-designed flexible pavements can perform for extended periods of time (Mahoney, 2001; Harvey et al., 2004). Many of these pavements in the past forty years were the products of full-depth or deep-strength asphalt pavement designs, and both have design philosophies that have been shown to provide adequate strength over extended life cycles (APA, 2002). It is significant that these pavements have endured an unprecedented amount of traffic growth. For instance, from 1970 to 1998, the average daily ton-miles of freight increased by 580 percent, and the average freight loading continues to increase 2.7 percent per year (D'Angelo et al., 2004). As the demand on existing pavements in the U.S. increases with potentially minimal funding for expansion and rehabilitation, efficient design of new and rehabilitated sections through Perpetual Pavement design will become increasingly important. Congestion on the existing system is at a point that requires pavements that can be maintained with minimal disruption of traffic.

Full-depth pavements are constructed by placing asphalt layers on modified or unmodified soil or subgrade material. Deep-strength pavements consist of asphalt layers on top of a thin granular base. Both of these design scenarios allow pavement engineers to employ a thinner total pavement section than if a thick granular base were used. By reducing the potential for fatigue cracking and confining cracking to the upper removable/replaceable layers, many of these pavements have far exceeded their design life of 20 years with minimal rehabilitation; therefore, they are considered to be superior pavements (APA, 2002).

Pavements which are either under-designed or poorly constructed exhibit structural distresses, such as fatigue cracking and rutting (Mahoney, 2001), before their design life is achieved. The successes seen in the full-depth and deep-strength pavements are the results of designing and constructing pavements that resist these detriments to the pavement's structure. In recent years, pavement engineers have begun to adopt a methodology of designing pavements to resist bottom-up fatigue cracking and deep structural rutting, the two most devastating pavement distresses, and through this change in thinking the idea of Perpetual Pavements or long-lasting pavements has evolved.

The approach to the design of long-life or Perpetual Pavements requires a different strategy than that which has normally been applied to pavement design in the past. Empirical pavement design must rely on relationships between observations of pavement performance, a scale that represents traffic, some gross indicator of material quality such as a structural coefficient, and the thickness of the layers. For a given level of material quality, the thickness of the pavement increases with increasing traffic. However, there comes a point beyond which the thickness of the pavement is more than adequate for the heaviest loads expected and any additional pavement results in an overly-conservative cross section and an unnecessary added cost. In addition to being extravagant from a cost standpoint, such an overuse of resources does not fit within an environmental sustainability framework. As a case in point, Huber et al. (2009) found that the 1993 AASHTO pavement design guide (AASHTO, 1993) typically over-designed pavements in Indiana by 1.5 to 4.5 inches which amounts to approximately 600 to 1,800 tons of material per lane-mile beyond what is needed.

A better approach to the design of Perpetual Pavements is the mechanistic-empirical method. This approach uses the elements of a rational engineering analysis of the reaction of the pavement in terms of stresses, strains, and displacements in the context of the pavement's expected life. A flowchart showing a typical mechanistic-empirical design approach is shown in Figure 5 - 9. This is an iterative approach in which the pavement response in terms of stresses, strains, or deflections is used to estimate the allowable number of loads to failure (Nf) for a given condition of loading and material properties. The actual number of anticipated traffic loads (n) is divided by Nf to define the degree of damage (D). The point at which the damage equals one is considered failure. This was originally defined by Miner (1959) as a way of describing metal fatigue. In many cases, engineers consider pavement failure to occur at either 20% fatigue cracking in the wheelpath or 0.5 inches of rutting (Von Quintus, 2001a). Currently, there are existing M-E pavement design methodologies (AI, 1982; Monismith, 1992; KTC, 2007; Timm et al., 1998), but as the new M-E Pavement Design Guide (MEPDG) is being completed and implemented, more attention is being spent on proper material and pavement response characterization (Timm and Priest, 2006). In Perpetual Pavement design, there are limiting strains below which damage does not occur, and thus damage is not accumulated. This concept is illustrated in Figure 5-10.

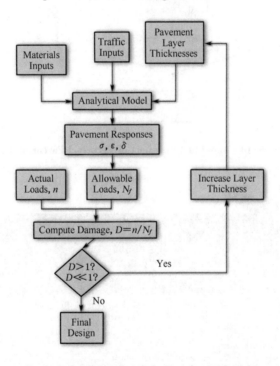

Figure 5-9   Simplified flowchart for M-E design

Most pavement engineers in the U.S. approach the idea of Perpetual Pavements with a 50-year structural design life in mind. However, while the structural integrity of the pavement should be intact during the entirety of the pavement's life, periodic resurfacing generally needs to occur within 20 years to improve friction, reduce noise, and mitigate surface cracking (Newcomb et al., 2001). The basic concept of a Perpetual Pavement is illustrated in Figure 5-11. While the importance of proper design for a long-lasting pavement must be recognized, one must also understand that design life is a function of the design requirements, material characteristics, construction practices, layer thicknesses, maintenance activities, and the failure criterion.

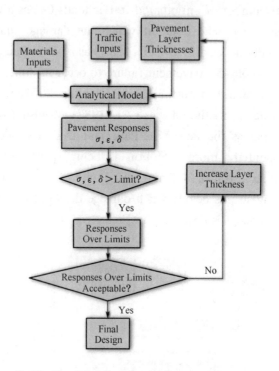

Figure 5-10 Simplified flowchart of Perpetual Pavement design

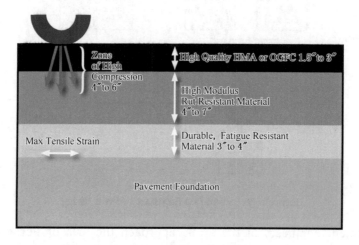

Figure 5-11 Perpetual Pavement design concept

Ferne (2006) expanded upon this idea by saying a "long-life pavement is a well-designed and constructed pavement that could last indefinitely without deterioration in the structural elements provided it is not overlooked and the appropriate maintenance is carried out." Pavement performance is more than a function of design. Traffic, climate, subgrade and pavement parameters (such as modulus), pavement materials, construction, and maintenance levels all contribute to how a pavement will perform over the course of its life (Von Quintus, 2001a; Walubita et al., 2008).

Assuming that pavements will be constructed adequately, engineers approach designing Perpetual Pavements using the following philosophy (Walubita et al., 2008; Merrill et al., 2006):

1) Perpetual Pavements must have enough structural integrity and thickness to preclude distresses such as fatigue cracking, permanent deformation, and structural rutting.

2) Perpetual Pavements must be durable enough to resist damage from traffic (such as abrasion) and the environment.

While one might think pavements designed to last longer would incur more or have higher initial costs than pavements with shorter life-cycles, it has been shown that Perpetual Pavements have the following benefits (Timm and Newcomb, 2006):

1) They provide a more efficient design, eliminating costly overly conservative pavement sections.

2) They eliminate reconstruction costs by not exceeding a pavement's structural capacity.

3) They lower rehabilitation-induced user delay costs.

4) They reduce use of non-renewable resources like aggregates and asphalt.

5) They diminish energy costs while the pavement is in service.

6) They reduce the life-cycle costs of the pavement network.

In order to provide the above advantages, it is necessary to know what thickness of pavement section will support the heaviest anticipated traffic loads without grossly over-designing the pavement. Research has shown that this can be identified mechanistically by identifying the stresses, strains, or displacements in a structure which are low enough to avoid the initiation of cracking or rutting deep in the pavement structure. These thresholds are often referred to as limiting pavement responses.

Text C   Cool Pavements

1. Introduction

Pavements are critical to transportation in all of its aspects — walking, riding in passenger vehicles, carrying goods in commercial vehicles, providing mobile services, and parking. They account for a significant percentage of the land surface in an urban area. Analyses in cities such as Chicago, Houston, Sacramento, and Salt Lake City have shown that pavements for both travel and parking can account for 29 to 39 percent of the land surface in an urban area. A large portion of this is due to parking; in the Houston, Texas, metropolitan area, the parking facilities account for approximately 60 percent of the transportation land use.

By altering landcover, pavements have important localized environmental effects in urban areas. This report focuses on the contributions pavements make to the urban heat island. As with roofing materials, paving materials can reach 150° F in daytime, radiating away this excess heat during both day and night into the air in the urban canopy layer (as well heating stormwater that reaches the pavement surface). Due to the large area covered by pavements in urban areas, they are important elements to consider in heat island mitigation.

This contribution can be reduced by using "cool pavements". Cool pavements can be achieved with existing paving technologies and do not require new materials. Their "cool" nature comes about by the attention given to the choice of materials and engineering design. The use of cool pavements is meant to reduce pavement temperature by increasing pavement reflectivity or controlling temperature by other means, with the selected technique (s) applied as appropriate throughout the urban area. Specific pavement technologies with cool attributes will not be appropriate for all uses; some may be better suited to light traffic areas, for instance; others to areas where noise management is considered crucial. In addition, certain paving technologies may not always be appropriate or feasible in a particular region of the country — whether technically, economically, organizationally, or institutionally — and local pavement engineers and owner agencies may not be sufficiently familiar with cool pavements to apply them confidently.

This report gives additional information on cool pavement technology and options for implementation. It describes the types of pavements now in use throughout the United States, the candidates for cool pavements within this context, and some of the elements that go into decisions on pavement selection at the state and local levels. Pavement type selection, design, and construction is influenced by a number of technical, economic, organizational, and institutional factors. Understanding how these factors are perceived by public- and private-sector facility owners can help those wishing to develop a more effective approach to implementing cool pavement policies, long-range plans, programs, and projects.

2. Benefits of Cool Pavements

As part of a heat island reduction strategy, cool pavements contribute to the general benefits of heat island mitigation, including increased comfort, decreased energy use, and likely improved air quality. Cool pavements also can be one component of a larger sustainable pavements program, or a "green" transportation infrastructure.

Cool pavements can contribute to local as well as regional comfort improvements. For instance, they help make large paved areas such as parking lots more comfortable for users. Shopping centers may feel this enhances the shopping experience.

Quantifying the heat island mitigation benefits of cool pavements is complicated by several factors in a real urban setting. The reflectivity of pavement surfaces changes over time; buildings, trees, and vehicles cast shadows; some of the reflected light could be reabsorbed by surrounding structures, negating the effect of the cooler pavement; and the degree of cooling afforded by permeable pavements is not well quantified. There may be offsetting effects or tradeoffs in the several mechanisms at work, all complicating the estimate of the benefits that cool pavements can yield.

The benefits of cool pavements are not limited to heat island reduction. There also are a number of ancillary benefits that can be gained from the use of cool pavement technologies, which can make their use worthwhile in their own right or as additional factors contributing to sustainable or green pavement initiatives. These additional benefits of cool pavements include:

1) Water quality. Cool pavements can create improvements in water quality in two ways:

① Permeable roadway pavements and especially parking facilities of all types (asphalt, concrete, and reinforced grass and gravel paving systems) can address water quality problems by reducing the percentage of land covered by impervious surfaces. When combined with water treatment wetlands, these pavements help to act as filters, improving water quality and providing greater groundwater protection. These improvements can translate into savings for urban areas by reducing the need to construct separate sewers or expanded water treatment facilities.

② Both permeable and non-permeable cool pavements can help water quality through reduced heating of runoff. Laboratory tests with permeable pavers have shown reductions in runoff temperatures of two to four degrees Celsius in comparison to conventional asphalt paving.

2) Noise. The open pores of permeable pavements have been shown to significantly reduce tire noise.

3) Safety. Permeable roadway pavements can enhance safety by reducing water spray from moving vehicles and increasing traction through better water drainage.

4) Nighttime illumination. More reflective pavements can enhance visibility at night, potentially reducing lighting requirements and saving both money and energy. European road designers often take pavement color into account when planning lighting needs. Better illumination from lighter pavements is sometimes considered valuable at private establishments as well, for security or customer appeal. Some sources cite nighttime illumination enhancements of 10 to 30 percent with more reflective pavements.

In designing a cool pavement policy, it is important to consider the differing characteristics of pavements in selecting the appropriate one for each situation. For instance, high-albedo pavements, which reflect away more solar radiation, will absorb less heat than darker pavements and thus stay cooler. However, they may not be appropriate in places where people will be uncomfortably exposed to the reflected radiation for long periods, as in a children's playground. In contrast, other pavements may take longer to heat during the day, but release excess heat at night — effectively transferring some of the day's heat to the evening. This may be appropriate in situations where the main concern is daytime heat or air pollution (such as ozone formation). In some sense, one can think of this as deciding how best to "manage" the urban climate. The information provided in this report can provide some of the background necessary to begin considering these issues.

3. Types of Cool Pavements

Several candidate technologies for cool pavement have been investigated in past research, including options for new pavement construction, reconstruction, and maintenance and rehabilitation activities. Brief descriptions of these technologies are given below, including additional suggestions discussed in interviews with paving industry representatives. It is clear from these interviews that the industry associations will be responsive in supporting development and application of cool pavement technologies if their clients demand these options. This review is not intended to be exhaustive, but rather to outline the basic technologies involved.

Both the asphalt and the concrete industries stress the importance of competent design, construction, and inspection of paving projects, including those that are using technologies consistent for cool pavements. This guidance is a matter of good practice generally and of being responsive to client and customer needs, and is not limited to cool pavements alone. All of these technologies depend upon the proper proportioning of cement, aggregate, and other constituents, and good control over the characteristics of these materials (e. g., in the size distribution of the aggregates). Lack of care in specifying and enforcing materials properties can lead to poor performance, which — if associated with a local cool pavement initiative — could contribute to a reluctance to consider future cool pavements and a loss of credibility for the initiative.

1) Conventional Portland Cement Concrete Pavement

Conventional PCC pavement has been proposed as a cool pavement because of its light color and reflectivity. It is used in new construction and reconstruction. The degree of surface reflectivity is affected by both the color of the cement and the type and color of aggregate (particularly as the cement surface becomes worn and the aggregate is exposed).

2) Concrete Additives

Additives are routinely used in concrete to enhance its placement during construction or its performance during its service life.

One example is the use of slag cement in combination with Portland cement. Slag is a byproduct of processing iron or copper ore in a blast furnace. It is obtained from the surface of the molten ore, granulated, and ground to produce a cement that replaces a portion of the portland cement in a concrete mixture. It has several benefits to concrete workability and performance (e. g., improved strength, resistance to aggressive chemicals, better ability to place concrete in hot weather). Among these benefits is a lighter color, which can enhance reflectivity of the finished pavement. In addition, by reducing the amount of portland cement used, slag cement reduces greenhouse gas emissions and energy use in the production of concrete.

Fly ash also can be used as a replacement for portland cement in concrete mixtures. Fly ash is a powdery byproduct of burning coal. While fly ash provides several benefits to concrete (e. g. , improved workability, strength, durability, resistance to chemical attack), there is no mention in the reviewed literature of special characteristics that would favor cool pavements. In particular, fly ash color varies considerably, ranging from light tan to black depending on the source; any reflectivity benefits also would be source specific. It does have other environmental benefits; as with slag cement, the use of fly ash reduces greenhouse gas emissions and energy use. A potential drawback is that fly ash mixes initially gain strength more slowly than typical mixes; however, they generally reach higher final strengths. Federal agencies are required to allow the use of fly ash in construction projects; in addition, the California Department of Transportation (Caltrans) mandates a minimum of 25 percent fly ash.

3) Whitetopping and Ultra-Thin Whitetopping

Whitetopping consists of a concrete pavement applied over an existing asphalt pavement as a form of maintenance or resurfacing. Conventional whitetopping is more than four inches thick. UTW is a newer form of this process in which a two- to four-inch thickness of concrete, usually high strength and fiber reinforced, is placed over an asphalt surface that has been milled. The UTW is different from conventional whitetopping in that it relies on bonding with the asphalt surface for strength, and joint spacing is much shorter (typically 2 to 6 feet for UTW, in comparison to 5 to 25 feet for conventional whitetopping). As a potential cool paving technology, UTW provides the color and reflectance of concrete over an existing asphalt surface. It has been used on a number of projects across the country for resurfacing road segments, intersections, and parking lots.

4) Roller-Compacted Concrete Pavement

Roller-compacted concrete is a specially mixed and placed form of concrete. It employs a very stiff mix that is placed with techniques and equipment much like that used for asphalt pavement. While it results in a strong pavement, its surface is not finished or textured, as is conventional concrete pavement. It is used for heavy hauling roads where speed is not a factor, bulk commodity storage areas, intermodal container facilities, automotive manufacturing plant parking areas, military facilities, and warehouse floors. Parts of its surface may become abraded over time. However, it is economical, with initial cost lower than that of conventional concrete, and competitive with asphalt concrete.

5) Light Aggregate in Asphalt Concrete Pavement

The reflectance of ACP can be increased by using light-colored aggregate such as limestone. This type of aggregate is available naturally in parts of the country (e. g. , Houston area, Florida) and is used in conventional pavement construction and reconstruction. In these locations, the incremental cost of this technology is nil. In other locations, however, the cost to transport such aggregate to the job site is prohibitively expensive.

6) Chip Seals with Light Aggregate

Chip seals are a frequently used preventive maintenance technique on asphalt pavements. Use of light-colored aggregate in these seals could increase the albedo of asphalt-paved surfaces. The cost of such treatment depends on the local availability of suitable aggregate, as noted earlier for ACP. Chip seals are traditionally associated with roads carrying light traffic volumes because of the tendency of the chips (stones) to loosen and be propelled by the action of moving vehicles toward other vehicles, potentially resulting in windshield damage. While Texas has had experience in applying chip seals to high-volume highways (including interstate highways), the trend now is to use another preventive maintenance treatment (microsurfacing) in lieu of chip seals on these high-volume roads.

7) Porous Pavement and Surfaces

Porous pavement and permeable surfaces have been investigated as mechanisms for stormwater discharge control and ground water management in urbanized areas. Both concrete and asphalt pavements can be built as porous surfaces on roads and parking lots.

A porous asphalt surface can improve skid resistance and reduce traffic noise, rutting, and splash due to ponded water on the surface. Noise reduction benefits may decline over time, and there may be reduced strength and durability in comparison to conventional surfaces. The extent to which they can be used on high-speed, high-volume roadways needs more investigation. Permeable friction courses laid on top of an impermeable base have been used successfully in Texas on interstates (e.g., I-35 in San Antonio) to improve traction and visibility in wet weather, as well as reduce noise. Pervious concrete may be used where reductions in stormwater runoff are desirable; it is appropriate for low-speed traffic (less than 35 miles per hour).

Opinions differ on whether porous pavements present a problem in winter. As long as the pavement remains free draining, there should be no problem due to freezing and thawing. However, if water becomes trapped in the pavement voids, expansion during freezing may degrade the pavement layer. Because road sand may clog the pavement pores, other methods of snow and ice control (e.g., use of chemicals) may be needed with porous pavements.

Other types of permeable surfaces can be built using plastic grids or masonry blocks with grass or gravel filled in. This type of "unbound" surface can be used in several applications: driveways, shoulders, parking lots, bicycle trails, pedestrian and golf paths, equestrian trails, and slope stabilization.

8) Color Pigments and Seals

Pigments and seals are available to change the color of an asphalt surface to make it lighter. However, these products are expensive, and tend to be used only in special situations where color is a dominant paving criterion. (Pigments also are available for concrete pavements; however, because concrete pavements are already light-colored, pigments are unlikely to improve their "coolness".)

9) Rubberized Asphalt

A composite pavement design with a rubberized asphalt layer over a PCC base is now being used in the Phoenix, Arizona, area. The primary purpose of these pavements is to reduce tire noise, but current research indicates that they are cooler at night than adjacent PCC. To determine their potential benefits in reducing the heat island, these pavements will be instrumented through an ongoing program by ASU. However, preliminary indications from satellite photos of different types of pavements in the Phoenix metropolitan area suggest promising results from a cool pavement perspective. Because the program has just gotten underway, detailed data are not yet available.

Integrated Skills: Writing a Report

Please investigate road distresses in your city or on your university campus, and write an investigating report, and then communicate it with your classmates. The contents of your investigating report should include types, patterns, causes, prevention and treatment of road distresses. When preparing your investigation, you may refer to the typical road distresses shown in Figure 5-12 to Figure 5-28.

Figure 5-12   Subgrade depression induced pavement cracking

Figure 5-13   Rutting

Figure 5-14　Upheaval

Figure 5-15　Shoving

Figure 5-16　Bleeding

Figure 5-17　Longitudinal and transverse cracking

Figure 5-18　Potholes

Figure 5-19　Fatigue cracking

Figure 5-20　Coner break

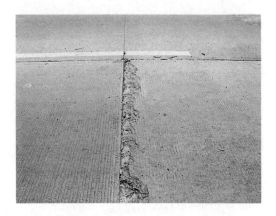

Figure 5-21　Joint destruction

Figure 5-22　Pavement slab break

Figure 5-23　Pumping

Figure 5-24　Patched failure

Figure 5-25　Faulting

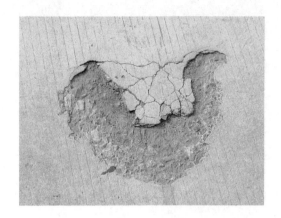

Figure 5-26   Spalling

Figure 5-27   Popouts

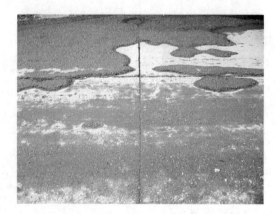

Figure 5-28   Failure of pavement overlay

Unit 6 History, Types and Structures of Bridges

Lead-in

1. What are bridges and why are they necessary in transportation?
2. How are bridges classified according to your knowledge?
3. Can you **match** the following bridges with their types?

_____ 1) Luoyang Bridge，China   a. the arch bridge

_____ 2) Zhaozhou Bridge，China   b. the beam bridge

_____ 3) Akashi-Kaikyō Bridge，Japan   c. the suspension bridge

_____ 4) Ikitsuki Bridge，Japan   d. the cable-stayed bridge

_____ 5. The Second Nanjing Yangtze River Bridge，China

                                            e. the truss bridge

Vocabulary Warm-up：Matching

_____ 1. beam bridge   a. 拱桥

_____ 2. cantilever bridge   b. 上承式拱桥

_____ 3. arch bridge   c. 悬臂桥

_____ 4. suspension bridge   d. 梁桥

_____ 5. truss bridge   e. 叠涩拱桥

_____ 6. corbel arch bridge   f. 斜拉桥

_____ 7. deck arch bridge   g. 桁架桥

_____ 8. through arch bridge   h. 系杆拱桥

_____ 9. tied arch bridge   i. 悬索桥；吊桥

_____ 10. cable-stayed bridge   j. 下承式拱桥

Text A Types of Bridges

A bridge is a structure providing passage over an obstacle such as a valley，road，railway，canal，river，without closing the way beneath. The required passage may be for road，railway，canal，pipeline，cycle track or pedestrians.

The branch of civil engineering which deals with the design，planning construction and maintenance of bridge is known as bridge engineering. Designs of bridges vary depending on the function of the bridge and the nature of the *terrain* where the bridge is constructed.

There are six main types of bridges：beam bridges，cantilever bridges，arch bridges， suspension bridges, cable-stayed bridges and truss bridges.

1. Beam Bridges

Beam bridges are *horizontal beams* supported at each end by *piers*. The earliest beam bridges were simple *logs* that sat across streams and similar simple structures. In modern times, beam bridges are large box steel *girder* bridges. Weight on top of the beam pushes straight down on the piers at either end of the bridge. They are made up mostly of wood or metal.

The beam bridge, also known as a girder bridge, is a firm structure that is the simplest of all the bridge shapes. Both strong and economical, it is a solid structure comprised of a horizontal beam, being supported at each end by piers that endure the weight of the bridge and the vehicular traffic. *Compressive* and tensile *forces* act on a beam bridge, due to which a strong beam is essential to resist *bending* and *twisting* because of the heavy loads on the bridge. When traffic moves on a beam bridge, the load applied on the beam is transferred to the piers. The top portion of the bridge, being under compression, is shortened; while the bottom portion, being under tension, is consequently stretched and lengthened. Trusses made of steel are used to support a beam, enabling dissipation of the compressive and tensile forces. In spite of the reinforcement by trusses, length is a limitation of a beam bridge due to the heavy bridge and truss weight. The *span* of a beam bridge is controlled by the beam size since the additional material used in tall beams can assist in the dissipation of tension and compression.

Extensive research is being conducted by several private enterprises and the state agencies to improve the construction techniques and materials used for the beam bridges. The beam bridge design is oriented towards the achievement of light, strong, and long-lasting materials like reformulated concrete with high performance characteristics, fiber reinforced *composite materials*, *electro-chemical corrosion* protection systems, and more precise study of materials. Modern beam bridges use *prestressed concrete* beams that combine the high *tensile strength* of steel and the superior compression properties of concrete, thus creating a strong and durable beam bridge. *Box girders* are being used that are better designed to undertake twisting forces, and can make the spans longer, which is otherwise a limitation of beam bridges. The modern technique of the *finite element analysis* is used to obtain a better beam bridge design, with a meticulous analysis of the stress distribution, and the twisting and bending forces that may cause failure.

2. Cantilever Bridges

A cantilever bridge is a bridge built using cantilevers, structures that project horizontally into space, supported on only one end. For small *footbridges*, the cantilevers may be simple beams; however, large cantilever bridges designed to handle road or rail traffic use trusses built from structural steel, or box girders built from prestressed concrete. The steel truss cantilever bridge was a major engineering breakthrough when first put into practice, as it can span distances of over 460 m, and can be more easily constructed at difficult crossings by virtue of using little or no *falsework*.

A simple cantilever span is formed by two cantilever arms extending from opposite sides of the obstacle to be crossed, meeting at the center. In a common variant, the *suspended span*, the cantilever arms do not meet in the center; instead, they support a central truss bridge which rests on the ends of the cantilever arms. The suspended span may be built off-site and lifted into place, or constructed in place using special traveling supports.

A common way to construct steel truss and prestressed concrete cantilever spans is to counterbalance each cantilever arm with another cantilever arm projecting the opposite direction, forming a balanced cantilever; when they attach to a solid foundation, the counterbalancing arms are called *anchor* arms. Thus, in a bridge built on two foundation piers, there are four cantilever arms: two which span the obstacle, and two anchor arms which extend away from the obstacle. Because of the need for more strength at the balanced cantilever's supports, the bridge superstructure often takes the form of towers above the foundation piers. The Commodore Barry Bridge is an example of this type of cantilever bridge.

Steel truss cantilevers support loads by tension of the upper members and compression of the lower ones. Commonly, the structure distributes the tension via the anchor arms to the outermost supports, while the compression is carried to the foundations beneath the central towers. Many truss cantilever bridges use pinned joints and are therefore statically determinate with no members carrying mixed loads.

Prestressed concrete balanced cantilever bridges are often built using *segmental construction*. Some steel arch bridges are built using pure cantilever spans from each side, with neither falsework below nor temporary supporting towers and *cables* above. These are then joined with a pin, usually after forcing the union point apart, and when *jacks* are removed and the bridge *decking* is added the bridge becomes a truss arch bridge. Such unsupported construction is only possible where appropriate rock is available to support the tension in the upper chord of the span during construction, usually limiting this method to the spanning of narrow *canyons*.

3. Arch Bridges

An arch bridge is a bridge with *abutments* at each end shaped as a curved arch. Arch bridges work by transferring the weight of the bridge and its loads partially into a horizontal thrust restrained by the abutments at either side. A *viaduct* may be made from a series of arches, although other more economical structures are typically used today.

There are some variations of arch bridges:

1) Corbel Arch Bridges

The corbel arch bridge is a *masonry* or stone bridge where each successively higher course cantilevers slightly more than the previous course. The steps of the masonry may be trimmed to make the arch have a rounded shape. The corbel arch does not produce thrust, or outward pressure at the bottom of the arch, and is not considered a true arch. It is more stable than a true arch because it does not have this thrust. The disadvantage is that this type of arch is not suitable for large spans.

2) **Aqueducts** and Canal Viaducts

In some locations it is necessary to span a wide gap at a relatively high elevation, such as when a canal or water supply must span a valley. Rather than building extremely large arches, or very tall supporting **columns**, a series of arched structures are built one atop another, with wider structures at the base. Roman civil engineers developed the design and constructed highly refined structures using only simple materials, equipment, and mathematics. This type is still used in canal viaducts and roadways as it has a pleasing shape, particularly when spanning water, as the reflections of the arches form a visual impression of circles or ellipses.

3) Deck Arch Bridges

This type of bridge comprises an arch where the deck is completely above the arch. The area between the arch and the deck is known as the **spandrel**. If the spandrel is solid, usually the case in a masonry or stone arch bridge, it is called a closed-spandrel arch bridge. If the deck is supported by a number of vertical columns rising from the arch, it is known as an **open-spandrel arch bridge**. The Alexander Hamilton Bridge is an example of an open-spandrel arch bridge.

4) Through Arch Bridges

This type of bridge comprises an arch which supports the deck by means of **suspension cables** or **tie bars**. The Sydney Harbour Bridge is a through arch bridge which uses a truss type arch. These through arch bridges are in contrast to suspension bridges which use the catenary in tension to which the aforementioned cables or tie bars are attached and suspended.

5) Tied Arch Bridges

Also known as a **bowstring arch**, this type of arch bridge incorporates a tie between two opposite ends of the arch. The tie is capable of withstanding the horizontal **thrust forces** which would normally be exerted on the abutments of an arch bridge.

4. Suspension Bridges

A suspension bridge is a type of bridge in which the deck is hung below suspension cables on vertical **suspenders**. This type of bridge dates from the early 19th century, while bridges without vertical suspenders have a long history in many mountainous parts of the world.

This type of bridge has cables suspended between towers, plus vertical suspender cables that carry the weight of the deck below, upon which traffic crosses. This arrangement allows the deck to be level or to arc upward for additional **clearance**. The suspension bridge is often constructed without falsework.

The suspension cables must be anchored at each end of the bridge, since any load applied to the bridge is transformed into a tension in these *main cables*. The main cables continue beyond the *pillars* to deck-level supports, and further continue to connections with anchors in the ground. The roadway is supported by vertical suspender cables or *rods*, called *hangers*. In some circumstances the towers may sit on a *bluff* or canyon edge where the road may proceed directly to the *main span*, otherwise the bridge will usually have two smaller spans, running between either pair of pillars and the highway, which may be supported by suspender cables or may use a truss bridge to make this connection. In the latter case there will be very little arc in the outboard main cables.

The main forces in a suspension bridge of any type are tension in the cables and compression in the pillars. Since almost all the force on the pillars are vertically downwards and they are also stabilized by the main cables, the pillars can be made quite slender, as on the Severn Bridge, near Bristol, England. In a suspended deck bridge, cables suspended via towers hold up the road deck. The weight is transferred by the cables to the towers, which in turn transfer the weight to the ground.

Assuming a negligible weight as compared to the weight of the deck and vehicles being supported, the main cables of a suspension bridge will form a *parabola*. One can see the shape from the constant increase of the gradient of the cable with linear distance, this increase in gradient at each connection with the deck providing a net upward support force. Combined with the relatively simple constraints placed upon the actual deck, this makes the suspension bridge much simpler to design and analyze than a cable-stayed bridge, where the deck is in compression.

The suspension bridge has some advantages over other bridge types:

1) Longer main spans are achievable than any other types of bridges;

2) Less material may be required than other bridge types, even at spans they can achieve, leading to a reduced construction cost;

3) Except for installation of the initial temporary cables, little or no access from below is required during construction, for example allowing a *waterway* to remain open while the bridge is built above;

4) May be better able to withstand earthquake movements than heavier and more rigid bridges can.

However, the suspension bridge has some disadvantages too, compared with other bridge types:

1) Considerable *stiffness* or *aerodynamic* profiling may be required to prevent the bridge deck vibrating under high winds;

2) The relatively low deck stiffness compared to other (non-suspension) types of bridges makes it more difficult to carry heavy rail traffic where high concentrated *live loads* occur;

3) Some access below may be required during construction, to lift the initial cables or to lift deck units. This access can often be avoided in cable-stayed bridge construction.

5. Cable-stayed Bridges

A cable-stayed bridge is a bridge that consists of one or more columns, with cables supporting the bridge deck. There are two major classes of cable-stayed bridges: In a harp design, the cables are made nearly parallel by attaching cables to various points on the tower so that the height of attachment of each cable on the tower is similar to the distance from the tower along the roadway to its lower attachment. In a fan design, the cables all connect to or pass over the top of the towers.

Compared with other bridge types, the cable-stayed is optimal for spans longer than typically seen in cantilever bridges and shorter than those typically requiring a suspension bridge. This is the range in which cantilever spans would rapidly grow heavier if they were lengthened, and in which suspension cabling does not get more economical were the span to be shortened.

A *multiple-tower cable-stayed bridge* may appear similar to a suspension bridge but in fact they are very different in principle and in the method of construction. In the suspension bridge, a large cable is made up by "spinning" small diameter *wires* between two towers, and at each end to *anchorages* into the ground or to a massive structure. These cables form the primary load-bearing structure for the bridge deck. Before the deck is installed, the cables are under tension from only their own weight. Smaller cables or rods are then suspended from the main cable, and used to support the load of the bridge deck, which is lifted in sections and attached to the suspender cables. As this is done the tension in the cables increases, as it does with the live load of vehicles or persons crossing the bridge. The tension on the cables must be transferred to the earth by the anchorages, which are sometimes difficult to construct due to poor soil conditions.

In the cable-stayed bridge, the towers form the primary load-bearing structure. A cantilever approach is often used for support of the bridge deck near the towers, but areas further from them are supported by cables running directly to the towers. This has the disadvantage, compared with the suspension bridge, that the cables pull to the sides as opposed to directly up, requiring the bridge deck to be stronger to resist the resulting horizontal compression loads; but has the advantage of not requiring firm anchorages to resist a horizontal pull of the cables, as in the suspension bridge. All static horizontal forces are balanced so that the supporting tower does not tend to *tilt* or slide, needing only to resist such forces from the live loads.

Key advantages of the cable-stayed form are as follows:

1) much greater stiffness than the suspension bridge, so that deformations of the deck under live loads are reduced;

2) can be constructed by cantilevering out from the tower; and the cables act both as temporary and permanent supports to the bridge deck;

3) for a symmetrical bridge, the horizontal forces balance and large ground anchorages are not required;

A further advantage of the cable-stayed bridge is that any number of towers may be used. This bridge form can be as easily built with a single tower, as with a pair of towers. However, a suspension bridge is usually built only with a pair of towers.

6. Truss Bridges

A truss bridge is a bridge composed of connected elements which may be stressed from tension, compression, or sometimes both in response to dynamic loads. Truss bridges are one of the oldest types of modern bridges. A truss bridge is economical to construct owing to its efficient use of materials.

Truss girders, *lattice girders* or *open web girders* are efficient and economical structural systems, since the members experience essentially axial forces and hence the material is fully utilized. Members of the truss girder bridges can be classified as chord members and *web* members. Generally, the chord members resist overall *bending moment* in the form of direct tension and compression and web members carry the shear force in the form of direct tension or compression. Due to their efficiency, truss bridges are built over wide range of spans. Truss bridges compete against *plate girders* for shorter spans, against box girders for medium spans and cable-stayed bridges for *long spans*.

For short and medium spans it is economical to use *parallel chord trusses* such as Warren truss, Pratt truss, Howe truss, etc. to minimize fabrication and *erection* costs. Especially for shorter spans the Warren truss is more economical as it requires less material than either the Pratt or Howe truss. However, for longer spans, a greater depth is required at the centre and variable depth trusses are adopted for economy. In case of truss bridges that are continuous over many supports, the depth of the truss is usually larger at the supports and smaller at mid-span.

Glossary

Complete the glossary using words in the text.

1. _____	地形,地面,地域,地带		
2. _____	水平梁	3. _____	桥墩;支柱
4. _____	原木,木材,木料		
5. _____	主梁,大梁,纵梁;桁架		
6. _____	压力	7. _____	弯曲,弯折;挠度
8. _____	扭曲;翘曲	9. _____	跨距,跨径
10. _____	复合材料	11. _____	电化学腐蚀
12. _____	预应力混凝土		
13. _____	拉伸强度,抗拉强度		
14. _____	箱梁	15. _____	有限元分析

16.	_____	人行桥	17.	_____	临时支架；脚手架

16. _____ 人行桥　　17. _____ 临时支架；脚手架

18. _____ 挂孔,悬孔　　19. _____ 锚固,锚碇,固定

20. _____ 预制段拼装施工,节段施工

21. _____ 缆索,钢丝绳　　22. _____ 千斤顶,起重器

23. _____ 作桥面,桥面　　24. _____ 峡谷

25. _____ 桥台,拱座　　26. _____ 高架跨线桥,栈道

27. _____ 砖石,砌体,圬工　28. _____ 渡槽

29. _____ 圆柱；柱形物　　30. _____ 拱肩,拱腹

31. _____ 空腹拱桥　　32. _____ 悬索

33. _____ 系杆　　34. _____ 系杆拱

35. _____ 推力　　36. _____ 吊杆,悬杆

37. _____ 净空,间距　　38. _____ 主缆

39. _____ 索塔柱　　40. _____ 杆,拉杆

41. _____ 吊杆,悬架　　42. _____ 悬崖,陡岸

43. _____ 主跨　　44. _____ 抛物线

45. _____ 航道；排水渠　　46. _____ 刚度,刚性

47. _____ 空气动力学的　　48. _____ 活载,动载

49. _____ 多塔斜拉桥

50. _____ 钢丝,钢索,金属线

51. _____ 锚碇　　52. _____ 倾斜

53. _____ 桁架梁　　54. _____ 井字梁

55. _____ 空腹梁　　56. _____ 腹板

57. _____ 弯矩　　58. _____ 板梁

59. _____ 大跨度　　60. _____ 平行弦桁架

61. _____ 建造

Exercises

1. Decide whether the following statements are true（T）or false（F）according to the text.

_____1）The earliest beam bridges were constructed by stones.

_____2）Arch bridges work by transferring the weight of the bridge and its loads partially into a horizontal thrust restrained by the abutments at either side.

_____3）A cable-stayed bridge is a type of bridge in which the deck is hung below suspension cables on vertical suspenders.

_____4）A cable-stayed bridge can achieve longer main span than any other types of bridges.

_____5）For bridges with short or medium spans，it is economical to use parallel chord trusses to minimize fabrication and erection costs.

2. Answer the following questions.

_____1）What is the difference between pier and abutment?

_____ 2) What are the variations of arch bridges?

_____ 3) What are the advantages of cable-stayed bridges compared with suspension bridges?

3. Write down the words with the following prefix or suffix.

eletro-_____

semi-_____

para-_____

-duct _____

-ment _____

4. Translate the following paragraph into Chinese.

Extensive research is being conducted by several private enterprises and the state agencies to improve the construction techniques and materials used for the beam bridges. The beam bridge design is oriented towards the achievement of light, strong, and long-lasting materials like reformulated concrete with high performance characteristics, fiber reinforced composite materials, electro-chemical corrosion protection systems, and more precise study of materials. Modern beam bridges useprestressed concrete beams that combine the high tensile strength of steel and the superior compression properties of concrete, thus creating a strong and durable beam bridge. Box girders are being used that are better designed to undertake twisting forces, and can make the spans longer, which is otherwise a limitation of beam bridges. The modern technique of the finite element analysis is used to obtain a better beam bridge design, with a meticulous analysis of the stress distribution, and the twisting and bending forces that may cause failure.

5. Translate the following sentences into English.

1) 通常小型人行桥中的悬臂梁采用简单的梁即可;但对于设计承受道路或铁路交通荷载的大型悬臂梁桥而言,则应采用结构钢制成的钢桁架或是预应力混凝土箱梁。

2) 虽然多塔斜拉桥在外形上与悬索桥很相似,但实际上两者在基本原理与建造方法上存在很大的不同。

3) 由于桁架梁、井字梁、空腹梁构件主要承受轴向力,因而材料特性可以得到充分利用,故它们属于高效、经济的结构体系。

Text B   Substructure and Foundations

1. *Substructure*（下部结构）

The portion of the bridge structure below the level of the *bearing*（支座,支承）and above the foundation is generally referred to as substructure. Thus for a river bridge with well foundation, the substructure will consist of the piers, the abutments and the *bed blocks*（墩帽）over the piers and abutments.

1) Piers

The general shape and features of a pier depend to a large extent on the type, size and dimensions of the superstructure and also on the environment in which the pier is located. Piers can be solid, cellular, trestle or hammerhead types. Solid and cellular piers for river bridges should be provided with semicircular cut waters to facilitate streamlined flow and to reduce scour. Other designs such as reinforced concrete framed type have also been used. *Solid piers*（实体桥墩）are of masonry or mass concrete. It is permissible to use stone masonry for the exposed portions and to fill the interior with *lean concrete*（贫混凝土）. This would save expenses on *shuttering*（模板）and would also enhance appearance. Cellular, trestle and hammerhead types use reinforced concrete. The cellular type permits saving in the quantity of concrete, but usually requires difficult shuttering and additional labor in placing reinforcements. The trestle type consists of columns（usually circular or octagonal）with a *bent cap*（排架帽）at the top. In some recent designs, concrete hinges have been introduced between the top of columns and the bent cap in order to avoid moments being transferred from deck to the columns. For tall trestles as in *flyovers*（立交桥，高架道路）and elevated roads, connecting *diaphragms*（横隔板）between the columns may also be provided. The hammerhead type provides slender substructure and is normally suitable for *elevated roadways*（高架路）. When used for a river bridge, this design leads to minimum restriction of waterway.

The top width of pier depends on the size of the *bearing plates*（支座垫板）on which the superstructure rests. It is usually kept at a minimum of 600 mm more than the out-to-out dimension of the bearing plates, measured along the longitudinal axis of the superstructure.

The length of pier at the top should be not less than 1.2 m in excess of the out-to-out dimension of the bearing plates measured perpendicular to the axis of the superstructure. The bearing plates are so dimensioned that the bearing stress due to dead and live loads does not excess 4.2 MPa.

The bottom width of pier is usually larger than the top width so as to restrict the net stresses within the permissible values. It is normally sufficient to provide a batter of 1 in 25 on all sides for the portion of the pier between the bottom of the bed block and the top of the well or *pile cap*（承台）or foundation footing, as the case may be.

Reinforced concrete framed type of piers have been used in recent years. The main advantage in their use is due to reduced effective span lengths for girders on either side of the centre line of the pier leading to economy in the cost of superstructure.

However, the author would suggest caution in their wide adoption. First, such framework would be conducive to accumulation of debris and especially floating trees if used in rivers subjected to sudden floods near hills and forests. Second, such designs call for two *expansion joints*（伸缩缝,胀缝）at close intervals of about 1 to 2 m on each pier, resulting in riding discomfort beside maintenance problems. If this type is to be adopted, the author would suggest that the ends of the decking on either side of the pier centre line be cantilevered beyond the bearings so that one expansion joint would be adequate.

2）Abutments

An abutment is the substructure which supports one terminus of the superstructure of a bridge and, at some time, laterally supports the embankment which serves as an *approach* （进路,引桥） to the bridge. For a river bridge, the abutment also protects the embankment from scour of the stream. Bridge abutments can be made of masonry, plain concrete or reinforced concrete.

An abutment generally consists of the following three distinct structural elements: ① the *breast wall* （胸墙） which directly supports the dead and live loads of the superstructure, and retains the filling of the embankment in its rear; ② the *wing walls* （翼墙,侧墙）, which act as extensions of the breast wall in retaining the fill though not taking any loads from superstructure; ③ the *back wall* （背墙）, which is a small retaining wall just behind the bridge seat, preventing the flow of material from the fill on to the bridge seat.

The design of an abutment consists in assuming preliminary dimensions depending on the type of the superstructure and foundation, and checking the stresses at the *sill* （基础,底面） level. The front face of the breast wall should have a batter of not less than 1 in 25, preferably of 1 in 12. The rear batter is adjusted to get the width required to restrict the net pressure within the prescribed limits.

The wing walls of typical forms of reinforced concrete abutments have been cantilevered without extending the base of breast wall for support, as would have been necessary for masonry abutment. The slope of the bottom edge of the wing should be such as to have this edge below the level of the *revetment* （护坡,护岸工程,挡土墙） of the embankment.

A bridge abutment may fail in several ways as below, and the final design should be checked to avoid these failures. The breast wall may fail by tensile cracks, crushing or shear. The wall may tilt forward caused by excessive overturning moment due to earth pressure. The wall may slide forward due to earth pressure if the vertical forces are inadequate. Though the wall may be structurally strong, failures may occur along a curved surface by *rupture* （破裂,断裂） of the soil due to inadequate shear resistance.

2. Foundations

The design of foundations is an important part of the overall design for a bridge and affects to a considerable extent the aesthetics, the safety and the economy of the bridge. The design demands a detailed knowledge of hydraulics, soil mechanics and structural analysis. The engineer will have to gather the data on *soil profile* （土剖面） at the site and on the waterway characteristics. Since the soil at any place is not of one uniform type, the design of a suitable foundation would involve the exercise of considerable judgment. A young engineer would do well to study critically many case studies from the past practice and try to develop a feel for the subject.

In order to design a foundation for a bridge, the designer must determine the followings reasonably and accurately：① the maximum likely scour depth；② the minimum *grip length*（握固长度）required；③ the soil pressures at the base；④ the stresses in the structure constituting the foundation.

1）Scour at Abutments and Piers

The pattern of scour occurring at a bridge across a river depends on many factors including discharge，bed slope，bed material，direction of flow，alignment of piers，their shape and their size. Hence the prediction of scour depth is difficult. It has been noticed that the maximum scour in the case of an abutment occurs at *upstream*（上游的，逆流的）corner，while in the case of a pier，it occurs at the *downstream*（下游的，顺流的）end. The scour will be further aggravated if the pier is not aligned in the direction of flow.

2）Grip Length

Unless the foundations are rested on rock，adequate grip length below the maximum scour level should be provided. The minimum required grip length is specified as one-third the maximum scour depth for road bridges and one-half of the maximum scour depth for railway bridges. The purpose of the grip length is to ensure stability under heavy flood conditions and to facilitate mobilization of passive pressure against horizontal forces.

3）Types of Foundations

The foundations used in bridge structures may be broadly classified as shallow foundation and deep foundation. A deep foundation is sometimes defined as one whose depth is greater than its width. For the purpose of discussion，a deep foundation would refer to one which cannot be prepared by *open excavation*（明挖）.

Deep foundations could be further classified as piled foundations and *caisson*（沉箱，船坞闸门）foundations. A *pile*（桩；打桩）is defined as a column-support type of foundation which may be precast or formed at site. Piled foundations may be divided into two groups：First，foundations with friction piles；second，foundations with point *bearing piles*（承压桩，承重桩）. As the name suggests，a friction pile develops bearing capacity，to a major extent by *skin friction*（表面摩擦），i.e.，it transfers the load to the adjacent soil by friction along the *embedded length*（埋置深度）of the pile. Friction piles are driven in ground whose strength does not increase appreciably with depth. A point bearing pile transfers practically all of its load by end bearing to a hard stratum on which it rests. It should be ensured that the strata beneath the *bearing layer*（承重层）are not too weak to carry the additional loads. The load carrying capacity of the pile is the lesser of the values of its structural capacity and the capacity of the supporting soil to carry the load. For an economical design，the dimensions of the pile should be so chosen that the structural capacity of the pile is nearly equal to the estimated supporting capacity of the soil for the pile. The pile tip should be rested on hard strata having a thickness of about twelve times the diameter or side of the pile. Since the soil conditions are generally non-uniform even at one site of a major bridge, considerable study and exercise of engineering judgment are necessary to design a proper *pile foundation*（桩基础）.

On the other hand, a caisson is a member with a hollow portion, which after installing in place by any means is filled with concrete or other material. Caisson foundations can be classified as *open caissons*（沉井,开口沉箱）and *pneumatic caissons*（气压沉箱）. An open caisson is one that has no top or bottom cover during its sinking. It is more popularly known as well foundation. A pneumatic caisson is a caisson with a permanent or temporary roof near the bottom so arranged that men can work in the *compression chamber*（压气室）under it. Pneumatic caisson can be used for a depth of about 30 m below water level, beyond which piled foundation would have to be resorted to.

The selection of the foundation system for a particular site would depend on many considerations, including the nature of subsoil, the presence or otherwise in the subsoil of the boulders, buried tree trunks, etc., and the availability of expertise and equipment with the contractors operating in the region where the bridge work is located. Generally, piles would be suitable when a thick stratum of soft soil overlays a hard soil. Caissons are preferred in sandy soils. It is not uncommon to come across cases where a contractor, owning *pile driving machinery*（打桩机）would quote lower for precast piling, even when the client has specified open caissons.

Text C   Ancient Chinese Bridges

China is a great country with a written history of about 5,000 years. She has a vast territory, topographically higher in the northwest and lower in the southeast. Networked with rivers, she has the best-known valleys of the Yangtze River, the Yellow River and the Pearl River, which are the cradle of the Chinese nation and her brilliant culture. Throughout history, the Chinese nation has erected thousands of bridges, which form an important part of her culture.

Ancient Chinese bridges are universally acknowledged and have enjoyed high prestige in the bridge history of both the East and the West.

Ancient Chinese bridges can be classified under four categories: the beam, arch, cable suspension and floating bridges. Nowadays, the first three categories of ancient bridges can still be seen, and they are introduced as follows:

1. Beam Bridges

The earliest reference to the beam bridge in Chinese history is the Ju Bridge dating from the Shang Dynasty. King Wu of the Zhou Dynasty launched a campaign against King Zhou, and having captured Zhaoge. At the Ju Bridge, he ordered a hoard of millet distributed to the relief of the poor. From the Zhou Dynasty through to the Qin and Han Dynasties, bridges with timber beams and stone piers were predominant.

During the Song Dynasty, a large number of stone-pier and stone-beam bridges were constructed. In Quanzhou alone, as recorded in ancient books, 110 bridges were erected during the two centuries, including ten well-known ones. For example, the 362-span Anping Bridge was known for its length of 2,223 m, a national record for over 700 years. Its construction lasted 16 years, from the 8th year to the 21st year of the reign of Shaoxing of the Song Dynasty. Another famous one is the 47-span Wan'an Bridge, situated at the outlet of the Luoyang River to the sea, better known as Luoyang Bridge. It is about 890 m long and 3.7 m wide. The construction began in the 5th year of the reign of Huangyou and ended in the 4th year of the reign of Jiayu of the Song Dynasty. Both bridges are included in the list of major cultural relics under state protection.

The Jiangdong Bridge in Zhangzhou, Fujian Province boasts the largest stone beams. In the first year of the reign of Jiaxi of the Song Dynasty, the timber beams of this bridge were replaced by stone ones. The bridge had 15 spans, each consisting of 3 slices of stone beams. But today only 5 spans remain. The largest stone beam, 23.7 m in length, 1.7 m in width and 1.9 m in height, weighs 2,000 kn. It seems incredible that such an arduous task could be performed then as there was no heavy-duty craning equipment to quarry the stone and to haul to the site and set in position such enormous stone beams.

To elongate the span, either the timber beams or the stone ones were placed horizontally on top of each other, the upper layer cantilevering over the lower one, thus supporting the simple beam in the middle. That kind of stone beam is called Diese (overlapping beam), which, however, could not extend long; while the timber cantilever beam, called flying bridge or extended arm bridge, could reach as far as 20 m. The earliest record of the timber cantilever beam dates as far back as the 4th century B.C.. The extant single-span timber cantilever bridge, the Yinping Bridge at Wenxian, Gansu Province, which was rebuilt in the Qing Dynasty, has a span of more than 60 m with covered housings on it.

It was common practice to build bridge housings or galleries on timber beam bridges, and a case in point is the Fengyu bridge (all-weather bridge) built by the Dong people. Situated at Sanjiang Dong Autonomous County, Guangxi Zhuangzu Autonomous Region, the Chengyang Bridge across the Yongji River built in 1916, is a 644 m 4-span timber cantilever covered bridge. Each of its five piers is crowned by a pavilion and the decks are roofed by a spacious gallery, which joins the pavilions. The pavilions not only perform the function of balance, but have added to the charm and elegance of the bridge as well.

2. Arch Bridges

There are different views on the origin of arches. Some believe the first arch was a natural formation over the caverns, others claim that it was brought into being by the piling of the collapsed stones, and still others hold that it was evolved from the false arch which was formed by the openings in the walls. However, a study of the tombs and the extant old arches in China indicates that the joint of the beam and sides evolved gradually into isometric trilateral, pentalateral and septilateral arches and finally into semicircular arch. The span, too, was gradually elongated, from 2 m or 3 m up to 37.02 m.

The oldest arch bridge in China, which is still surviving and well-preserved, is the Anji Bridge. It is also known as the Zhaozhou Bridge, at Zhouxian, Hebei Province, built in the Sui Dynasty. It is a single segmental stone arch, composed of 28 individual arches bonded transversely, 37.02 m in span and rising 7.23 m above the chord line. Narrower in the upper part and wider in the lower, the bridge averages 9 m in width. The main *arch ring* (拱圈) is 1.03 m thick with protective arch stones on it. Each of its spandrels is perforated by two small arches, 3.8 m and 2.85 m respectively in *clear span* (桥涵净跨径), so that flood water can be drained and the bridge weight is lightened as well. The Anji Bridge has a segmental deck and the parapets are engraved with dragons and other animals. Its construction started in the 15th year of the reign of Kaihuang and was completed in the 1st year of Daye's reign of the Sui Dynasty. Up till now it has survived for 1,387 years. The bridge, exquisite in workmanship, unique in structure, well-proportioned and graceful in shape, and meticulous yet lively in engraving, has been regarded as one of the greatest achievements in China. Great attention and protection have been given to it through successive dynasties. In 1991, the Anji Bridge was named among the world cultural relics.

Stone arches in China vary in structure in accordance with different land transport as well as different natures between the north and south waterways. In the north, what prevails is the flat-deck bridge with solid spandrels, thick piers and arch rings, whereas in the south crisscrossed with rivers, the hump-shaped bridge with thin piers and shell arches prevails.

The Lugou Bridge across the Yongding River is located at Wanping County. The project began in the 28th year of the reign of Dading and was completed in the 3rd year of the reign of Mingchang of the Jin Dynasty. It is 212.2 m long, 9.3 m wide and has 11 semicircular arches, ranging from 11.4 m to 13.45 m in span. The piers are from 6.5 m to 7.9 m wide; their pointed cutwaters upstream are inlaid with triangular iron bars, while the downstream sides are square in shape but without two angles. The parapets are divided into 269 sections with columns in-between, each column crowned with a carved lion. When the bridge was first erected in the Jin Dynasty, all the lions were alike and very simple, but through the ages they were replaced each time by better ones, more delicately carved and different in style. Now, each lion has its individual posture. And more fascinating are the lion cubs. They are playing around their parents, clinging to the breast, squatting on the shoulder, nestling at the feet, or licking the face. These exquisite sculptures on the bridge and on the ornamental columns, which show the practical application of the aesthetic principle of unity and variation, have become a scene of attraction. The bridge has long since been included in the historical relics under state protection.

In the southern part of China, say, Jiangsu and Zhejiang Province, networked with navigable rivers, boats were the main means of transportation. As bridges were to be built over tidal waters and their foundations laid in soft soil, even the stone arch bridge had to be built with thin piers and shell arches in order that its weight could be lessened as much as possible. The spans ranged in number from one to as many as eighty-five, for example, the Chuihong Bridge in Wujiang, Jiangsu Province, but the bridge has collapsed and now only eight ruined spans remain.

The bridge in classical Chinese gardens not only serves as a passage, but is an integral part of the garden scenery, either as a highlight or a setoff. One masterpiece of this kind is the Wuting (five pavilions) Bridge, also known as the Lianhua Bridge in the Shouxi Lake in Yangzhou, Jiangsu Province.

The timber arch bridge in China dates back to the Song Dynasty. In the panorama *Riverside Scene on Qingming Festival* by Zhang Zeduan of the Song Dynasty, there is portrayed a timber arch bridge, known as the Hong Bridge spanning the Bian River in Bianjing, capital of the Song Dynasty. It was built following the pattern of guanmu (the interlocking of logs), put forward by a prison guard during the reign of Mingdao of the Song Dynasty. Planks were interdigitated to form a kind of arch, thus obviating the necessity for piers which would usually get in the way of navigation. The span of this bridge was around 18.5 m in length, 4.2 m in height above the chord line, and the deck averaged 9.6 m in width. The Hong Bridge fell into ruin between the Jin and Yuan Dynasties. For hundreds of years it had been considered to be peerless. However, investigations during the recent decades have shown that in the mountain areas of Zhejiang and Fujian Province, there were found dozens of ancient timber arch bridges, similar to the Hong Bridge in structure, but with some improvements, their spans as long as 35 m or so. Another example is the Xidong Bridge spanning the Sixi Stream at Taishun County, Zhejiang Province. The bridge is 41.7 m long, 4.86 m wide, with a 25.7 m span rising 5.85 m above the chord line. On the deck there is an exquisite gallery, both sides of which are paneled with overlapping boards for protection. It was constructed in the 4th year of the reign of Longqing of the Ming Dynasty. Amazingly enough, the Yeshuyang Bridge, another of this kind in the same county, has survived for 511 years.

The timber arch of such bridges as the Hong Bridge is a unique achievement of ancient China. Besides, there are still other kinds of peculiar structures, like bamboo arch bridges, which are also unparalleled in the world.

3. Cable Suspension Bridges

Cable suspension bridges vary in kinds according to the material of which the cables are made: rattan, bamboo, leather and iron chain. Li Bin of the Qin State superintended the establishment of seven bridges in Gaizhou (now Chengdu, Sichuan

Province), one of which was built of bamboo cables.

The iron-chain bridge is said to date as far back as the early time of the Western Han Dynasty. Senior General Fan Kuai superintended the construction of the Fanhe Bridge on the ancient plank road in Maocheng County, Shanxi Province; and the bridge is believed to be of iron chain. In historical records there is a clear reference to the iron chains across the Yangtze River. In A.D. 280, the army of Kingdom Wu, when attacked by the Western Jin, threw several iron chains across the mouth of the Xilin Gorge, one of the three gorges of the Yangtze River, to block away the enemy ships.

The Jihong Bridge at Yongping County, Yunnan Province is the oldest and broadest bridge with the most iron chains in China today. Spanning the Lancang River, it is 113. 4 m long, 4.1 m wide and 57.3 m in clear span. There are 16 bottom chains and a hand rail chain on each side. The bridge is situated on the ancient road leading to India and Burma.

The Luding Iron-chain Bridge in Sichuan Province, the most exquisite of the extant bridges of the same type, spans the Dadu River and has served as an important link between Sichuan Province and Tibet. The bridge was believed to be set up by bridge-builders who were noted for making iron chains. Its erection began in the 4th year of the reign of Kangxi of the Qing Dynasty and was completed in the following year. It is 100 m in clear span, 2.8 m in width, with boards laid on the bottom chains. There are nine bottom chains, each about 128 m long, and two hand-rail chains on each side. On each bank, there is a stone abutment, whose dead weight balances the pulling force of the iron chains. The Luding Bridge is included in the first group of major cultural relics under state protection.

Integrated Skills: Writing a Letter

Please write a letter in English to your foreign friend who has the same major as you, and introduce the ancient, modern and contemporary bridges in your hometown or in the city where you are studying.

Unit 7 Design, Construction, Maintenance and Rehabilitation of Bridges

Lead-in

1. What can be the major challenges during the rehabilitation of bridges?
2. What is the chief factor to decide the type of a bridge?

Vocabulary Warm-up: Matching

Can you write down the full forms of the following abbreviations?

_____1. GPS：   a. 荷载抗力系数设计法

_____2. LRFD：   b. 纤维增强复合材料

_____3. FRP：   c. 全球定位系统

Text A   Bridge Construction

In the 20th century, bridge construction technology evolved and was fueled by the Industrial Revolution. At the turn of the century, steel bridges were *riveted* together, not *bolted*; concrete bridges were cast in place, not *precast*; and large bridge members were built from *lacing bars* and smaller sections, not *rolled* in one piece. Plastic had not yet been invented. Construction techniques such as *post-tensioning*, *slurry walls*, soil freezing, and *reinforced earth walls* had not yet been conceived. Surveying was performed mechanically since infrared, optical technology was still 75 years away.

Bridge construction is changing as the new millennium begins. New construction techniques and new materials are emerging. There are also new issues facing the bridge building industry relative to the research needs associated with these new techniques and materials.

1. Long-Span Bridges

1）Suspension Bridges

While suspension bridge building was conducted at a modest pace throughout the 20th century, an unprecedented number of spans of remarkable record lengths were built in the Far East and Denmark. Both the Akashi Kaikyo Bridge in Japan and the Great Belt Bridge in Denmark were completed in 1998. The Akashi Kaikyo Bridge is the largest suspension bridge in the world, with a span of 1,991 m, and the Great Belt Bridge is the second largest, with a span of 1,624 m.

While span lengths have increased nearly fivefold during the course of this century, they may have reached their physical limits with today's materials. Research will be necessary to develop the new, *ultra-high-strength* steel wire or carbon fiber wire required to build the longer main suspension cables that will make it possible to increase span lengths to beyond 2,000 m.

As we enter the new millennium, *rehabilitation* and ongoing *maintenance* of the existing suspension bridges must continue as well. Recent rehabilitation measures for the main cables and suspension systems of these bridges have uncovered *degradation* through *corrosion* and *hydrogen embrittlement*. Research is needed to determine the remaining useful *service life* of suspension bridge cables and what measures can be taken to slow or halt the degradation process.

Other components of *long-span bridge*s, existing and new, are being revolutionized as technology moves forward. Advances in deck technology are producing stronger, lighter decks. *Orthotropic* decks are becoming increasingly popular on long-span structures as a means of reducing *dead load*. Bearings, joint systems, and *seismic retrofitting* components are becoming increasingly efficient as more large-scale testing facilities are built.

2) Cable-Stayed Bridges

Cable-stayed bridges are a phenomenon of the latter half of this century, and have become more efficient and longer in span as a result of advances in computer technology during the 1980s and 1990s. Cable-stayed bridges represent an efficient alternative to suspension bridges, particularly when spans are in the 300~500 m range. However, record spans, such as the Tatara crossing in Japan, are now in the 900 m range.

Perhaps one of the greatest areas of concern and debate with regard to cable-stayed bridge construction is protection measures for the stay *tendons*. The debate is focused on the use of *epoxy* versus *grout*, encased or nonencased stay cables, and a variety of issues regarding the protection of these key bridge elements. There has been a modest amount of research to date on this topic, but more research will be essential as these unique structures age. Other stay cable research topics should include the effects of *wind-induced vibrations*, *anchorage* details, and determination of *in-service* stay tensions.

2. Short- and Medium- Span Bridges

Changes can be expected as well for short and medium span bridges, which represent the vast majority of bridges to be built. To meet public demand for minimal traffic disruption, construction time for these bridges will have to decrease, and *traffic flow* will need to be maintained during construction, often within a few feet of workers and equipment. Consequently, public agencies and contractors will seek new materials and methods that enable shorter construction time without compromising safety. The need to make work conditions safer and more efficient will also continue to be emphasized.

There will be advances in the construction of these bridges in many areas. New technology will enable better quality control of the positioning of bridges and members.

The design-build approach, which takes into account actual soil and environmental conditions, will also become more widely accepted. Problems in construction will be dealt with immediately to keep the work moving. More experienced people that know how to approach problems and devise solutions will be involved.

The Global Positioning System (GPS) will be used to locate bridge working points more accurately and quickly with fewer people. Each succeeding unit of segmental structures will be located by GPS, and adjustments to *elevation* and *plan locations* will be made instantly. Construction equipment will be run by computers that are directed by GPS. One person will monitor multiple units from a remote location via video cameras and computers.

Precast foundation, abutment, pier, and *superstructure* units will eliminate costly field *formwork*. Quality control of plant-cast units will minimize variations in size and strength, as well as extend seasonal construction time into fall and spring. Delivery systems will grow in number and size to handle complex units. Labor economies will help drive the development and use of segmental production craftsmen directed by a few highly skilled individuals. Precast arch-like structures, such as Bebo Arch, will continue to gain acceptance. Use of these structures will result in an aesthetically pleasing bridge without the associated deterioration of exposed concrete decks.

As skilled technicians who can monitor bridge construction become less available, local governments will increasingly turn to the use of wood-panel and similar types of bridges. Components that are easily assembled and inspected will gain favor because they are viewable, and do not require *inspectors* with extensive experience and skills in such areas as the performance of concrete air and *slump tests*.

A recent development is the resurgence in the construction of post-tensioned concrete box girder bridges built on placed and *compacted* fills. Traffic is maintained on *bypass*es, and fill from the bridge construction is reused to build the bridge approaches. This type of construction allows forms and workers to operate efficiently at grade, instead of being suspended above the ground and traffic. Contractors that are equipped to move fill material will take advantage of this method.

New materials, such as plastics, *polymer concretes*, and *high-performance concretes*, will be used for the construction and rehabilitation of bridges. As *bonded fiber-reinforced composites* become temperature tolerant, they will increasingly be used to rehabilitate bridges and make them capable of carrying the latest truck loads. To bring deteriorated concrete beams and other components up to capacity, carbon reinforcement bars will be used to keep the existing uncoated and epoxy-coated bars from developing corrosion hot spots. High-strength polymers and concretes may be used to extend span lengths through reductions in dead loads. Use of larger-sized steel strands in tensioned members, along with higher-strength concretes, will become common practice.

Full acceptance of plastic chairs and form ties will keep corrosion from progressing through the steel to the concrete. Plastic and aluminum stay-in-place forms will be used to reduce form weights and costs. The ability to leave a form in place will reduce removal as well as placement costs.

Before the 1970s, most bridge designs had deck *expansion joints* combined with fixed and expansion support bearings. The idea was to allow the structure to expand and contract. However, expansion joints tend to fill with dirt and debris, and bearings deteriorate over time. Thus, the structure stiffens with age, and maintenance needs increase more rapidly with time.

In the early 1970s, bridge engineers investigated ways of minimizing this problem. The solution found was to eliminate costly expansion devices and support bearings. The result was the development of the *jointless bridge*. A jointless bridge is built on a flexible substructure so the structure can expand and contract with minimal distress. This type of structure becomes more flexible as it ages.

Today jointless structures have become a viable alternative for short and medium span bridges. However, many questions remain unanswered. For example, should there be a span length limitation for this type of structure? Should length limitations be different for steel and concrete bridges? Are *thermal stresses* more important to these bridges than to conventional bridges? What are the most effective methods for connecting abutments to the superstructure? Research on such issues will be required to support the use of jointless bridges for short and medium span bridges.

Short and medium span bridges using the new load and resistance factor design (LRFD) codes will benefit from the safety factors related to dead load. However, this benefit will be offset to some degree by heavier truck loads and associated safety concerns.

The LRFD codes represent an attempt to keep live-load conditions and values in line with load factor design. This may be a rather moot point, however, as final LRFD moment and shear values are comparable to existing load factor design values.

Challenges in the construction and repair of new short and medium span bridges will become more complex, but can be met through better planning and equipment. Designers must learn to adapt to changing conditions and technology to provide a better product for the public.

3. Vision for the Future

In the 21st century, new technology will meet changing needs and provide alternatives that will lead to new standards in engineering and construction worldwide. The future economic impact of bridge construction will revolve around ways of implementing simple design and construction solutions through innovative thinking. Improved interaction among bridge design, construction, maintenance and field performance will be essential in providing the most economical infrastructure.

Cutting-edge research in new enhanced materials, advanced smart sensing, and life-cycle management will provide the technology needed to construct more durable and maintenance free structures, and to rejuvenate and extend the life of older structures. Nondestructive evaluation techniques suitable for in-field construction, quality control, and structural integrity assessment of large structures will be developed.

Reliable, inexpensive, rapid, automated inspection techniques will emerge as well. To optimize the safe operation of civil structures with a minimum of expense, it will be necessary to develop innovative, built-in remote monitoring systems for highway bridges. Such systems will yield complementary data on anomalies, bridge traffic, and project-life estimation, as well as management and maintenance planning for the structure. The data will be read out remotely on demand and transmitted to a central monitoring station.

Advanced composites made of *resin-impregnated* strong fibers [fiber-reinforced plastic (FRP)] could become prominent construction materials in the 21st century. This class of light, durable materials could provide sizable benefits to the global infrastructure. The speed and ease of installation of composites as compared with conventional materials will make use of FRP particularly competitive with respect to construction costs. FRP composites can be used as stand-alone structural members, as reinforcement for prestressed and nonprestressed concrete, or in combination with other structural materials for new construction or repair and rehabilitation. Composite bridges made with new materials will afford the opportunity to use embedded sensors and actuators.

Finally, education and training programs in bridge inspection, evaluation, and design incorporating cutting-edge technology and promoting lifelong learning are likely to be created. These programs will link universities, industry, and departments of transportation worldwide using satellite television and the World Wide Web.

Glossary

Complete the glossary using words in the text.

1. _____ 铆接，固定   2. _____ 螺栓连接
3. _____ 预先浇注，预制   4. _____ 缀条
5. _____ 轧制   6. _____ 后张法
7. _____ 地下连续墙   8. _____ 加筋土挡土墙
9. _____ 超高强度的   10. _____ 修复，复原，改善
11. _____ 养护，维修   12. _____ 腐蚀，剥蚀
13. _____ 腐蚀，锈蚀   14. _____ 氢脆
15. _____ 使用寿命   16. _____ 大跨径桥梁
17. _____ 正交各向异性的   18. _____ 静荷载
19. _____ 地震的，抗震的   20. _____ 加固，翻新

21. _____ 钢丝束,钢绞线束,受力筋束

22. _____ 环氧树脂(的) 23. _____ 水泥浆

24. _____ 风振 25. _____ 锚碇,锚固

26. _____ 在役的,服役的 27. _____ 交通流

28. _____ 高程,高度 29. _____ 平面位置

30. _____ 上部结构 31. _____ 模板

32. _____ 监测;监理 33. _____ 坍落度试验

34. _____ 密实 35. _____ 便道,旁道

36. _____ 聚合物混凝土 37. _____ 高性能混凝土

38. _____ 黏结

39. _____ 纤维增强复合材料

40. _____ 伸缩接头 41. _____ 无伸缩缝桥梁

42. _____ 温度应力,热应力

43. _____ 浸渍在树脂中的

Exercises

1. Write down the words with the following prefix or suffix.

ultra- _____

poly- _____

post- _____

mini- _____

-ance _____

2. Decide whether the following statements are true (T) or false (F) according to the text.

_____ 1) In the 20th century, concrete bridges were precast, not cast in place.

_____ 2) Although span lengths of suspension bridges in the 20th century increased greatly, they may have reached their physical limits with today's materials.

_____ 3) Before the 1970s, jointless structures have become a viable alternative for short and medium span bridges.

_____ 4) Compared with conventional materials, the speed and ease of installation of FRP will make its use particularly competitive with respect to construction costs.

3. Translate the following paragraphs into Chinese.

_____ 1) Other components of long-span bridges, existing and new, are being revolutionized as technology moves forward. Advances in deck technology are producing stronger, lighter decks. Orthotropic decks are becoming increasingly popular on long-span structures as a means of reducing dead load. Bearings, joint systems, and seismic retrofitting components are becoming increasingly efficient as more large-scale testing facilities are built.

_____ 2) Precast foundation, abutment, pier, and superstructure units will eliminate costly field formwork. Quality control of plant-cast units will minimize variations in size and strength, as well as extend seasonal construction time into fall and spring. Delivery systems will grow in number and size to handle complex units. Labor economies will help drive the development and use of segmental production craftsmen directed by a few highly skilled individuals.

4. Translate the following sentences into English.

1) 20 世纪之前,诸如后张法、地下连续墙、土壤冻结、加筋土挡土墙等施工技术尚未出现。

2) 关于斜拉索研究的另一些课题包括:风振影响、锚固细节以及在役拉索的拉力确定。

3) 考虑实际土壤与环境条件的"设计—建造一体化"方法将被更为广泛地接受。

Text B   Bridge Design and Alignment

1. Bridge Design

The design of bridges requires the collection of extensive data and from this the selection of possible options. From such a review the choice is narrowed down to a shortlist of potential bridge designs. A sensible work plan should be devised for the marshalling and deployment of information throughout the project from conception to completion. Such a checklist will vary from project to project but a typical example might be drawn up on the following lines.

1) Selection of Bridge Type

The chief factors in deciding whether a bridge will be built as girder, cantilever, truss, arch, suspension or some other type are: ①location, for example, across a river; ② purposes, for example, a bridge for carrying motor vehicles; ③ span length; ④strength of available materials; ⑤cost; ⑥beauty and harmony with the location.

Each type of bridge is most effective and economical only within a certain range of span lengths. In some cases, alternative preliminary designs are prepared for several types of bridge in order to have a better basis for making the final selection.

2) Selection of Materials

The bridge designer can select from a number of modern high-strength materials, including concrete, steel and a wide variety of corrosion-resistant alloy steels.

For the Verrazano-Narrows Bridge, for example, the designer used at least seven different kinds of alloy steel, one of which has a yield strength of 345 N/mm^2 and does not need to be painted because fin oxide coating forms on its surface and inhibits corrosion. The designer also can select steel wires for suspension cables that have tensile strengths up to 1,724 N/mm^2.

Concrete with compressive strength as high as 55 N/mm^2 can now be produced for use in bridges, and it can be given high durability against chipping and weathering by the addition of special chemical agents and control of the hardening process. Concrete that has been prestressed and reinforced with steel wires has a tensile strength of 1, 724

N/mm^2.

Other useful materials for bridges include aluminum alloys and wood. Modern structural aluminum alloys have yield strength exceeding 276 N/mm^2. Laminated strips of wood glued together can be made into beams with strength twice that of natural timbers; glue-laminated southern pine, for example, can bear working stresses approaching 21 N/mm^2.

3) Analysis of Forces

A bridge must resist a complex combination of tension, compression, bending, shear, and torsion force. In addition, the structure must provide a safety factor as insurance against failure. The calculation of the precise nature of the individual stresses and strains in the structure, called analysis, is perhaps the most technically complex aspect of bridge building. The goal of analysis is to determine all of the forces that may act on each structural member.

The forces that act on bridge structural members are produced by two kinds of loads — static and dynamic. The static load — the dead weight of the bridge structure itself — is usually the greatest load. The dynamic, or live load, has components, including vehicles carried by the bridge, wind forces, and accumulations of ice and snow.

Although the total weight of the vehicles moving over a bridge at any time is generally a small fraction of the static and dynamic load, it presents special problems to the bridge designer because of the vibration and impact stresses created by moving vehicles. For example, the severe impacts caused by irregularities of vehicle motion or bumps in the roadway may momentarily double the effect of the live load on the bridge.

Wind exerts force on a bridge both directly by striking the bridge structure and indirectly by striking vehicles that are crossing the bridge. If the wind induces aeroelastic vibration, as in the case of the Tacoma Narrows Bridge, its effect may be greatly amplified. Because of this danger, the bridge designer makes provisions for the strongest winds that may occur at the bridge location.

Other forces that may act on the bridge, such as stresses created by earthquake tremors must also be provided for.

Special attention must often be given to the design of bridge piers, since heavy loads may be imposed on them by currents, waves, and floating ice and debris. Occasionally a pier may even be hit by a passing ship.

Electronic computers are playing an ever-increasing role in assisting bridge designers in the analysis of forces. The use of precise model testing, particularly for studying the dynamic behavior of bridges, also helps designers. A scaled-down model of the bridge is constructed, and various gauges to measure strains, accelerations, and deformations are placed on the model. The model bridge is then subjected to various scaled-down loads or dynamic conditions to find out what will happen. Wind tunnel tests may also be made to ensure that nothing like the Tacoma Narrows Bridge failure can occur. With modern technological aids, there is much less chance of bridge failure than in the past.

2. Bridge Alignment

Formerly, the bridge dictated the alignment of the road. If possible, the bridge was built at right angles to the river or railroad, even when the road had then to be linked up with curves. In elevation too, steep ramps were chosen with a small radius of summit. Nowadays, the alignment of the road governs the design of the bridge and the bridge engineer is well advised to accept this subordination, provided that the alignment is good. This change of attitude has arisen as a result of the requirements of high speed driving and by the challenge to achieve a steady, good looking road in the landscape.

To design a good alignment for a road nowadays has become a great art requiring the consideration of many factors. One has learned, for example, that long straight sections may lull drivers to sleep and are therefore dangerous. One has learned that curves in plan must be seen in space in combination with curves and slopes in elevation. Easement curve was introduced, whereby curvature changes steadily from a straight line to the desired radius. The integration of a planned alignment into the environment is usually checked by perspective computer drawings as seen from different sight points.

Bridges are a part of the alignment and it is evident these days that the bridge must be adapted to it. This means that bridges often lie within plan curvature and that they follow the slopes or curvature in elevation. The radii of curvature are generally large in order to achieve adequate range of vision. Such large radii are also desirable for aesthetic reasons.

1) Elevation and Vertical Alignment

For overpass bridges over motorways in flat country, the vertical curve should be extended into the approach ramps. For bridges crossing rivers in plains, it is desirable to extend the curve over the total length of the bridge, even when a very large radius results. For bridges across the River Rhine with main spans of 200 m and more, the bridge looks better if the curvature continues over the central span, rather than having a central horizontal section between curves over the side spans.

For bridges crossing valleys, the vertical alignment along the bridge depends upon the further alignment of the road; does it continue without much slope, or does the road descend from higher ground and ascend again after crossing the valley? An example of the first situation is the Sulzbach viaduct of the autobahn Stuttgart-Ulm which was one of the first viaducts built for the autobahn system (1934). It was designed with great care and the spans of the continuous beam decrease with the slope. For the view from the valley a slight summit curvature looks better than a stiff straight line, and thus the bridge was built with profile. As soon as the autobahn was finished, the hump of the bridge, as the car-driver saw it, was criticized and the bridge — destroyed in the war — was later rebuilt without the hump but with a straight steady upper line of the beam.

A similar situation in which car-drivers objected to the vertical alignment was experienced with the bridge crossing the Werra valley near Hedemtinden, where the autobahn ascends from both ends of the bridge towards the mountains. In 1936 it was considered unacceptable to build a bridge with a hanging trough-shaped vertical alignment and thus the bridge was built with a horizontal beam between short vertical curves leading to the slopes. Seen from the autobahn, the bridge looked like a stiff board swung from the side slopes. Here also the reconstruction of the bridge, destroyed during the war, enabled a correction to be made. It was rebuilt in 1952 with a smooth sag curve over the total length of the bridge. As seen from the valley, the bridge looks quite natural and nobody senses the hanging profile of the continuous beam as unpleasant; most people do not even notice this peculiarity.

The experience gained from these two bridges has become history and both teaches and proves that the bridge must be subordinated to the alignment and that is must be integrated into the steady line of the road both in plan and elevation.

2) Plan-layout

For alignment in plan, one tries to make the angle between the crossing of the new road and another traffic route, river or valley as close to 90° as possible, at the same time assuring a steady alignment. The rectangular or almost rectangular crossing leads of course to the shortest and cheapest bridge and gives the best chance for a pleasing design.

However, in densely populated areas or in mountainous country one has often to accept skew crossings as unavoidable.

For narrow bridges, skew angles down to about 60° can be achieved with a rectangularly structured bridge by means of small rectangular abutments at the top of the embankments and slender columns in the axis of the bridge providing all necessary intermediate supports.

When skew bridges are wide there is only one good solution: all lines and surfaces of transverse members must be parallel to the direction of the river or valley. This complies with the guideline of good order, limiting the directions of lines to as few as possible. In rivers, the piers must in any case be placed parallel to the stream's direction for hydraulic reasons, in order to reduce scour action. On river banks, also, abutments parallel to the river look better than if placed rectangularly to the crossing road.

On steep slopes of valleys, wide abutments or piers will cut into the slope at a skew inclination if designed rectangularly and this will not only look bad, but also make foundation work more difficult.

Of course the structure can be designed rectangularly for river bridges and viaducts, if it can be carried by column-like piers in the axis and if the abutments are raised up to the crown of the embankment in order to reduce their size.

If a good alignment requires a curved bridge — over a part or the total length, then all external longitudinal lines or edges of the structure should be parallel to the curved axis, thereby following again the guideline of good order.

The transverse axis of piers or groups of columns should be rectangular (radial) to the curved axis, unless skew crossings over roads or rivers enforce other directions.

The requirements of traffic design result occasionally in very acute angles or in level branching which cause difficulties for the bridge engineer to find pleasing solutions for the bridges.

Text C　Experience of Bridge Maintenance and Reconstruction in Latvian

1. Summary

This report reviews the experience of maintaining and reconstructing the bridges under the responsibility of the Latvian Road Administration (LRA).

Bridges as in many other countries are very important part of the road network in Latvia, as they connect parts of the cities or countries, or countries themselves. If bridges are out of order, the infrastructure is not able to function sufficiently well. If safety is concerned the road network may be even dangerous for road users if the bridge maintenance quality is poor.

2. Introduction

Bridges and culverts are built in order to lead roads over obstacles: other roads, rivers, ravines, etc. These structures considerably change the natural regimes of rivers and other nature processes.

Modified natural materials are used in bridge and culvert structures. However, the influence of the environment tends to turn every modified material into its natural initial state.

The task of bridge and culvert owners is to maintain and preserve these structures so that they could serve the needs of the publics in long term and with sufficient safety.

3. Bridge Characteristics

The Latvian Road Administration is responsible for 928 bridges in the state road network and 740 large ($l \geqslant 2$ m) culverts.

Bridge division according to construction materials is done taking into consideration the main bearing elements of bridge superstructure: timber, stone, steel and reinforced concrete bridges.

Historically the bridge construction in Latvia developed depending on the public needs, economical capacities and level of bridge construction technologies. It was greatly influenced by the wars and the political situation.

In the end of the 19th century and the beginning of the 20th century mostly timber bridges were built in Latvia; proportion of stone masonry, steel and reinforced concrete bridges was smaller.

The World War I and the World War II had a great influence on the composition of the Latvian bridges. During the World War II approximately 90% of the bridges were fully or partially destroyed. After the war less damaged capital bridges were repaired and other bridges were replaced by unimpregnated timber structures. Their lifetime was short, approximately 15~20 years.

Timber bridges gradually were replaced by cast in situ reinforced concrete structures, and since 1955 only bridges from prefabricated reinforced concrete elements were built. A considerable number of bridges were built in the conditions of general deficit of construction materials, production capacities, appropriate structures and mounting equipment and without any motivation for quality assurance.

4. Analysis of the Experience

After analyzing the drawbacks and the found defects of the existing bridges the following main causes of drawbacks and defects may be divided in the following categories:

1) Design defects:

① Normative documents, which were used by the bridge designers, allowed the use of extremely thin concrete covering layer in reinforced concrete structures.

② Simple supported beams with expansion joint on each pier were used in bridges with many spans (many sensitive elements).

③ Insufficiently durable expansion joints.

④ Thin reinforced concrete structures used in roadway structures.

⑤ Concrete covering layer is envisaged on water insulation.

⑥ Steel structures in guardrails and parapets were coated with insufficiently durable paint.

⑦ Sufficient bridge span and appropriate reinforcement of fill ends is not always chosen.

2) Drawbacks and defects in the execution of construction works:

① Structures from prefabricated reinforced concrete elements were constructed with poor quality: Insufficient concrete density; Positioning of the reinforcement does not conform the design; Insufficient concrete frost resistance.

② Defects of cast in situ reinforced concrete: Insufficient concrete density; Positioning of the reinforcement does not conform the design; Insufficient concrete frost resistance; Honeycombs in the concrete.

③ Defects of steel structures: Poor preparation of structure surfaces prior to painting; Poor quality painting.

④ Defects of expansion joints: Use of inappropriate material; Poor technology of construction works.

3) Drawbacks of maintenance works:

① Only routine maintenance is executed;

② Appropriate periodic maintenance is not implemented;

③ Defects are repaired with materials of poor quality.

5. Bridge Maintenance

1) General Rules for Bridge Maintenance

After analyzing the existing situation and acquainting with the experience of foreign countries it was stated that in order to improve bridge management it was necessary to assess the bridge assets.

In order to implement appropriate system for bridge management the Latvian Road Administration in co-operation with the Norwegian Public Roads Administration elaborated a Bridge Management System (BMS). Initially bridge management guidelines were elaborated which provided the description of the main tasks and detailed duties.

Bridge maintenance purpose is to preserve the planned level of traffic safety and quality during the whole lifetime of the bridge. Maintenance works have to be performed after such intervals, which would allow minimizing the costs and at the same time ensuring the necessary safety and traffic possibilities. In addition to that no harm to the environment caused by the bridges may be allowed.

Bridge maintenance starts with a good design containing as less details causing problems during the bridge operation, as possible. Quality assurance and control in construction works is also crucial. During the bridge operation timely and qualitative bridge inspections have to be ensured.

In Latvian BMS the bridge maintenance is divided into: Routine maintenance; Periodic maintenance; Rehabilitation.

2) Routine Maintenance Works

The purpose of routine maintenance works is to provide the same traffic safety and driving conditions, which exist on the road where bridge is located, as well as, to preserve esthetical outlook of the structure and ensure the functioning of the bridge as specified in the design. The works are implemented by construction companies.

Routine maintenance works include: Cleaning of bridge elements; Cleaning of the area under the bridge; Elimination of scours in the soil; Temporary elimination of accident consequences; Elimination of small structural defects.

3) Periodic Maintenance Works

The purpose of routine maintenance works is to eliminate the defects, which could influence the traffic and bridge bearing capacity, as well as, traffic safety, environment and bridge lifetime. Maintenance works have to be executed after such intervals, which would allow the minimization of the costs. The works are executed by specialized construction companies according to the specifications and in the amounts specified in the bills of quantities.

Periodic maintenance works include: Renewal of coatings on the surfaces of steel structures; Renewal of coatings on the surfaces of reinforced concrete structures; Renewal of coatings on the surfaces of timber structures; Renewal of worn elements in roadway structure and water insulation.

4) Rehabilitation

The purpose of rehabilitation works is to renew the functionality of a damaged bridge element without its replacement and without limiting the traffic possibilities, traffic safety and bridge lifetime. The works are executed by specialized construction companies according to specially elaborated construction designs.

Rehabilitation works include: Repairs of the damages in pier and foundation concrete above and under the water; Repairs of the damages of steel, stone, timber or other structures in bridge substructure or superstructure; Replacement of damaged parts of structural elements.

6. Reconstruction

Bridge reconstruction is divided into strengthening and rebuilding.

1) Strengthening

The purpose of the works is to increase the bearing capacity of the bridge or its elements disregarding the presence or absence of any defects up to a level which is higher than the initial bearing capacity.

2) Rebuilding

The purpose of the works is to modify the function, use or standard of the bridge or its elements. They may include bridge widening or increase of clearance under the bridge, replacement of the whole superstructure or bridge deck, adding of pedestrian sidewalk, etc.

Strengthening or rebuilding works are executed by specialized construction companies according to construction designs elaborated particularly for this purpose.

The financing is allocated according to the following scheme in Figure 7-1:

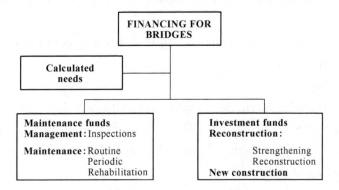

Figure 7-1   Financing for bridges

Bridge maintenance and reconstruction works have to be implemented in optimal intervals and in the necessary amount. Bridge inspection and routine maintenance works are practically implemented in the whole necessary amount.

However, insufficient financing may cause the delay of periodic maintenance or reconstruction works on several bridges. In such cases priorities are determined so that the delay of the works would have a minimal negative result.

7. Alternative Strategies

If the results of main or special bridge inspections show that the costs of the necessary repair works exceed 20% of the bridge replacement value, alternative strategies have to be studied. At least two different strategies have to be evaluated from the cost benefit point of view. The calculations in addition to maintenances costs have to include also all the costs, which arise to the road users and the publics due to the implementation of the chosen strategy.

The following strategies may be considered:

1) Temporary action: execution of small repairs in order to postpone the main works or bridge replacement until later moment.

2) Mayor action: execution of large repairs in a short period of time in order to increase the remaining bridge lifetime substantially.

3) New element or bridge: No repairs are executed; the bridge or its element is replaced at the end of its lifetime.

Different technical solutions may be considered for each strategy.

If the maintenance costs exceed 50% of the value of bridge replacement, the 3rd strategy is to be considered. Special attention has to be paid to bridge elements, as well.

When choosing the optimal strategy the following factors, which usually are not included in cost calculations, have to be considered, as well: the age of the bridge and its remaining lifetime; bridge bearing capacity; bridge width and road curvature; clearance under the bridge; traffic safety; traffic capacity; bridge use in the future; esthetical aspects; historical value.

If the financing for implementing the optimal strategy is not allocated, the bridge priorities have to be determined.

8. Reconstruction

Bridge reconstruction usually is substantiated with the choice of the 2nd or the 3rd strategy. With the increase of daily traffic loads according to the EU norms the bridge strengthening or the replacement of main bearing elements becomes topical. Some examples may be mentioned:

1) Bridge over the Venta in Skrunda. The bridge bearing capacity was increased and roadway was widened by replacing the span superstructure with lighter steel beams and reinforced concrete deck.

2) Bridge over the Gauja in Carnikava. The bridge bearing capacity was increased and roadway was widened by strengthening and connecting the separate span beams in a continuous girder. Particularly light sidewalk structure with steel grid deck was installed.

3) Bridge over the Malta in Vilani. The bridge bearing capacity was increased and roadway was widened by strengthening the existing reinforced concrete structures and reconstructing the bridge deck.

9. Conclusion

Bridge maintenance in Latvia historically has gone through all development phases: from the false idea about the eternity of reinforced concrete structures to planned and preventive maintenance works today.

To provide the necessary bridge bearing capacity and traffic safety in the conditions with limited financing the bridge engineers have to find balance between bridge preservation and reconstruction. The Bridge Management System implemented in Latvia is a modern tool for the solution of such bridge problems.

Integrated Skills: Writing a Scientific Essay

By means of the field research, Internet search or the review of special literatures, you should search for some technological information on a bridge that you are interested in. The information includes basic facts, design, structures, materials, construction, maintenance and rehabilitation of this bridge. Then you should write a scientific essay based on one of the above aspects, and finally present your essay in class and exchange your ideas with others.

Unit 8   Rail Transit Engineering

Lead-in

1. What are the advantages and disadvantages of light rail?
2. Could you draw a distinction between light rail and streetcar? What's that?

Vocabulary Warm-up: Matching

_____1. aerial cableway/ropeway/tramway	a. 缆索铁路
_____2. aerial railway / elevated railroad	b. 地下铁道,地铁
_____3. cable/rope railway/railroad	c. 磁悬浮列车
_____4. tram	d. 架空索道,缆车道
_____5. tunnel	e. 有轨电车
_____6. metro / subway / tube / underground	f. 高架铁路
_____7. maglev (magnetically levitated) train	g. 大众运输工具
_____8. people mover	h. 隧道;地道

Text A   Light Rail

Light rail or light rail transit (LRT) is a particular class of urban and suburban passenger railway that uses equipment and infrastructure that is generally less massive than that used for rapid transit systems, with modern light rail vehicles usually running along the system.

Light rail is the *successor* term to streetcar, trolley and tram in many *locales*, although the term is most *consistently* applied to modern tram or trolley operations employing features more generally associated with metro or subway operations, including *exclusive* rights-of-way, multiple unit train configuration and signal control of operations.

The term "light rail" is derived from the British English term "light railway" long used to distinguish *tram* operations from steam railway lines, and also from its usually lighter infrastructure.

Light rail systems are almost universally operated by electricity delivered through overhead lines, though several systems are powered through different means, such as the JFK Air Train, which uses a standard third for its electrical power, and trams in Bordeaux which use a special third-rail configuration in which the rail is only powered while a tram is on top of it. A few unusual systems like the River Line in New Jersey and the O-Train in Ottawa use diesel-powered trains, though this is sometimes intended as an *interim* measure until the funds to install electric power become available.

1. History of the Light Rail

From the mid-19th century, horse-drawn trams (or horse cars) were used in many cities around the world. In the late 1880s electrically-powered street railways became technically feasible after the invention of a trolley system of collecting current by American inventor Frank J. Sprague who installed the first successful system at Richmond, Virginia. They became popular because roads were then poorly surfaced, and before the invention of the internal combustion engine and the advent of motorbuses, they were the only practical means of public transport around cities.

The light rail systems built in the 19th and early 20th centuries generally only ran in single-car setups. Some rail lines experimented with multiple unit configurations, where streetcars were joined together to make short trains, but this didn't become common until later. When lines were built over longer distances (typically with a single track) before good roads were common, they were usually called interurban streetcars in North America or radial railways in Ontario.

In North America, many of these original light-rail systems were decommissioned in the 1950s and onward as the popularity of the automobile increased. Though some traditional trolley or tram systems still exist to this day, the term "light rail" has come to mean a different type of rail system. Beginning in the 1980s, some cities began reintroducing light-rail systems that more like subway or metro systems that operate at street level. These light-rail systems include modern, multi-car trains that can only be reached at stations that spaced anywhere from a couple blocks to a mile or more apart. Some of these systems operate within roadways alongside automobile traffic, and others operate on their own separate right-of-way.

As with other rail systems, the rail gauge has a lot of variations, but today standard gauge is dominant. Narrow gauge was common in many earlier systems, though as systems merged or died out, old lines were often upgraded, removed, or replaced. Some systems still use other track gauges, however.

2. Definition

Most rail technologies, including high-speed, freight, commuter/regional, and metro/subway are considered to be "heavy rail" in comparison. A few systems such as *people movers* and personal rapid transit could be considered as even "lighter", at least in terms of how many passengers are moved per vehicle and the speed at which they travel. Monorails are also considered to be a separate technology. Light rail systems can handle steeper inclines than heavy rail, and curves sharp enough to fit within street *intersections*. They are generally built in urban areas, providing frequent service with small, light trains or single cars.

The most difficult distinction to draw is that between light rail and streetcar or tram systems. There is a significant amount of *overlap* between the technologies, and it is usual to classify streetcar/trams as a subtype of light rail instead of as a distinct type of transportation. The two common versions are:

1) The traditional type, where the tracks and trains run along the streets and share space with road traffic. Stops tend to be very frequent, but little effort is made to set up special station. Because space is shared, the tracks are not usually visible.

2) A more modern variation, where the trains tend to run along their own right-of-way and are often separated from road traffic. Stops are usually less frequent, and the vehicles are often got on from a platform. Tracks are highly visible, and in some cases significant effort is used to keep traffic away through the use of special signaling and even grade crossings with gate arms. At the highest degree of separation, it can be difficult to draw the line between light rail and metros, as in the case of London's Docklands Light Railway, which would likely not be considered "light" compared with London Underground.

Many light rail systems have a combination of the two, with both on road and off road sections. In some countries, only the latter is described as light rail. In those places, trams running on mixed right-of-way are not regarded as light rail, but considered distinctly as streetcars or trams.

Light rail is usually powered by electricity, generally by means of overhead wires, but sometimes by a live rail, also called third (a high voltage bar alongside the track), requiring safety measures and warning to the public not to touch it. In some cases, especially when initial funds are limited, diesel-powered versions have been used, but it is not a preferred option. Some systems, such as the JFK Air Train in New York City, are automatic without a driver; however, such systems are not what are usually thought of as light rail. Automatic operation is more common in smaller people mover systems than in light rail systems, where the possibility of grade crossings and street running make driverless operation of the latter inappropriate.

3. Advantages of Light Rail

Light rail systems are usually cheaper to build than heavy rail, since the infrastructure does not need to be considerable, and tunnels are usually not required as most metro systems. In addition, the ability to handle sharp curves and steep gradients can reduce the amount of work required.

Traditional streetcar systems and also newer light rail systems are used in many cities around the world because they generally can carry a larger number of people than any bus-based public transport system. They are also cleaner, quieter, more comfortable, and in many cases faster than buses. In an emergency, light rail trains are easier to evacuate than monorail or elevated rapid rail trains.

Many modern light rail projects re-use parts of old rail networks, such as abandoned industrial rail lines.

4. Disadvantages of Light Rail

Like all modes of rail transport, light rail tends to be safest when operating in dedicated right-of-way with complete grade separations. Nevertheless, grade separations are not always financially or physically feasible.

In California, the development of light rail systems in Los Angeles and San Jose caused a high rate of collisions between automobiles and trolleys during the 1990s. The most common cause was that many senior citizens were unfamiliar with light rail trolleys and often mistook the trolley "T" signal lights for left-turn signal lights. They would then make a left turn, right into the path of a trolley. The same high crash rate problem existed when the Metro Rail was first set up in Houston, Texas.

To reduce such collisions, brighter lights and louder warning klaxons have been added to many at-grade crossings. However, consequently, many people do not like to live next to light rail crossings because the noise makes them impossible to sleep. A more effective means of reducing or preventing automobile-light rail collisions has been the installation of *quad* crossing gates at gate crossings. These gates block both lanes of a street when the gate closes. These prevent those driving automobiles from driving around the gates when they are lowered.

Monorail supporters like to point out that light rail trolleys are heavier per pound of cargo carried than heavy rail cars or monorail cars, because they must be designed to avoid collisions with automobiles.

Glossary

Complete the glossary using words in the text.

1. _____	接替的人或事物	2. _____	现场，场地
3. _____	一致地,相符	4. _____	独有的;独占的
5. _____	有轨的电车	6. _____	暂时的;临时的
7. _____	大众运输工具	8. _____	交叉
9. _____	重叠,重复,重合	10. _____	四边形

Exercises

1. Answer the following questions according to the text.

1) What's the definition of light rail?

2) Where is the term "light rail" derived?

3) What are the powers for light rail systems?

4) What is considered to be "heavy rail"?

5) What could be considered as even "lighter"?

6) What advantages does light rail have over heavy rail?

2. Choose the best answer to each question.

1) Most rail technologies, including high-speed, freight, commuter/regional, and metro/subway are considered to be _____.

A. light rail   B. heavy rail

C. personal rapid transit   D. people movers

2) Light rail is generally powered by _____.

A. steam   B. diesel

C. electricity   D. electromagnet

3) Light rail systems are generally _____ to build than heavy rail.

A. cheap   B. expensive

C. more expensive   D. cheaper

4) In an emergency, light rail trains are _____ to evacuate than monorail.

A. easier   B. difficult

C. easy   D. more difficult

5) Quad crossing gates prevent those driving automobiles from driving around the gates when they are _____.

A. lifted   B. lowered

C. reduced   D. increased

3. Write down the words with the following prefix or suffix.

over-   _____

auto-   _____

mono-   _____

-sist   _____

-ify   _____

4. Translate the following paragraph into Chinese.

Most rail technologies, including high-speed, freight, commuter/regional, and metro/subway are considered to be "heavy rail" in comparison. A few systems such as people movers and personal rapid transit could be considered as even "lighter", at least in terms of how many passengers are moved per vehicle and the speed at which they travel. Monorails are also considered to be a separate technology. Light rail systems can handle steeper inclines than heavy rail, and curves sharp enough to fit within street intersections. They are generally built in urban areas, providing frequent service with small, light trains or single cars.

Text B   Rapid Transit

A rapid transit, underground, subway, elevated, or metro system is a railway system, generally in an urban area, that generally has high capacity and frequency, with large trains and total or near total grade separation from other traffic.

1. Definitions and Nomenclature

There is no single term in English that all speakers would use for all rapid transit or metro systems. This fact reflects variations not only in national and regional usage, but in what characteristics are considered essential.

One definition of a metro system is as follows: an urban, electric mass transit railway system totally independent from other traffic with high service frequency.

But those who prefer the American term "subway" or the British "underground" would additionally specify that the tracks and stations must be located below street level so that pedestrians and road users see the street exactly as it would be without the subway; or at least that must be true for the most important, central parts of the system. On the contrary, those who prefer the American "rapid transit" or the newer term "metro" tend to regard this as a less important characteristic and are pleased to include systems that are completely elevated or at ground level (at grade) as long as the other criteria are met. A rapid transit system that is generally above street level may be called an "elevated" system (often shortened to "el" or, in Chicago, "L"). In some cities the word "subway" applies to the entire system, in others only to those parts that actually are underground; and analogously for "el".

Germanic languages usually names meaning "underground railway" (such as "subway" or "U-Bahn"), while many others use "metro".

2. Train Size and Motive Power

Some urban rail lines are built to the full size of main-line railways; others use smaller tunnels, limiting the size and sometimes the shape of the trains (in the London Underground the informal term tube train is commonly used). Some lines use light rail rolling stock, perhaps surface cars merely routed into a tunnel for all or part of their route. In many cities, such as London and Boston's MBTA, lines using different types of vehicles are organized into a single unified system.

Although the initial lines of what became the London Underground used steam engines, most metro trains, both now and historically, are electric multiple units, with steel wheels running on two steel rails. Power is usually supplied by means of a single live third rail (as in New York) at 600 to 750 volts, but some systems use two live rails (noticeably London) and thus eliminate the return current from the running rails. Overhead wires, allowing higher voltages, are more likely to be used on metro systems without much length in tunnel, as in Amsterdam; but they also exist on some that are underground, as in Madrid. Boston's Green Line trains derive power from an overhead wire, both while traveling in a tunnel in the central city and at street level in the suburban areas.

Systems usually use DC power instead of AC, even if this requires large rectifiers for the power supply. DC motors were formerly more efficient for railway applications, and once a DC system is in place, converting it to AC is usually considered too large a project to contemplate.

3. Tracks

Most rapid transit systems use conventional railway tracks, though since tracks in subway tunnels are not exposed to wet weather, they are often fixed to the floor instead of resting on ballast. The rapid transit system in San Diego, California operates tracks on former railroad rights of way that were acquired by the governing entity.

Another technology using rubber tires on narrow concrete or steel rollways was pioneered on the Paris Metro, and the first complete system to use it was in Montreal. Additional horizontal wheels are required for guidance, and a conventional track is often provided in case of flat tires and for switching. Advocates of this system note that it is much quieter than conventional steel-wheeled trains, and allows for greater inclines given the increased traction allowed by the rubber tires.

Some cities with steep hills incorporate mountain railway technologies into their metros. The Lyon Metro includes a section of tack (cog) railway, while the Carmelit in Haifa is an underground funicular.

For elevated lines, still another alternative is the monorail. Supported or "straddle" monorails, with a single rail below the train, include the Tokyo Monorail; the Schwebebahn in Wuppertal is a suspended monorail, where the train body hangs below the wheels and rail. Monorails have never gained wide acceptance except for Japan, although Seattle has a short one, which it hopes to replace with a new, larger system, and one has lately been built in Las Vegas. One of the first monorail systems in the United States was installed at Anaheim's Disneyland in 1959 and connects the amusement park to a nearby hotel. Disneyland's builder, animator and filmmaker Walt Disney, offered to build a similar system between Anaheim and Los Angeles.

4. Crew Size and Automation

Early underground trails often carried an attendant on each car to operate the doors or gates, in addition to a driver. The introduction of powered doors around 1920 permitted crew sizes to be decreased, and trains in many cities are now operated by a single person, where the operator would not be able to see the whole side of the train to tell whether the doors can be safely closed, mirrors or closed-circuit TV monitors are often provided for that purpose.

An alternative to human drivers became available in the 1960s, as automated systems were developed that could start a train, accelerate to the correct speed, and stop automatically at the next station, also taking into account the information that a human driver would obtain from line side or cab signals. The first complete line to use this technology was London's Victoria Line, in 1968. In usual operation the one crew member sits in the driver's position at the front, but just closes the doors at each station; the train then starts automatically. This style of system has become widespread. A variant is seen on London's Docklands Light Railway, opened in 1987, where the "passenger service agent" (formerly "train captain") rides with the passengers instead of sitting at the front as a driver would. The same technology would have allowed trains to operate completely automatically with no crew, just as most elevators do; and as the cost of automation has decreased, this has become financially attractive. But a countervailing argument is that of possible emergency situations. A crew member on board the train may be able to prevent the emergency in the first place, drive a partly failed train to the next station, assist with an evacuation if needed, or call for the correct emergency services (police, fire, or ambulance) and help direct them.

In some cities the same reasons are considered to justify a crew of two instead of one; one person drives from the front of the train, while the other operates the doors from a position farther back, and is more conveniently able to help passengers in the rear cars. The crew members may exchange roles on the reverse trip (as in Toronto) or not (as in New York).

Completely crewless trains are accepted on newer system where there are no existing crews to be removed, and especially on light rail lines. Thus the first such system was the VAL (véhicule automatique léger or "automated light vehicle") of Lille, France, inaugurated in 1983. Additional VAL lines have been built in other cities. In Canada, the Vancouver Sky Train carried no crew members, while Toronto's Scarborough RT, opening the same year (1985) with otherwise similar trains, uses human operators.

These systems generally use platform-edge doors (PEDs), in order to improve safety and ensure passenger confidence, but this is not universal: for example, the Vancouver Sky Train does not. (And on the contrary, some lines which retain drivers, however, still use PEDs, noticeably London's Jubilee Line Extension. MTR of Hong Kong also uses platform screen doors, the first to install PEDs on an already operating system.)

With regard to larger trains, the Paris Metro has human drivers on most lines, but runs crewless trains on its newest line, Line 14, which opened in 1998. Singapore's North East MRT Line (2003) claims to be the world's first completely automated underground urban heavy rail line. The Disneyland Resort Line of Hong Kong MTR is also automated.

5. Tunnel Construction

The construction of an underground metro is an expensive project, often carried out over many years. There are several different methods of building underground lines.

In one usual method, known as cut-and-cover, the city streets are excavated and a tunnel structure strong enough to support the road above is built at the trench, which is then filled in and the roadway rebuilt. This method often involves extensive relocation of the utilities usually buried not far below city streets — especially power and telephone wiring, water and gas mains, and sewers. The structures are generally made of concrete, perhaps with structural columns of steel; in the oldest systems, brick and cast iron were used. Cut-and-cover construction can take so long that it is often necessary to build a temporary roadbed while construction is going on underneath in order to avoid closing main streets for long periods of time; in Toronto, a temporary surface on Yonge Street supported cars and streetcar tracks for several years while the Yonge subway was built.

Some American cities, like Newark, Cincinnati and Rochester, were originally built around canals. When the railways took the place of canals, they were able to bury a subway in the disused canal's trench, without rerouting other utilities, or acquiring a fight of way piecemeal.

Another common way is to start with a vertical shaft and then dig the tunnels horizontally from there, often with a tunneling shield, thus avoiding almost any disturbance to existing streets, buildings, and utilities. But problems with ground water are more likely, and tunneling through native bedrock may require blasting. (The first city to extensively use deep tunneling was London, where a thick sedimentary layer of clay largely avoids both problems.) The confined space in the tunnel also restricts the machinery that can be used, but specialized tunnel-boring machines are now available to overcome this challenge. One disadvantage with this, nevertheless, is that the cost of tunneling is much higher than building systems cut-and-cover, at-grade or elevated. Early tunneling machines could not make tunnels large enough for conventional railway equipment, necessitating special low, round trains, such as are still used by most of the London Underground, which cannot fix air conditioning on most of its lines because the amount of empty space between the trains and tunnel walls is so small.

The deepest metro system in the world was built in St. Petersburg, Russia. In this city, built in the marshland, stable soil starts more than 50 meters deep. Above that level the soil is mostly made up of water-bearing finely dispersed sand. As a result of this, only three stations out of nearly 60 are built near the ground level and three more above the ground. Some stations and tunnels lie as deep as $100\sim120$ meters below the surface.

One advantage of deep tunnels is that they can dip in a basin-like profile between stations, without incurring significant extra costs owing to having to dig deeper. This technique, also referred to as putting stations "on humps", allows gravity to help the trains as they accelerate from one station and brake at the next. It was used as early as 1890 on parts of the City and South London Railway, and has been used many times since.

Text C   Tram

A tram (tramcar, trolley, or streetcar) is a rail borne vehicle, lighter than a train, designed for the transport of passengers (and/or, very occasionally, freight) within, close to, or between villages, towns and/or cities. Trams are distinguished from other forms of railway systems in that they travel wholly or partly along tracks laid down in streets, usually on track reserved for the tram system. A cable car is a special type of tram.

Tram systems are common throughout Europe and were common throughout the western world in the early 20th century. Although they disappeared from many cities for many years in the mid 20th century, in recent years they have made a comeback.

The terms "tram" and "tramway" were originally Scottish and Northern English words for the type of truck used in coal mines and the tracks on which these trucks ran — probably derived from a North Sea Germanic word of unknown origin meaning the "beam or shaft of a barrow or sledge", also "a barrow or truck body". The sense of "streetcar" is first recorded in 1860.

1. History

Appearing in the first half of the 19th century, trams were at first pulled by horses.

The first trams, known as streetcars or horsecars, were built in the US, and developed from city stagecoach lines and omnibus lines that picked up and dropped off passengers on a regular route and without the need to be pre-hired. These first lines operated in Baltimore, Maryland in 1828, in 1832 on the New York and Harlem Railroad in New York City, and in 1834 in New Orleans. At first the rails protruded above street level, causing accidents and major trouble for pedestrians. They were supplanted in 1852 by grooved rails, invented by Alphonse Loubat. The first tram in France was inaugurated in 1853 for the World's Fair, where a test line was presented along the Cours de la Reine, in the 8th Arrondissement. Trams were first regularly used in Europe in Sarajevo, starting in 1885.

These streetcars were an animal railway, usually using horses and sometimes mules to haul the cars, usually two as a team. Rarely other animals were tried, including humans in emergencies.

One of the advantages over earlier forms of transit was the low rolling resistance of metal wheels on steel rails, allowing the animals to haul a greater load for a given effort. Problems included the fact that any given animal could only work so many hours on a given day, had to be housed, groomed, fed and cared for day in and day out, and produced prodigious amounts of manure, which the streetcar company was charged with disposing of. Since a typical horse pulled a car for perhaps a dozen miles a day and worked for four or five hours, many system needed ten or more horses in stable for each horsecar. New York City had the last regular horsecar lines in the U. S., closing in 1914. A mule-powered line in Celaya, Mexico operated until 1956. Horse-drawn trams still operate in Douglas, Isle of Man.

The tram developed after that in numerous cities of Europe (London, Berlin, Paris, etc.). Faster and more comfortable than the omnibus, trams had a high cost of operation because they were pulled by horses. That is why mechanical drives were rapidly developed: with steam power in 1873, and electrical after 1881, when Siemens AG presented the electric drive at the International Electricity Exhibition in Paris.

The convenience and economy of electricity resulted in its rapid adoption once the technical problems of production and transmission of electricity were solved. The first electric tram opened in Berlin in 1881.

2. Cable Pulled Cars

The next type of streetcar was the cable car, which sought to reduce labor costs and the hardship on animals. Cable cars are pulled along a rail track by a continuously moving cable running at a constant speed on which individual cars stop and start by releasing and gripping this cable as required. The power to move the cable is provided at a site away from the actual operation. The first cable car line in the United States was tested in San Francisco, California in 1873.

Cable cars suffered from high infrastructure costs, since a vast and expensive system of cables, pulleys, stationary engines and vault structures between the rails had to be provided. They also require strength and skill to operate, to avoid obstructions and other cable cars. The cable had to be dropped at particular locations and the cars coast, for example when crossing another cable line. After the development of electrically-powered streetcars, the more costly cable systems declined rapidly.

Cable cars were especially useful in hilly cities, partially explaining their survival in San Francisco, though the most extensive cable system in the U. S. was in Chicago, Illinois, a flat city. The San Francisco cable cars continue to perform a regular transportation function, in addition to being a tourist attraction.

3. Electric Trams (Trolley Cars)

Electric-powered trams (trolley cars, so called for the trolley pole used to gather power from an unshielded overhead wire), were first successfully tested in service in Richmond, Virginia in 1888, in an installation by Frank J. Sprague. There were earlier commercial installations of electric streetcars, including one in Berlin, Germany, as early as 1881 by Werner von Siemens and the company that still bears his name, and also one in St. Petersburg, Russia, invented and tested by Fiodor A. Pirotskiy in 1880. The earlier installations, however, proved difficult and/or unreliable. Siemens' line, for example, provided power through a live rail and a return rail, like a model train setup, limiting the voltage that could be used, and providing unwanted excitement to people and animals crossing the tracks. Siemens later designed his own method of current collection, this time from an overhead wire, called the bow collector. Once this had been developed his cars became equal to, if not better, than any of Sprague's cars.

Since Sprague's installation was the first to prove successful in all conditions, he was credited with being the inventor of the trolley car.

A rare but significant variant of the trolley car was the conduit car, which drew its power from an underground third rail.

4. Golden Age

Trams experienced a rapid expansion at the start of the 20th century until the period between the two world wars. There was a rapid increase in the number of lines and increase in the number of riders; indeed, it became the primary mode of urban transportation. Horse-drawn transport virtually disappeared in all European, American and Indian cities by 1910. Buses were still in a development phase at this time, gaining in mechanical reliability, but remaining behind compared to the benefits offered by trams;

the automobile was still reserved for the well-to-do.

5. A Temporary Disappearance from Many Cities

In several countries the advent of personal motor vehicles caused the rapid disappearance of the tram from most Western and Asian countries by the end of the 1950s. The technical progress of the bus rendered it more reliable, and it became a serious competitor to the tram because it did not require the construction of costly infrastructure.

In many cases buses also provided a smoother ride and a faster journey than the older trams. For example, the tram network survived in Budapest but for a considerable period of time bus fares were higher to recognize the superior quality of the buses.

Governments thus put investment principally into bus networks. Indeed, infrastructure for roads and highways meant for the automobile were perceived as a mark of progress. The priority given to roads was illustrated in the proposal of French president Georges Pompidou who declared in 1971 that "the city must adapt to the car".

Tram networks were no longer maintained or modernized, a state of affairs that served to discredit them in the eyes of the public. Old lines, considered archaic, were then bit by bit replaced by buses.

Tram networks has disappeared almost completely in North America, France, the UK, India, Turkey and Spain. On the other hand, they were maintained or modernized in Switzerland, Germany, Poland, Finland, Romania, Austria, Italy, Belgium, the Netherlands, Scandinavia, Japan and Eastern Europe. In France and the UK, only the networks in Lille, Saint-Etienne, Marseille, and Blackpool survive from this period, but they are each reduced to a single line. Australian tram networks disappeared by the 1970s, with the exception of the extensive system in Melbourne and the Glenelg line in Adelaide.

6. Return to Grace

The priority given to personal vehicles and notably to the automobile led to a loss in quality of life, particularly in large cities where smog, traffic congestion, sound pollution and parking became problematic. Acknowledging this, some authorities saw fit to redefine their transport policies. The bus had shown its limits on account of its low capacity and its difficult coexistence with automobile traffic, which made it slow both on the road and commercially. Subways required a heavy investment and presented problems in terms of subterranean spaces that required constant security. For subways, the investment was mainly in underground construction, which made it impossible in some cities (with underground water reserves, archaeological remains, etc.). Subway construction thus was not a universal panacea.

The advantages of the tram thus became more visible. At the end of the 1970s, some governments studied, and then built new tram lines. In France, Nantes and Grenoble lead the way in terms of the modern tram, and new systems were inaugurated in 1985 and 1988. In 1994 Strasbourg opened a system with novel British-built trams, specified by the city, with the goal of breaking with the archaic conceptual image that was held by the public.

The public, who realized with each installation of tram lines their benefits in urban flexibility and redistribution and the reduction in downtown automobile traffic, encouraged numerous city governments to so equip their streets. Many cities already equipped with trams have extended their lines and built new ones.

A great example of this shift in ideology is the city of Munich, which began replacing its tram network with a metro a few years before the 1972 Summer Olympics. When the metro network was finished in the 1990s the city began to tear out the tram network (which had become rather old and decrepit), but now faced opposition from many citizens who enjoyed the enhanced mobility of the mixed network — the metro lines deviated from the tram lines to a significant degree. New rolling stock was purchased and the system was modernized, and a new line was proposed in 2003.

7. Technical Developments

Later, trans were attached to a moving cable underneath the road. The cable would be pulled by a steam engine at a powerhouse. The Monongahela and Duquesne Inclines in Pittsburgh, Pennsylvania, USA, had some of the appearance of trams, but were more accurately funiculars. Modern trams generally used overhead electric cables, from which they draw current through a pantograph, a bow collector (less commonly) or the now-rare trolley pole (the first was most common and used on most new tram designs). The first operational electric street railway was started in Scranton, Pennsylvania, but the first large-scale electric street railway was built in Richmond, Virginia in January, 1888. By 1890 over 100 such systems had been installed or were planned.

There are other methods of powering electric trams, sometimes preferred for aesthetic reasons since poles and overhead wires are not required. The old tram systems in London, Manhattan (New York City), and Washington D. C. used live rails, like those on third-rail electrified railways, but in a conduit underneath the road, from which they drew power through a plough. Washington's was the last of these to close, in 1962. Today, no commercial tramway uses this system. More recently, a modern equivalent has been developed which allows for the safe installation of a third rail on city streets, which is known as surface current collection or ground level power supply; the main example of this is the new tramway in Bordeaux.

In narrow situations double-track tram lines sometimes reduce to single track, or, to avoid switches, have the tracks interlaced, e. g. in the Leidsestraat in Amsterdam on three short stretches; this is known as interlaced or gauntlet track.

Traditionally trams had high floors, requiring passengers to climb several steps in order to board, but since the 1990s this design has been largely replaced by low-floor trams, or occasionally by high-floor trams with level boarding platforms, as in

Manchester's Metrolink and some parts of Cologne's network, which allow passengers in wheelchairs or with perambulators to access vehicles more easily.

8. Tram-Train

Tram-train operation uses vehicles such as the Flexity Link and Regio-Citadis which are suited for use on urban tram lines, but also meet necessary indication, power, and resistance requirements to be certified for operation on main line railways. This allows passengers to travel from suburban areas into city-centre destinations without having to change from a train to a tram when they arrive at the central station.

It has been primarily developed in Germanic countries, in particular Germany and Switzerland. Karlsruhe is a notable pioneer of the tram-train. This system should be brought into service in the Paris area in 2005.

Integrated Skills: Summary

In no more than 100 words give an account of the passage about light rail. Use your own words as far as possible. Finally present your essay in class and exchange your ideas with others.

Unit 9   Tunnel Engineering

Lead-in

1. When does an engineer decide to construct a tunnel?
2. When does an engineer give up considering building a traffic tunnel?
3. What has made the construction of tunnels easier?

Vocabulary Warm-up: Matching

_____ 1. tunnel blasting		a. 盾构
_____ 2. tunnel diversion		b. 隧道衬砌
_____ 3. tunnel muck hauling		c. 隧洞导流
_____ 4. tunnel shield		d. 隧道挖掘机
_____ 5. tunnel support		e. 隧洞爆破
_____ 6. tunnel lining		f. 隧洞支撑
_____ 7. tunnel drill		g. 隧道凿岩机
_____ 8. tunnel excavator		h. 出渣

Text A   Traffic Tunnel Construction

Whenever the proposed path of a road or railway is *obstructed* by a hillside, a *waterway* or some form of construction, the engineer designing the project has to decide whether or not it is practical to construct a *tunnel* through or under the obstacle. In making his decision, he has not only to consider the economic aspect, but also weigh up all the constructional advantages and disadvantages of both tunneling and the *alternative* method of either passing around or over the obstruction. In practice, it is often found that tunneling, although costly, proves to be less expensive in the long run than any alternative system. There are instances, however, where the engineer has no choice but to pass around or over the *obstacle*. For example, *intensely* hard rock might well make tunneling extremely difficult and far too costly to even contemplate.

The inability of existing road systems of many large towns to cope with modern traffic requirements has made tunnel construction a proposition well worth serious consideration. Up to a point, *flyovers* and *underpasses*, being the first really decisive steps to speed up and *divert* traffic, have eased the situation, but a considerable contribution towards a satisfactory solution can be made by underground railway networks.

Many of the larger cities of the world have been successfully served by underground railways for years, and most of them are still extending their networks. In the meantime, other cities have introduced them for the first time in an effort to reduce traffic congestion. In the broadest sense, there are two fundamentally different types of construction employed for underground systems. They are sub-pavement tunnels and "*tube*s", the latter being situated at a far greater depth. The technique in both forms has improved dramatically over the years.

The decision of city planning authorities as to whether a sub-pavement or "tube" system should be adopted depends a great deal on what type of public transport system already exists. If a perfectly efficient surface *tramway* system is functioning at the time, it is often found advantageous to extend it as a sub-pavement system. Normally, it is relatively straightforward to provide a *ramp*ed connection between the existing street level system and the underground extension. On the other hand, it may prove more satisfactory to re-organize the public transport system and provide an entirely separate tube system, working independently from the service at steel level.

Modern equipment and ingenuity have enabled the construction of tunnels to be far less laborious than hitherto. In applying the method known as the "Berlin Building System", firstly, the open *excavation* is taken down to a depth of between 12 and 25 meters. The sides of the excavation are supported by lines of driven *pile*s and timbering. Generally, this type of preliminary construction is also required to carry temporary bridges for both vehicles and pedestrians.

Supporting open excavations are often employed. For example, steel sheet piling and the more modern methods of either "Benoto-Veder" pile walling or "I. C. O. S. - Veder" *diaphragm* reinforced concrete walling are all in common use today, because in each no *lateral* supports are necessary.

The "Benoto-Veder" pile walling consists of concrete piles accurately *interlock*ed one with the other to form a complete *vertical* support to the sides of the excavation. The use of *bentonite* mud is the basis of the "I. C. O. S.-Veder" process. As the deep *trench* is being excavated by means of a special *grab* operated by an electric-powered *winch*, the mud is poured in. In this way, the excavation is stabilized and held open until the diaphragm walling has been placed in position in the trench. The steel reinforcement is then *prefabricate*d into cages, which are lowered by crane into position through the bentonite mud. Finally, the concrete is placed around the reinforcement by *tremie* and the "*slurry*" is *displace*d as the concrete rises to fill the trench. The diaphragm walling is usually *cast* in sections not exceeding 6 meters in length. Vertical construction joints between sections are formed by temporarily placing a steel pipe having an outside diameter equal to the width of walling, in the trench at the end of each section. The pipe acts as end *shuttering* for the concrete, and on being removed; the half-round end formed on one section allows a neat and fairly *watertight* joint to be made with the next section. The *span* between two such diaphragm walls can easily be bridged with precast reinforced concrete *beam*s or a cast *in-situ* reinforced concrete slab to complete the main *shell* of the tunnel.

Glossary

Complete the glossary using words in the text.

1. _____ 阻隔;阻塞　　2. _____ 航道;水路;排水沟
3. _____ 隧道;地下通道　4. _____ 可选择的;替换的
5. _____ 障碍　　　　　6. _____ 密集地
7. _____ 立交桥;天桥
8. _____ 地下通道;下穿交叉道路
9. _____ 转移;改道　　10. _____ 地铁
11. _____ 电车;电车轨道　12. _____ 斜面,斜坡
13. _____ 挖掘;被挖掘地带
14. _____ 桩
15. _____ 横隔墙;横隔板;横隔梁
16. _____ 侧的;横向的　17. _____ 连结;结合
18. _____ 垂直的,直立的
19. _____ 膨润土,斑脱土;膨土岩
20. _____ 沟槽,沟渠;深沟
21. _____ 抓具;抓斗挖土机
22. _____ 绞车;卷扬机
23. _____ 预先建造;预制构件
24. _____ 混凝土导管;漏斗管
25. _____ 泥浆　　　　26. _____ 移置;移走
27. _____ 浇筑;筑造　28. _____ 模板;支板
29. _____ 防水的;不透水的
30. _____ 跨距;跨度　31. _____ 横梁
32. _____ 原位;现场　33. _____ 外壳

Exercises

1. Write down the words with the following prefix or suffix.

inter- _____
under- _____
dis- _____
-vert _____
-struct _____

2. Answer the following questions according to the passage.

1) What are the first really decisive steps taken to speed up and divert traffic?

2) What are the two fundamentally different types of construction employed for an underground system?

3) What is the method to connect the existing street level system with the underground extension?

4) What are the two most commonly used methods for supporting open excavations?

5) How is the concrete placed around the reinforcement?

6) How are the vertical construction joints formed between sections?

7) What is the function of a steel pipe in forming the vertical construction joints between sections?

3. Translate the following into Chinese.

1) In practice, it is often found that tunneling, although costly, proves to be less expensive in the long run.

2) The inability of existing road systems of many large towns to cope with modern traffic requirements has made tunnel construction a proposition well worth serious consideration.

3) Many of the large cities of the world have been successfully served by underground railways for years, and most of them are still extending their networks.

4) Modern equipment and ingenuity have enabled the construction of tunnels to be far less laborious than hitherto.

5) Vertical construction joints between sections are formed by temporarily placing a steel pipe, having an outside diameter equal to the width of walling, in the trench at the end of each section.

Text B   Shields

A tunnel shield is a structural system, normally constructed of steel, used during the face excavation process. The shield has an outside configuration which matches the shape of the tunnel. The shield provides protection for the men and equipment and also furnishes initial ground support until structural supports can be installed within the tail section of the shield. The shield also provides a reaction base for the breast-board system used to control face movement. The shield may have either an open or closed bottom. In a closed-bottom shield, the shield structure and skin provide 360-degree ground contact and the weight of the shield rests upon the invert section of the shield skin. The open shield has no bottom section and requires some additional provision to support its weight and the superimposed weight of ground pressure bearing on the skin; normally this provision is a pair of side drifts driven in advance of shield excavation. Rails or skid tracks are installed within these side drifts to provide bearing support for the shield.

Shield length generally varies from 1/2 to 3/4 of the tunnel diameter. The front face of the shield is generally hooded so that the top of the shield protrudes forward further than the invert portion; this provides additional protection for the men working at the face and also eases pressure on the breast-boards. The steel skin of the shield may vary from 0.5 to 4 in (1.3 to 10 cm) in thickness, depending on the expected ground pressures. The type of steel used in the shield is the subject of many arguments within the tunneling fraternity. Some prefer mild steel in the A36 category because of its ductility and ease of welding in the underground environment where precision work is difficult. Others prefer a high-strength steel such as T-1 because of its higher strength/weight ratio. Shield weight may range from 5 to 500 tons. Most of the heaviest shields are found in the former Soviet Union because of their preference for cast iron in both structural and skin elements.

Propulsion for the shield is provided by a series of hydraulic jacks installed in the tail of the shield that thrust against the last steel set that has been installed. The total required thrust will vary with skin area and ground pressure. Several shields have been constructed with total thrust capabilities in excess of 10,000 tons. Hydraulic systems are usually self-contained, air-motor powered, and mounted on the shield. Working pressures in the hydraulic system may range from 3,000 to 10,000 psi (20,000 to 70,000 kN/m^2). To resist the thrust of the shield jacks, a horizontal structure member (collar brace) must be installed opposite each jack location and between the flanges of the steel set. In addition, some structural provision must be made for transferring this thrust load into the tunnel walls. Without this provision the thrust will extend through the collar braces to the tunnel portal.

An Englishman, Marc Brunel, was credited with inventing the shield. Brunel supposedly got his idea by studying the action of the Teredo navalis, a highly destructive woodworm, when he was working at the Chatham dock yard. In 1818 Brunel obtained an English patent for his rectangular shield which was subsequently used to construct the first tunnel under the River Thames in London. In 1869 the first circular shield was devised by Barlow and Greathead in London and is referred to as the Greathead-type shield. Later that same year, Beach in New York City produced a similar shield. The first use of the circular shield came during 1869 when Barlow and Greathead employed their device in the construction of the 7 ft (2.1 m) diameter Tower Subway under the River Thames. Despite the name of the tunnel, it was used only for pedestrian traffic. Beach also put his circular shield to work in 1869 to construct a demonstration project for a proposed New York City subway system. The project consisted of an 8 ft (2 m) diameter tunnel, 294 ft (90 m) long, which was used to experiment with a subway car propelled by air pressure.

Shields are most commonly used in ground conditions where adequate stand-up time does not exist. The advantage of the shield in this type of ground, in addition to the protection afforded men and equipment, is the time available to install steel ribs, liner plates, or precast concrete segments under the tail segment of the shield before ground pressure and movement become adverse factors.

One of the principal problems associated with shield use is steering. Non-uniform ground pressure acting on the skin tends to force the shield off line and grade. This problem is particularly acute with closed-bottom shields that do not ride on rails or skid tracks. Steering is accomplished by varying the hydraulic pressure in individual thrust jacks. If the shield is trying to drive, additional pressure on the invert jacks will resist this tendency. It is not unusual to find shields wandering several feet from the required line and grade. Although lasers are frequently used to provide continuous line and grade data to the operator, once the shield wanders off its course, its sheer bulk resists efforts to bring it back. Heterogeneous ground conditions, such as a clay with random boulders, also present steering problems.

One theoretical disadvantage of the shield is the annular space left between the support system and the ground surface. When the support system is installed within the tail section of the shield, the individual support members are separated from the ground surface by the thickness of the tail skin. When steel ribs are used, the annular space is filled with timber blocking as the forward motion of the shield exposes the individual ribs. A continuous support system presents a different problem. In this case, a filler material, such as pea gravel or grout, is pumped behind the support system to fill the void between it and the ground surface.

The main enemy of the shield is ground pressure. As ground pressure begins to build, two things happen: more thrust is required for shield propulsion and stress increases in the structural members of the shield. Shields are designed to withstand and function under a preselected ground pressure. Designers will select this pressure as a percentage of the maximum ground pressure contemplated by the permanent tunnel design. In some cases, unfortunately, the shield just gets built without specific consideration of the ground pressures it might encounter. When ground pressure exceeds the design limit, the shield gets "stuck". The friction component of the ground pressure on the skin becomes greater than the thrust capability of the jacks. Several methods, including pumping bentonite slurry into the skin-ground interface, pushing with heavy equipment, and bumping with dynamite, have been applied to stuck shields with occasional success.

Because ground pressure tends to increase with time, the cardinal rule of operation is "keep moving". This accounts for the frantic activity when a shield has suffered a temporary mechanical failure. As ground pressure continues to build on the nonmoving shield, the load finally exceeds its structural limit and bucking begins. An example of shield destruction occurred in California in 1968 when two shields being used to drive the Carly V. Porter Tunnel were caught by excessive ground pressure and deformed beyond repair. One of the Porter Tunnel shields was brought to a halt in reasonably good ground by a water-bearing ground fault that required full breast-boards. While the contractor was trying to bring the face under control, skin pressure began to increase. With the face condition finally stabilized, the contractor prepared to resume operations and discovered the shield was stuck. No combination of methods was able to move it, and the increasing ground pressure destroyed the shield.

To offset the ground pressure effect, a standard provision in design is a cutting edge radius several inches greater than the main body radius. This allows a certain degree of ground movement before pressure can come to bear on the shield skin. Another approach, considered in theory but not yet put into practice, is the "watermelon seed" design. The theory calls for a continuous taper in the shield configuration; maximum radius at the cutting edge and the minimum radius at the trailing edge of the tail. With this configuration, any amount of forward movement would create a drop in skin pressure.

Working decks, spaced 8 to 10 ft (2.4 to 3.0 m) vertically, are provided inside the shield. These working decks enable the miners to excavate the face, drill and load blast holes, if necessary, and adjust the breast-board system. The hydraulic jacks for the breast-boards are mounted on the underside of the working decks. Blast doors are sometimes installed as an integral part of the work decks if a substantial amount of blasting is expected.

Some form of mechanical equipment is provided on the rear end of the working decks to assist the miners in handling and placing the elements of the support system. In larger tunnels, these individual support elements can weigh several tons and mechanical assistance becomes essential sufficient. Vertical clearance must be provided between the invert and the first working deck to permit access to the face by the loading equipment.

Text C   The Channel Tunnel: The Dream Becomes Reality

The idea of constructing a fixed link between France and Great Britain took root in the imagination of engineers and geologists as early as 1751: it has taken nearly 250 years for this dream to be accomplished.

On 20 January 1986, French President Francois Mitterand and British Prime Minister Margaret Thatcher announced that the "Transmanche" Project, proposed by the Channel Tunnel Group for the U. K. side and the France Manehe for the French side, had been selected by both countries.

The Franco-British Treaty concerning the Channel Tunnel was signed on 12 January 1986 by the Foreign Ministers in Canterbury Cathedral, and the Concession Agreement was signed in March 1986.

The largest private sector project of the century was given the go-ahead and the works commenced at the end of 1986. On 1 December 1990, the historic junction was made between France and Great Britain.

On 6 May 1994, Queen Elizabeth and President Mitterand inaugurated the Channel Tunnel.

1. Scope of the Contract

The contract that was awarded to the TML Joint Venture covered the design, procurement, construction, and testing of the fixed link between France and Great Britain. It included:

1) The design and construction of two running tunnels and a service tunnel, with communication networks, communication passage, piston relief ducts, and pumping stations constructed under the sea and under land over a total distance of 50 km.

2) The services areas and buildings of the terminal, including access roads and railway links, on an area of 700 ha at Coquelles in France; and an area of 140 ha at Cheriton in the U.K.

3) The design, procurement and installation of railway lines, catenaries, signaling equipment, ventilation, drainage, fire detection and fire fighting installations, electrical supply, control and communications systems, etc.

4) The design and procurement of the rolling stock, i. e., locomotives and wagons for the transport of heavy goods vehicles (H. G. V.), automobiles and buses on single deck or double deck shuttles.

5) The testing of these systems, testing of the interfaces between these systems, and demonstration of the performance of the complete system.

2. A Few Statistics on the Channel Tunnel

Below are noted a few salient statistics on the Channel Tunnel, chosen from among hundreds that could be listed as:

Eleven Tunnel Boring Machines (TBMs) — five in France and six in the U. K. — were sent into battle on the 150 km of tunnels.

The project involved 554 transverse passages; communication passages, piston relief ducts and technical rooms were constructed.

To create the terminals, 840 ha of land were developed. This work involved: 16 million m^3 of earth-moving; 62,000 m^2 of buildings; 1,200,000 m^2 of roadworks; 190 km of railway track, with 176 switch points to permit more than 500 different routings; 8,000 km of cable and 500 km of pipework installed for the mechanical and electrical services.

The data transmission system handles 26,000 items of technical data and 15,000 control points for management of the rail traffic, the application software counting 250,000 lines of programme.

The 238 km fibre optic network handles 700 million items of data per second.

Nearly 1,000 partial acceptance tests and 230 system acceptance tests were carried out.

43 locomotives and 525 wagons were designed and built to comprise the tounists and H. G. V. shuttles, each nearly 800 m long and weighing 2,600 t.

Nearly 13,000 people (5,600 in France) worked directly on the construction of the

Channel Tunnel, corresponding to more than 100 million hours of work.

More than 2,000 subcontractors, suppliers and consultant practices participated in the construction of the project.

3. Driving the Tunnel: A Technical Challenge

The challenge of the Channel Fixed Link lay primarily in the capacity to drive more than 150 km of tunnels in an extremely short time-scale — 3.5 years — without an intermediate means of access for the undersea portion, which represented more than 37 km per tunnel.

4. The Geology and Choice of TBMs

The ground investigation geological reports and the limit criteria imposed for inclines and curves to permit passage of high-speed trains in the tunnels led to the definition of an alignment which on the U. K side of the project, passed within the stratum of chalk marl — a chalk of low permeability and few faults, and therefore possessing low water-bearing characteristics.

The choice of TBMs by our British colleagues was therefore for traditional shields operating in open mode.

The muck extracted from the cutter head was taken up directly by belt conveyors for evacuation by trains of muck wagons. The segments that composed the lining rings of the tunnels were placed directly in contact with the ground behind the TBMs.

The good lord must have been backing the British, since for us on the French side the ground had suffered important tectonic movement, giving abrupt reliefs. Because of this geology, the alignment encountered beds of heavily water-bearing chalk from extremely varied geological periods. Therefore, the French side chose to adopt TBMs that operated practically like submarines and were capable of resisting a 100 m load of water.

On the French side, muck from the cutter head was extracted via a watertight bulkhead. The muck was taken up by a shielded Archimedes screw, the pressure of which was reduced to atmosphere pressure by either a discharge pump or a twin screw system.

Water-sealing at the tail of the TBMs was ensured by using lining segments fitted with neoprene gaskets, assembled within the tailskin of the TBMs. The ingress of water was offset by metal brushes, injected continuously with mastic, between the ring and the tailskin of the TBMs.

5. Faster and Further

Scheduled to drive 560 m per month, the French TBMs managed to drive up to 1,232 m in 30 consecutive days, and more than 800 m per month on the running tunnels during the last six months. Their performance set an unequalled record for this type of TBM.

In view of the geological conditions, the French team should have driven 14.5 km on the marine service tunnel and 15.6 km on the marine running tunnels. In fact, our machines reached 15.6 km on the service tunnel; 20 km on the north running tunnel, thus penetrating on to British territory; and 19 km on the south running tunnel.

We could have passed the border twice, but the French are noted for diplomacy, aren't they?

On 28 June 1991, the final breakthrough of the tunnels was achieved — in advance of the contract programme.

6. Appropriate Logistics

From the start of the civil engineering design studies in 1986, the French construction team opted for industrialization and the use of advanced technology in the interest of mastering the tunnel driving programme, as well as to ensure the safety of personnel.

In order to achieve these goals, we decided to construct a shaft 55 m in diameter by 65 m deep, situated as close as possible to the French coast.

Sized to permit simultaneous operation of four TBMs, this shaft was a real nursemaid to the site. Its lifts and goods hoists, 40-to-400-tonne capacity overhead cranes, and traverse carriages permitted lowering down of personnel, locomotives and wagons, tunnel lining rings, pipework and cables or prepacked pallets. Its tippers, conveyors, crushers, slurry mixers and pumps handled the 5 million m^3 of muck extracted from the tunnels. And its sumps and pumps facilitated the drainage of water infiltrating the tunnels (capacity of 500 liter/sec per marine running tunnel).

Both tunnels, equipped with two 0.90 m temporary tracks divided into cantons 1 km long, were managed by a central traffic control post established in 1989. This temporary post was responsible for all train movements — more than 250 per day.

A separate radio channel for each tunnel permitted contact with the locomotives and man-riders. Each tunnel was managed by a controller, who was provided with a visual control panel showing the path of the programmed itineraries and the state of occupation of the cantons.

A temporary central electromechanical services central post allowed the controller to follow the state of the temporary installations at all times on control monitor. At any moment the controller could modify the equipment settings for electrical supply, ventilation, seepage water drainage, gas detection, rail signaling, etc. by remote control. Certain incidents were handled automatically by computers.

A 20,000 m^2 segment precasting factory was built a few hundred meters away from the shaft. Between 400 and 500 segments, of 24 different types, were manufactured daily at the factory. The reinforcement cages were manufactured in advance by a three-dimensional welding machine. The segments were transported to a stockyard by computer-controlled overhead cranes, to finish curling before being transferred in complete ring sets to the shaft by means of specially designed transporters.

These concrete segments have a crushing strength of 70 to 100 N/mm^2. As a comparison, the standard for the shell of a nuclear power station is 50 N/mm^2.

These sophisticated techniques allowed us not only to meet the challenge of driving the tunnels, but also to succeed in simultaneously executing the special works (technical rooms, communication passages, crossovers, pumping stations, etc) and the permanent electromechanical installations.

7. Special Works

While the TBMs were working the chalk in order to meet each other, a large number of other excavation works were in progress. These included the following special works:

1) The piston relief ducts, which straddle the service tunnel every 250 m. These 2.20 m-internal-diameter ducts linked the running tunnels and served as an outlet for the air pressure created by the passage of high-speed trains.

2) The communication passages, which linked the running tunnels to the service tunnel every 375 m. These passages had a 3.30 m internal diameter.

3) Three main pumping stations and numerous technological rooms and signaling rooms.

But the "gold medal" in the category of special works must go to the enormous underground caves of the railway crossovers, constructed to permit trains to change tracks for maintenance works. For this purpose, the running tunnels converged and the service tunnel was diverted sideways and underneath.

Sited at approximately 17 km from the tunnel portal, as for the U. K. crossover, the French crossover was a cylinder 162 m long, 12 m high and 19 m wide at the springing points, situated at a depth of 80 m below sea level.

The design and the method of construction of the French crossover were guided by the overriding requirements for safety of personnel and the works. They permitted the passage of the running tunnel TBMs at any time during the construction, while employing ample traditional methods.

In order to accomplish this, the access drifts and ramps, sloping at 10%, were carried out immediately after passage of the service tunnel, thus allowing 11 primary galleries to be excavated by Lynx road-header, and were then concreted to form the vault of the structure.

Once this vault had been constructed, the cove of the structures was excavated in successive phases, including demolition of the concrete lining of the running tunnels, which had by then passed through the structure, and the construction of a raft foundation in three phases.

This structure was then equipped with technical rooms and a metal partition door composed of two 90-tonne leaves.

8. The Terminals: Showcases of the Project

At each end of the tunnels, on which the attention of the media was fixed, were two gigantic terminals, which themselves alone represent the largest European civil engineering projects executed at the end of this century.

These terminals permit the shuttle trains to leave the tunnels and arrive at their respective platforms after making a 180 degree loop, before redeparting in the return direction. There, vehicles aboard the shuttles disembark, while others embark for the crossing under the channel. Mainline trains and goods trains alternate with the shuttles.

The French terminals at Coquelles required more than 12 million m^3 of earth-moving over a 700 ha area. Because a large part of the terminal site was located on compressible subsoils (peats and marshy slits), it was necessary to preload with fill on top of a sand drainage bed in order to stabilize the subsoil by dewatering. Drainage water was pumped via lifting stations and transferred to five retention basins, covering a total area of 100,000 m^2.

The structural works for the terminals comprised nearly 60,000 m^2 of bridges 5,000 m of platforms, and 40,000 m^2 of buildings. Four 300 m-long bridges gave access to the platforms via 24 ramps.

Maintenance for the Eurotunnel fixed link shuttle trains will be carried out at the French Terminal.

This terminal, now completed, is the architectural showcase of the project of the century.

9. The Channel Tunnel: A Human Gamble

The success of the Channel Tunnel depended on meeting a double challenge: 1) Implementing and mastering the exceptional techniques required for the works, and 2) maintaining good labor relations.

From the start of recruiting personnel, the management of French construction decided to carry out the works using local personnel to the maximum, since the rate of unemployment in the "Nord-Pas-de-Calais" region had reached 22%.

In order to do this, a unique training and tutorial programme was established with the help of the national and regional government authorities. The programme involved providing 183,000 hours of site access training and 560,000 hours of further on-the-job training so that people who initially had little experience and training in our profession could be qualified to work on the project.

The results speaked for themselves: 95% of the manual labor and 68% of the management and supervisory staff were recruited in the region.

Within the training function, safety occupied an important place. Five hundred senior staff attended a two-day safety conference; all supervisors and manual labor attended a one-day conference; and 700 individuals were trained in emergency and first aid procedures.

At the start of method studies for the execution of the structural works and the electromechanical installations, a risk analysis was undertaken. In addition, protective safety measures for incorporation in the machines were taken into account right at design stage, and were checked in partnership with safety organizations. More than 300 studies of this type were carried out.

Responsibility was delegated through the management chain in all fields with respect to safety, quality, cost and programme. These measures permitted reduction of the work accident rate to 50% that of the national average for our industry; and created an exceptional climate of labor relations that combined unity, confidence and exceptional motivation.

10. An Experience for the Future

Previous experiences in construction of long tunnels had been inconclusive. The Japanese had taken nearly 14 years to excavate a tunnel of under 50 km, albeit under extremely difficult working and geological conditions.

The Channel Tunnel has cleared away many uncertainties about undersea construction of megaprojects. The experience gained has permitted the vision, throughout the world, of other ambitious projects, not only in terms of excavating tunnels but also in the construction of complex railway systems.

History remembers also that the dream only has become a reality thanks to the people who constructed the Channel Tunnel, and to the trust that was placed in them.

Integrated Skills: Writing an E-mail

Please write something about your major to your classmates, teachers or friends by E-mail. You may simulate to write by referring the following two E-mails:

Example 1

From: Zhang Qiang<zq@sohu.com>

To: Wang Ming<wq@163.net>

Date: 15th March, 2010 09:15:30

Subject: Substructure Contract

Dear Mr. Wang,

With regard to the confirmation of the intent to award bid of the subject project, two sets of the latest drawings were issued to you on 12 March, 2019; and in addition, we enclose herewith a set of specifications.

In this connection, you are requested to review these drawings and specifications, and send your confirmation within 17 March 2019.

Your prompt action on this matter would be much appreciated.

Best regards,

Zhang Qiang

Example 2

　　From: Li Feng<lf@371. net>
　　To: Bill White<Bill@cmmail. com>
　　Date: 21 March, 2000 11:30:30
　　Subject: Queries of Specifications

Dear Mr. White,

　　Thank you for your response to and confirmation of the Specifications and Contract Drawings (letter dated 17 March 2019). And we would like to answer your queries of the same letter.

　　1) The ground floor slab should be included in the scope of the Substructure Contract. Please refer to clause 2 of the Letter of Acceptance.

　　2) The Substructure Contractor shall be responsible for the provision of various utility connections as shown on the Drawings and Specifications. The Substructure Contractor shall also provide attendance upon utility departments as stipulated in Clause 1.61 of the Preliminaries.

　　Regards,

Li Feng

Unit 10 Geotechnical Engineering and Underground Engineering

Lead-in

1. What methods may soils be classified by?
2. What is compressibility?
3. When did the science of soil mechanics begin its rapid growth?

Vocabulary Warm-up: Matching

_____1. three-phase systems	a. 黏聚力
_____2. tidal action	b. 三相系
_____3. bearing capacity	c. 压缩性
_____4. cohesion	d. 潮汐作用
_____5. compactness	e. 渗透性
_____6. compressibility	f. 承载力
_____7. degree of saturation	g. 密度
_____8. density	h. 密实度
_____9. permeability	i. 团结系数
_____10. coefficient of consolidation	j. 饱和度

Text A Soil

Loadings in buildings consist of the combined dead and imposed loads which exert a downward pressure upon the soil on which the structure is founded and this in turn promotes a reactive force in the form of an upward pressure from the soil. The structure is in effect sandwiched between these opposite pressures and the design of the building must be able to resist the resultant stresses set up within the structural members and the general building fabric. The supporting subsoil must be able to develop sufficient reactive force to give stability to the structure to prevent failure due to unequal settlement and to prevent failure of the subsoil due to shear. To enable a designer to select，design and detail a suitable foundation he（she）must have adequate data regarding the nature of the soil on which the structure will be founded and this is normally obtained from a planned soil investigation programmer.

Although they are closely related，soil mechanics and foundation engineering are not *synonymous*. Foundation engineering as a profession is a *subtle* combination of soil mechanics and engineering *geology* with the intuitive art of judgment and innovation where experience exercises an important role. Soil mechanics, on the other hand, is a study of the behavior of a material whose most important characteristic is its particulate composition. Since this characteristic is not unique to soils, the principles and techniques of soil mechanics may find application to a variety of problems in *geophysics*, materials processes, and most recently, **lunar exploration**.

The science of soil mechanics began its rapid growth with the pioneering studies of Karl Terzaghi during the early part of this century. Terzaghi developed many of the theories of soil mechanics out of the practical necessity of providing solutions to the many difficult foundation problems introduced by modern construction.

1. Characteristic of Soils

Soils are aggregates of **mineral** particles that cover extensive portions of the earth's surface. In soil engineering, the force applied to soil masses very frequently produces relative displacements between particles. Hence a study of soil mechanics requires an appreciation of the particulate nature of the material. Another characteristic of soils is that they are **three-phase systems**: they are composed of solid particles, water, and air. Since air is very **compressible**, and water may flow into or out of a soil, the relative proportions of three components change with time and load. Hence these components often form the basis for the **quantitative** description of soil behavior.

2. Structure of Soils

The structure of natural soils is the net product of the interaction between the forces of **sedimentation**, surface forces of the soil particles, and subsequent **geologic** forces. If particles of sand are allowed to settle from a suspension in water, the particles tend to take up stable positions to form a **single-grained structure**. Very loose sand or **silt** may have a **honeycomb** structure. If the fine particles consist of clay materials, the surface forces play an important part. If strong attractive forces exist between the edge or corner and the face of clay plates, a **flocculent** develops.

Otherwise, the clay plates may occupy nearly parallel positions as they settle from suspension. This is called a **dispersed structure.**

Soils with flocculent and honeycomb structures have large voids between soil particles and are held together by surface forces at the contact points. Such structures are generally not very stable. When a load is applied to the soil, the contacts may be broken and part of the structure destroyed, thus compressing the voids to form a more stable structure that can withstand the load. Some soil may be so unstable that the structure collapses with small disturbances. If the void space is filled with water, the soil-water mixture may lose all stability and flow as a viscous liquid. Occasionally very loose deposits of fine sand or silt have been observed to flow after small disturbances such as a **seismic tremor**, an adjacent slide, or even **tidal action**.

3. Classification of Soils

Many soil descriptions and classifications are based on the size of the soil particles.

This is the simplest criteria for soil description. Soils are commonly named gravel, sand, silt, and clay on the basis of the particle size. The dividing line between these categories is arbitrary, and, as is common to arbitrary definitions, there are several systems in current use. These systems were initiated independently by various agencies that worked with soils.

Soils may be classified by any of the following methods: ① Physical properties; ②Geological origin; ③Chemical composition; ④Particle size.

It has been established that the physical properties of soils can be closely associated with their particle size both of which are of importance to the foundation engineer, architect or designer. All soils can be defined as being coarse-grained or fine-grained each resulting in different properties.

Coarse-grained soils: these would include sands and grovels having a low proportion of voids, negligible cohesion when dry, high permeability and slight compressibility, which takes place almost immediately upon the application of load.

Fine-grained soils: these include the cohesive silts and clays having a high proportion of voids, high cohesion, very low permeability and high compressibility which takes place slowly over a long period of time.

There are of course soils which can be classified between the two extremes described above. BS 1377 deals with the methods of testing soils and divides particle sizes into follows:

Table 10-1　Classification of soils

Soil	Particle size
Clay particles	Less than 0.002 mm
Silt particles	Between 0.002 and 0.06 mm
Sand particles	Between 0.06 and 2 mm
Gravel particles	Between 2 and 60 mm
Cobbles	Between 60 and 200 mm

The silt, sand and gravel particles are also further subdivided into fine, medium and coarse with particle sizes lying between the extremes quoted above.

Grain size is the basis of soil mechanics, since it decides whether a soil is frictional or cohesive, a sand or a clay. Every large civil engineering job starts with a soil mechanics survey in its early stages.

4. Mechanical Property of Soils

The resistance which can be offered by a soil to the sliding of one portion over another or its shear strength is of importance to the designer since it can be used to calculate the bearing capacity of a soil and the pressure it can exert on such members as timbering in *excavations*. Resistance to shear in a soil under load depends mainly upon its particle composition. If a soil is *granular* in form, the frictional resistance between the particles increases with the load applied and consequently its shear strength also increases with the magnitude of the applied load. Conversely clay particles being small develop no frictional resistance and therefore its shear strength will remain constant whatever the magnitude of the applied load. Intermediate soils such as sandy clays normally give only a slight increase in shear strength as the load is applied.

Compressibility is the characteristic of the reduction of the volume of soils in the pressure (weight of additional stress or stress). Another important property of soils which must be ascertained before a final choice of foundation type and design can be made is compressibility, and two factors must be taken into account: ①rate at which compression takes place; ②total amount of compression when full load is applied.

When dealing with non-cohesive soils such as sands and grovels the rate of compression will keep pace with the construction of the building and therefore when the structure is complete there should be no further settlement if the soil remains in the same state. A soil is compressed when loaded by the expulsion of air and/or water from the voids and by the natural rearrangement of the particles. In cohesive soils the voids are very often completely *saturate*d with water which in itself is nearly incompressible and therefore compression of the soil can only take place by the water moving out of the voids thus allowing settlement of the particles. Expulsion of water from the voids within cohesive soils can occur but only at a very slow rate due mainly to the resistance offered by the plate-like particles of the soil through which it must flow. This gradual compressive movement of a soil is called *consolidation*. Uniform settlement will not normally cause undue damage to a structure but uneven settlement can cause progressive structural damage.

Glossary

Complete the glossary using words in the text.

1. _____	同义的,同义语的	2. _____	细微的,巧妙的
3. _____	地质学	4. _____	地球物理学
5. _____	月球勘探		
6. _____	矿物的,有矿物质的		
7. _____	三相系		
8. _____	可压缩的,压缩性的		
9. _____	数量的,定量的		
10. _____	沉积,沉积物,沉淀		
11. _____	地质的,地质学上的		
12. _____	单粒结构	13. _____	粉粒,粉砂土
14. _____	蜂窝,蜂窝状		

15. _____ 絮凝的,绒聚的,絮凝
16. _____ 分散结构　　17. _____ 地震震动
18. _____ 潮汐作用　　19. _____ 挖掘
20. _____ 颗粒状的
21. _____ 浸透,渗透,使充满
22. _____ 渗压,加强

Exercises

1. Decide whether the following statements are true（T）or false（F）according to the text.

1）Soil mechanics and foundation engineering are closely related, so they are synonymous.

2）Soils are three-phase systems, and they are composed of solid particles, water, and air.

3）Resistance to shear in a soil under load has no relation with its particle composition.

4）The shear strength of clay will remain constant whatever the magnitude of the applied load.

5）In cohesive soils, if the voids are very often completely saturated with water, compression of the soil can only take place by the water moving out of the voids thus allowing settlement of the particles.

2. Write down the words with the following prefix or suffix.

syn-/sym-　_____
-medi-　_____
-ward　_____
-ness　_____
-th　_____

3. Translate the following paragraph into Chinese.

When dealing with non-cohesive soils such as sands and grovels the rate of compression will keep pace with the construction of the building and therefore when the structure is complete there should be no further settlement if the soil remains in the same state.

4. Translate the following sentences into English.

1）土是广泛分布于地球表面的矿物颗粒堆积物。

2）大量关于土的描述和分类都基于土体颗粒的尺寸。

3）土体结构是在天然土沉积作用、土颗粒表面分子作用以及次生地质作用的共同作用下形成的网格式结构。

Text B　Ground Investigations

Engineers designing structures and machines normally choose materials and specify their strength and stiffness and they often combine materials to make composites (e. g. steel and concrete in reinforced concrete). Similarly, highway engineers can specify the soils and rocks to be used in the construction of roads. Geotechnical engineers, on the other hand, cannot choose and must work with the materials in the ground. They must therefore determine what there is in the ground and the engineering properties of the ground, and this is the purpose of ground investigations.

The basic techniques of ground investigation are drilling, sampling and testing, in situ and in the laboratory, but these must be complemented by geological information and a sound appreciation of the relevant soil mechanics principles. Consequently, it is in the area of ground investigation that geology and engineering combine and where engineering geologists and geotechnical engineers cooperate.

Ground investigation is, of course, far too big a topic. The detailed techniques vary from country to country, and from region to region, and depend both on the local ground conditions, on historical precedents, on contractual procedures and on the available equipment and expertise. As with laboratory testing, procedures for ground investigations are covered by national standards and codes of practice; in the United Kingdom this is BS 5930:1999. You should look up the standards covering the region where you work in to see what they contain. Detailed descriptions of the current practices in the United Kingdom are given by Clayton, Simons and Matthews (1995).

1. Objectives of Ground Investigations

When you look at the face of a cliff or an excavation you see a section of the ground and when you look at a site you have to imagine what an excavation would reveal. A major part of a ground investigation is to construct a three-dimensional picture of the positions of all the important soil and rock layers within the site that may be influenced by, or may influence, the proposed construction. Of equal importance is the necessity to sort out and identify the groundwater conditions. In distinguishing the important soil and rock layers engineering classifications based on the nature and state of the soils should be used rather than the geological classifications, which are based on age.

There is no simple answer to the problem of how many holes should be drilled and to what depths and how many tests should be carried out. Most of the standards and codes of practice make various recommendations, but really you should do enough investigation to satisfy everybody that safe and economical works can be designed and constructed. There will inevitably be uncertainties and these will require conservation in design which will lead to additional costs of construction. There is a balance to be struck between costs of more investigations and savings in construction: it is a matter of apportioning risk.

Fig. 10 - 1 illustrates a very simplified section along the centre-line of a road. (Notice that the horizontal and vertical scales are not the same.) The ground conditions revealed by drilling and other methods have been greatly idealized so that a number of characteristic layers have been identified and the boundaries between them drawn as smooth lines. The actual soils in the ground within any one layer are likely to be variable horizontally and vertically, and their boundaries irregular. Something like Figure 10-1 is about the best you can do with a reasonable investigation. Notice that Figure 10-1 is a section along the centre-line of the road and to complete the investigation you should be able to draw cross-sections and sections on either side of the road.

For each of the principal strata in Figure 10 - 1 you will need to determine representative parameters for strength, stiffness and water seepage flow (i. e. permeability). These will be selected from the results of laboratory and in situ tests. These parameters may be constant for a particular layer or they may vary with depth; generally we expect strength and stiffness to increase with depth. The question of which parameters to determine depends on the ground and on the structure to be designed and constructed. There are relationships between the engineering properties of soils, such as strength and stiffness, and their nature and state.

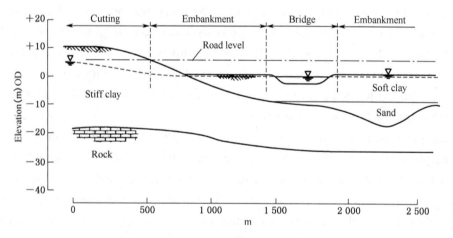

Figure 10-1　A simple geotechnical cross-section

After any ground investigation you should know the following for each of the principal strata:

1) Its engineering description and classification in terms of the nature (grading and plasticity) and state (stress and specific volume oroverconsolidation).

2) The positions of the boundaries between the different strata (i. e. you should be able to draw sections like that in Figure 10-1 in any direction).

3) The geological environment when the soil was deposited and the history of subsequent deposition, erosion, weathering and ageing.

4) Descriptions of visible features of structure and fabric (e. g. layering, fissuring and jointing).

5) Representative values for the parameters for strength, stiffness and permeability relevant to the design and construction of the works.

You should also be sure that you know all about the groundwater. A very experienced ground engineer once said to me that he would not start an excavation until he knew exactly what he was digging into and what the groundwater conditions were; this is a very good advice.

2. Planning and Doing Investigations

You cannot really plan an entire ground investigation because you do not know what is there before you start and SO you cannot select the best methods or decide how much to do. A ground investigation must, therefore, be carried out in stages; each stage can be planned with existing information and the knowledge gained from one stage will assist with planning the next. Currently in the United Kingdom a ground investigation is often let as a single contract with a specification and bill of quantities, which leads to major problems in planning the investigation and can often cause later difficulties.

There should be three principal stages in a ground investigation. (These are not rigid demarcations. There is often overlap between the stages; they need not be strictly sequential and one or other may have to be expanded later.)

1) Desk studies

This consists of study of all the information that you can find existing on paper or electronically. The major sources are topographical and geological maps and sections, geological reports and local authority records. Other sources include air photographs, historical archives and reports on earlier site investigations at the site or at nearby sites. Experienced geotechnical engineers and engineering geologists can often decipher the principal ground conditions from the desk study, so leading to well-planned later stages.

2) Preliminary investigations

Preliminary investigations are carried out at the site, rather than in the office, but they do not yet involve major expenditure on drilling, sampling and testing. The purposes are, firstly, to confirm or revise the findings of the desk study and, secondly, to add further information. This additional information will come from detailed engineering geological mapping, and this is best done by engineers and geologists working together or by experienced engineering geologists. Preliminary investigations may also involve some limited subsurface exploration by trial pits, probing or exploratory drilling and geophysical sensing using seismic, electrical resistivity and other methods.

3) Detailed investigations

Detailed investigations consist of drilling, sampling and laboratory and in situ testing. They may also involve more detailed geological mapping, groundwater and chemical studies and other appropriate investigations necessary for the works. This is where the bulk of the expenditure is incurred and planning of the detailed investigations should set out to discover the required facts in the most efficient way. This will require some foreknowledge which can be gained from the desk study and preliminary investigations.

3. Test Pitting, Drilling and Sampling

The standard method of ground investigation is excavation and sampling supplemented by in situ and laboratory testing. The excavations are usually done by drilling but also by opening test pits.

1) Test pitting

A test pit is an excavation that a geotechnical engineer or engineering geologist can enter to examine the soil profile in situ. Pits can be excavated by large drilling machines of the kind used for boring piles, by an excavator or by hand digging. Remember that any excavation in soil with vertical or steep sides is basically unstable and must be supported before anyone enters it.

2) Drilling

Drill holes can be advanced into the ground using a number of different techniques. Augers may be drilled to shallow depths by hand and large diameter augers can be drilled by machines used also for installation of bored piles. Wash boring is used in sands and gravels and rotary drilling is used mainly in rocks. Light percussion drilling is widely used in the United Kingdom and you can very often see the typical tripod rigs at work. In some soils, particularly stiff clays and in rocks, boreholes will remain open unsupported, but in soft clays and particularly in coarse-grained soils the hole will need to be cased to maintain stability. Boreholes should normally be kept full of water, or bentonite mud, to prevent disturbance below the bottom of the hole.

3) Sampling

Samples obtained from test pits or boreholes may be disturbed or intact. (Samples are often called disturbed or undisturbed but, as no soil sample is ever truly undisturbed, the word intact can be used for samples taken with minimum disturbance.) Disturbed samples are used principally for description and classification. They may be reconstituted to determine the properties of the soil. Intact samples may be cut from the base or sides of test pits using saws or knives or taken in tubes pushed into the bottom of a borehole.

Text C   Pile Foundation

Concrete piles came into general use during the last few years of the past century. Ordinarily concrete piles are divided into two principal types, cast-in-place and precast. The cast-in-place pile is formed in the ground in the position in which it is to be used in the foundation. The precast pile is made above the ground in a pile-casting yard, and it is then brought to the job site and driven or jetted into the ground just like a wood pile. The fields of usefulness of cast-in-place and precast concrete piles overlap to some extent, but usual conditions give one or the other a distinct advantage. In building-foundation work on land, the cast-in-place pile is more commonly used than the precast pile. The principal reasons for this are as follows: The cast-in-place pile foundation can be installed more rapidly because there is no delay while test piles are being driven to predetermine pile lengths, and there is no delay waiting for piles to cure sufficiently to withstand handling and driving stresses. Of particular importance in crowded locations is the fact that a cast-in-place pile job does not require a large open space for a pile-casting yard. The question of predetermining the required pile length is especially important on precast-pile work. On a cast-in-place pile job the contractor usually starts the work with equipment that will drive piles of widely varying lengths so that it is not necessary to know the exact required length before the job is started. On precast-pile work it is necessary to know the required lengths before the piles are cast, and time and money must usually be spent for driving test piles to determine the required length. If the probable pile length for a precast pile is not determined before the piles are cast, there is serious risk of a considerable excess cost on the work, because of the necessity for either adding to the length of precast piles that are too short or cutting off and throwing away lengths of precast piles that are too long.

1. Precast Piles

It is usually required that precast piles shall be stored and cured for 30 days before they are driven, but this curing period can often be reduced by the use of the high-early-strength cements that are now available. On all precast-pile work a good-sized area is required for a pile-casting yard. On some jobs there is space available so that the piles can be made close to the job site. In the downtown sections of large cities there is usually no vacant property available for a casting yard so that precast piles have to be made at some distance from the job site. This involves handling and transporting the pile from the casting yard to the job site, and it adds to the cost of the work. It is for these reasons that the cast-in-place pile is more commonly used than the precast pile on building-foundation work on land. In marine installations such as docks, piers, and bulkheads, the precast pile is used almost exclusively. This is due largely to the difficulty involved in placing cast-in-place piles in open water. For docks and bulkheads the cast-in-place pile is sometimes used in the anchorage system. On trestle type structures, such as highway viaducts, the precast pile is more commonly used. A portion of the pile often extends above the ground surface and serves as a column for the superstructure.

Precast concrete piles are always reinforced to withstand handling and driving stresses. They are usually cast in horizontal position and then picked up and delivered to the pile driver so that they are often subjected to severe bending stresses. They must be reinforced internally to take care of these stresses. The concrete in a cast-in-place pile is not subjected to handling and driving stresses since the pile is poured in place in the ground. When a cast-in-place pile is completely submerged in soil, it does not need to be reinforced internally. Both types of concrete piles are sometimes used as columns extending above the ground surface, and in such cases they must reinforced for column action. In some type of structures, either the cast-in-place or the precast pile may be subjected to horizontal force applied at the pile head. Under such conditions, the bending stress in the piles should be investigated and internal reinforcement should be provided as necessary to take care of these bending stresses.

Precast concrete piles may be divided into two kinds, tapered and parallel-sided. Precast piles are usually of square or octagonal cross section, although sometimes they are round. Since the piles are usually cast in a horizontal position on the ground, the round croaks section is not common because it is difficult to fill a cylindrical form with concrete when the form is horizontal. All precast piles are reinforced with longitudinal bars and transverse reinforcement, which may be separate hoops or a spiral. The pitch of the spiral reinforcement is always smaller for a length of 3 or 4 ft at each end of the pile than it is in the middle of the pile because the maximum stresses in the pile during driving occur near the ends of the pile.

Precast piles are usually manufactured in temporary casting yard on or near the site where they are to be used. Transportation and handling costs are comparatively high so that precast piles are not usually shipped over great distances from casting yard to job site.

When very hard driving conditions are encountered, precast piles are often equipped with steel driving shoes. These are pointed steel caps that fit over the point of the pile. They are available in various sites and are provided with straps point that extends into the concrete at the lower end of the pile so that the shoe can be made a part of the pile when the pile is cast.

Recently, there has been developed a centrifugally precast, prestressed, cylindrical, hollow concrete pile. These are cast in sections 16′ long, which are then fitted together into the number of sections required for the length of a pile. The sections, thus fitted together, are then prestressed by posttensioning wire cables that run through open holes cast longitudinally in the walls of each section. Special, but positive, methods have been developed for joining the sections, posttensioning the wire cables, and grouting the cables so as to form a homogeneous product of great strength and durability.

2. Cast-in-Place Piles

Cast-in-place piles are generally divided into two different kinds: ①those that are provided with permanent steel shells and ② those that are made without permanent shells. The purpose of the shell is to prevent mud and water from mixing with the fresh concrete and to provide a form to protect the concrete while it is setting. With the shell-less pile, the fresh concrete is in direct contact with the surrounding soil. Cast-in-place piles with permanent shells may be divided into three different kinds: ①tapered driven shells; ②cylindrical, dropped-in or driven shells; ③steel-pipe piles. Tapered driven shell pile is made in the following manners. A thin steel shell, closed at the bottom with a steel boot, is placed on the outside of a steel mandrel or core; the shell and the core are driven together into the ground. When sufficient driving resistance has been developed, driving is stopped and the core is withdrawn from the shell. This leaves a steel-lined hole in the ground that can be inspected with a lamp on a drop cord, and the concrete is then dumped into the shell from wheelbarrows. With this type of pile a relatively thin shell is used because the heavy steel-driving core supports the shell so that it will not collapse during driving.

Another type of driven shell has a scalloped cross section that is formed by a series of vertical flutings running the full length of the pile. These piles are tapered and are closed at the bottom by a steel boot. The shell is driven into the ground and is then filled with concrete to form a cast-in-place pile. Monotubes are driven without the aid of an internal driving core, and the shell must therefore be thick enough to be able to withstand the driving stresses and ground pressures. One important thing about both of these types of tapered driven shell piles is the fact that they maintain the driving resistance that is set up in the soil surrounding the pile during driving because the steel shells remain in place in the ground in the position in which they are driven. A third type of cast-in-place pile is a cylindrical, dropped-in driving apparatus. The driving apparatus consists of a piece of heavy steel-pipe that is usually about 14 to 16 in diameter and 1/2 in thick. A steel-driving core is fitted inside this pipe, and the two are driven together into the ground. When suitable driving resistance has been obtained, driving is stopped and the core is removed from the pipe. A thin corrugated-metal shell several inches smaller in diameter than the driving pipe is then dropped down inside the driving pipe. The thin corrugated shell is then pulled out of the ground. The pile that remains in the ground is therefore several inches smaller in diameter than the pile that was driven.

A cast-in-place pile with a permanent shell is the steel pipe pile. It is usually divided into two types: closed-end pipe pile and open-end pipe pile. The closed-end pipe pile is a piece of steel pipe closed at the bottom with a heavy boot. The pile is driven into the ground and filled with concrete. The uses of such piles are about the same as those of the other types of cast-in-place piles with driven shells. Sometimes these pipe piles are driven all in one piece and sometimes they are made of several pieces of pipe that are either welded together or fitted together with special internal sleeves.

The open-end steel-pipe pile is usually driven to bearing on rock. When the pipe is being driven, it is open at the bottom so that, when driving is stopped, the interior of the pipe is full of soil. This soil has to be removed from the pipe, and the usual procedure is to do this with air and water jets, although it used to be done occasionally with miniature orange peel dredging buckets. When the cleaning out is done by the jetting process, a long jet using water is worked down inside the driven pile in order to loosen up the soil. When the soil has been softened by the water, air is forced through the jet into the pile. This air jet is connected to a large compressed-air receiver, and a quick-throw valve is set in the airline between the receiver and the jet. This valve is thrown open suddenly so that the compressed air rushes into the bottom of the pipe almost like an explosion. The soil inside the pipe is blown out at the top. It is usually necessary to repeat this jetting and blowing process several times in order to get the pipe properly cleaned out. When this has been done, the pipe is redriven with the hammer to make sure that it is properly seated in the rock. The pile is then again blown out and is promptly filled with concrete. Open-end pipe piles installed in this manner are usually alloyed to carry relatively high working loads.

Integrated Skills: Making a PowerPoint

Shield tunneling method is an important construction method of tunnels in loose, non-cohesive or soft ground, and is generally used in construction of subway in civil cities. By means of Internet search and the review of special literatures, you should collect engineering pictures and technological information about shield, and make a relevant PowerPoint file. Then make a presentation in class, and report your ideas on shield.

Unit 11　Airport Engineering

Lead-in

1. What does the landside area of an airport include? What about the airside area?
2. How are airports usually named?
3. What environmental problems should airports concern?

Vocabulary Warm-up: Matching

_____1. runway	a. （靠近候机厅的）停机位,停机坪
_____2. taxiway	b. （机场）跑道
_____3. helipad	c. 飞机库
_____4. ramp / tarmac	d. 飞机滑行道
_____5. apron	e. 候机楼,航站楼
_____6. hangar	f. 直升机停机坪
_____7. terminal building	g. 候机厅
_____8. lounge / concourse	h. 停机坪（远离候机厅的）

Text A　Introduction to the Airport

An *airport* is a location where aircraft such as *fixed-wing aircraft*, *helicopters*, and *blimps take off* and *land*. Aircraft may be stored or maintained at an airport. An airport consists of at least one surface such as a runway, a helipad, or water for takeoffs and landings, and often includes buildings such as hangars and terminal buildings.

Larger airports may have fixed base operator services, *seaplane dock*s and ramps, air traffic control, passenger facilities such as restaurants and lounges, and emergency services. A *military airport* is known as an airbase or air station. Smaller or less-developed airports, which represent the vast majority, often have a single runway shorter than 1,000 m. Larger airports for airline flights generally have paved runways 2,000 m or longer. Many small airports have dirt, grass, or *gravel* runways, rather than asphalt or concrete.

As of 2006, there were approximately 49,000 airports around the world, including 14,858 in the United States. In the United States, the minimum dimensions for dry, hard *landing field*s are defined by the FAR Landing and Take off Field Lengths. These include considerations for safety margins during landing and takeoff. Heavier aircraft require longer runways. The longest public-use runway in the world is at Qamdo Bangda Airport in China, which has a length of 5,500 m. The world's widest paved runway is at

Ulyanovsk Vostochny Airport in Russia and is 105 m wide.

1. Airport Structures

Airports are divided into *landside* and *airside* areas. Landside areas include *parking lots*, public transportation train stations, tank farms and access roads. Airside areas include all areas accessible to aircraft, including runways, taxiways, ramps and tank farms. Access from landside areas to airside areas is tightly controlled at most airports. Passengers on commercial flights access airside areas through terminals, where they can purchase tickets, clear security, check or claim luggage and *board* aircraft through gates. The waiting areas which provide passenger access to aircraft are typically called concourses, although this term is often used interchangeably with terminal.

The area where aircraft park next to a terminal to load passengers and baggage is known as a ramp or tarmac. Parking areas for aircraft away from terminals are called aprons.

Airports can be towered or non-towered, depending on air traffic density and available funds. Due to their high capacity and busy *airspace*, many *international airports* have air traffic control located on site.

Airports with *international flights* have customs and immigration facilities. However, as some countries have agreements that allow travel between them without customs and immigrations, such facilities are not a definitive need for an international airport. International flights often require a higher level of physical security, although in recent years, many countries have adopted the same level of security for international and domestic travel.

2. Airport Access and Internal Transport

Many large airports are located near railway trunk routes for seamless connection of multi-modal transport. It is also common to connect an airport and a city with rapid transit, light rail lines or other non-road public transport systems. Such a connection lowers risk of missed flights due to traffic congestion. Large airports usually have access also through expressways from which motor vehicles enter either the departure loop or the arrival loop.

The distances passengers need to move within a large airport can be substantial. It is common for airports to provide moving walkways and buses. The Hartsfield-Jackson Atlanta International Airport has a tram that takes people through the concourses and baggage claim. Major airports with more than one terminal offer inter-terminal transportation, such as Mexico City International Airport, where the domestic building of Terminal 1 is connected by an *aerotren* to Terminal 2, on the other side of the airport.

3. Airport History and Development

The earliest aircraft takeoff and landing sites were grassy fields. The plane could approach at any angle that provided a favorable wind direction. A slight improvement was the dirt-only field, which eliminated the drag from grass. However, these only functioned well in dry conditions. Later, concrete surfaces would allow landings, whatever rain or shine, day or night.

The title of "world's oldest airport" is disputed, but College Park Airport in Maryland, US, established in 1909 by Wilbur Wright, is generally agreed to be the world's oldest continually operating airfield, although it serves only general aviation traffic. Another claim to the world's oldest airport is from Bisbee-Douglas International Airport Douglas, Arizona, USA, which had the first airplane in the state. In 1908, the Douglas Aeronautical Club was formed, starting with a glider made from mail order plans. This glider was pulled into the air by two horses and flew behind the Douglas YMCA building. In 1909, a motor and propeller were put on the airplane, making it the first powered airplane in Arizona. The airport's status as the first international airport in the USA is confirmed by a letter from President Roosevelt declaring it "the first international airport of the Americas". Furthermore, Albany International Airport is the oldest municipal airport in the United States.

Increased aircraft traffic during World War I led to the construction of landing fields. Aircraft had to approach these from certain directions and this led to the development of aids for directing the approach and landing slope.

Following the war, some of these military airfields added civil facilities for handling passenger traffic. One of the earliest such fields was Le Bourget Airport near Paris. The first international airport to open was the Croydon Airport, in South London, although an airport at Hounslow had been temporarily operating as such for nine months. In 1922, the first permanent airport and commercial terminal solely for commercial aviation was built at Königsberg, Germany. The airports of this era used a paved "apron", which permitted night flying as well as landing heavier aircraft.

The first lighting used on an airport was during the later part of the 1920s; in the 1930s *approach lighting* came into use. These indicated the proper direction and angle of descent. The colors and flash intervals of these lights became standardized under the International Civil Aviation Organization (ICAO). In the 1940s, the slope-line approach system was introduced. This consisted of two rows of lights that formed a funnel indicating an aircraft's position on the *glideslope*. Additional lights indicated incorrect altitude and direction.

Following World War II, airport design became more sophisticated. Passenger buildings were being grouped together in an island, with runways arranged in groups about the terminal. This arrangement permitted expansion of the facilities. But it also meant that passengers had to travel further to reach their plane.

An improvement in the landing field was the introduction of *grooves* in the concrete surface. These run *perpendicular* to the direction of the landing aircraft and serve to draw off excess water in rainy conditions that could build up in front of the plane's wheels.

Airport construction boomed during the 1960s with the increase in *jet aircraft* traffic. Runways were extended out to 3 km. The fields were constructed out of reinforced concrete using a *slip-form* machine that produces a continual slab with no disruptions along the length. The early 1960s also saw the introduction of jet bridge systems to modern airport terminals, an innovation which eliminated outdoor passenger boarding. These systems became commonplace in the United States by the 1970s.

Modern runways are thickest in the area where aircraft move slowly and are expected to have maximum load, i.e. runway ends. A common myth is that airplanes produce their greatest load during landing due to the impact of landing. This is untrue as much of the aircraft weight remains on the wings due to lift. Runways are constructed as smooth and level as possible.

4. Airport Naming

Airports are uniquely represented by their International Air Transport Association (IATA) airport *code* and ICAO airport code. An IATA airport code is often an abbreviation of the airport's common name, particularly older ones, such as PHL for Philadelphia International Airport. An airport sometimes retain their previous IATA code when its name, or even when its location is changed. Rafik Hariri International Airport in Beirut retains the IATA code BEY, from its former name of Beirut International Airport. Hong Kong International Airport retained both its name and its IATA code when moved from Kai Tak to Chek Lap Kok in 1998.

The name of the airport itself can be its location, such as San Francisco International Airport. It can be named after some public figure, commonly a politician, e.g. Paris-Charles de Gaulle Airport, or a person associated with the region it serves or prominent figures in aviation history, such as Norman Y. Mineta San Jose International Airport, Kingsford Smith International Airport and so on.

Airport names may include the word "International", reflecting their ability to handle international aviation traffic, although the airport may not actually operate any such flights. Some airports with international immigration facilities may also choose to drop the word from their airport names, such as Perth Airport and Singapore Changi Airport.

In the early years of the 21st century, low cost terminals, or even entire airports have been built to cater for discount airlines such as Ryanair.

5. Airport Environmental Concerns

Aircraft noise is major cause of noise disturbance to residents living near airports. Sleep can be affected if the airports operate night and early morning flights. Aircraft noise not only occurs from take-off and landings, but also ground operations including maintenance and testing of aircraft. Other noise and environmental concern are vehicle traffic causing noise and pollution on road leading the airport.

The construction of new airports or addition of runways to existing airports is often resisted by local residents because of the effect on countryside, historical sites, local flora and fauna. Due to the risk of collision between birds and airplanes, large airports undertake population control programs where they frighten or shoot birds.

The construction of airports has been known to change local weather patterns. For example, because they often flatten out large areas, they can be susceptible to fog in areas where fog rarely forms. In addition, they generally replace trees and grass with pavement; and change drainage patterns in agricultural areas, leading to more flooding, run-off and erosion in the surrounding land.

Some of the airport administrations are regulated to publish annual environmental reports in order to show how they consider these environmental concerns in airport management issues and how they protect environment from airport operations. These reports contain all environmental protection measures performed by airport administration in terms of water, air, soil and noise pollution, resource conservation and protection of natural life around the airport.

Glossary

Complete the glossary using words in the text.

1. _____	机场,航空港	2. _____	固定翼飞机
3. _____	直升机	4. _____	软式小飞机,飞艇
5. _____	起飞	6. _____	着陆,降落
7. _____	水上飞机码头	8. _____	军用机场
9. _____	碎石	10. _____	飞机场,停机坪
11. _____	陆侧	12. _____	空侧
13. _____	停车场	14. _____	登机
15. _____	空域	16. _____	国际机场
17. _____	国际航班	18. _____	机场轻轨
19. _____	着陆照明	20. _____	下滑道
21. _____	沟,槽	22. _____	垂直的,正交的
23. _____	喷气飞机	24. _____	滑模
25. _____	规范,法规,规则		

Exercises

1. Decide whether the following statements are true (T) or false (F) according to the text.

1) The landside areas of airports usually include runways, taxiways, ramps and tank farms.

2) Many large airports are located near railway trunk routes for seamless connection of multi-modal transport.

3) An airplane produces its greatest load during landing because of the impact of landing.

4) An airport, which has the ability to handle international aviation traffics but hasn't operated any such flights actually, shouldn't be named International Airport.

5) Airports often pollute environment for their aircraft noises, but they won't change local weather patterns.

2. Write down the words with the following prefix or suffix.

air- _____

aero- _____

-cess _____

-less _____

-ity _____

3. Translate the following paragraph into Chinese.

Airports are divided into landside and airside areas. Landside areas include parking lots, public transportation train stations, tank farms and access roads. Airside areas include all areas accessible to aircraft, including runways, taxiways, ramps and tank farms. Access from landside areas to airside areas is tightly controlled at most airports. Passengers on commercial flights access airside areas through terminals, where they can purchase tickets, clear security, check or claim luggage and board aircraft through gates. The waiting areas which provide passenger access to aircraft are typically called concourses, although this term is often used interchangeably with terminal.

4. Translate the following sentences into English.

1) 机场至少具有一个起降平面,如飞机跑道、直升机停机坪或是起降水域;此外,机场通常还有飞机库、航站楼等建筑物。

2) 配筋混凝土道面采用滑模摊铺机来施工,这样便在沿跑道长度方向形成无施工缝的连续面板。

3) 现代机场跑道在飞机低速运动、荷载最大的区域厚度最大,此区域即跑道端部。

Text B Airport Runway

A runway (RWY) is a strip of land at an airport on which aircraft can take off and land and forms part of the maneuvering area. Runways may be a man-made surface (often asphalt, concrete, or a mixture of both) or a natural surface (grass, dirt, gravel, or salt).

1. Runway Orientation and Dimensions

Runways are given a number between 01 and 36. This indicates the runway's heading: A runway with the number 36 points to the north (360°), runway 09 points east (90°), runway 18 is south (180°), and runway 27 points west (270°). Thus, the runway number is one tenth of the runway centerline's magnetic azimuth, measured clockwise from the magnetic declination.

A runway can be used in two directions, which means the runway has two names: "runway 33" and "runway 15". The two numbers always differ by 18(= 180°).

Runways in North America that lie within the Northern Domestic Airspace are, because of the magnetic north pole, usually numbered according to true north.

Runway designations change over time because the magnetic poles slowly drift on the Earth's surface and the magnetic bearing will change. Depending on the airport location and how much drift takes place, it may be necessary over time to change the runway designation. As runways are designated with headings rounded to the nearest 10 degrees, this will affect some runways more than others. For example, if the magnetic heading of a runway is 233 degrees, it would be designated Runway 23. If the magnetic heading changed downwards by 5 degrees, the Runway would still be Runway 23. If on the other hand the original magnetic heading was 226 (Runway 23), and the heading decreased by only 2 degrees to 224, the runway should become Runway 22. Because the drift itself is quite slow, runway designation changes are uncommon, and not welcomed, as they require an accompanying change in aeronautical charts and descriptive documents. When runway designations do change, especially at major airports, it is often changed overnight as taxiway signs need to be changed and the huge numbers at each end of the runway need to be repainted to the new runway designators. In July 2009 for example, London Stansted Airport in the United Kingdom changed its runway designations from 05/23 to 04/22 overnight.

If there is more than one runway pointing in the same direction (parallel runways), each runway is identified by appending Left (L), Center (C) and Right (R) to the number, for example, Runway One Five Left (15L), One Five Center (15C), and One Five Right (15R). Runway Zero Three Left (03L) becomes Runway Two One Right (21R) when used in the opposite direction.

Two runways pointing in the same direction are classed as dual or parallel runways depending on the separation distance. In some countries, flight rules mandate that only one runway may be used at a time under certain conditions (usually adverse weather) if the parallel runways are too close to each other.

At large airports with more than three parallel runways, some runway identifiers are shifted by 10 degrees to avoid the ambiguity that would result with more than three parallel runways. For example, in Los Angeles, this system results in Runways 6L, 6R, 7L and 7R, even though all four runways are exactly parallel (approximately 69 degrees). At Dallas-Fort Worth, there are five parallel runways, named 17L, 17C, 17R, 18L and 18R, all oriented at a heading of 175.4 degrees.

For fixed-wing aircraft it is advantageous to perform take-offs and landings into the wind to reduce takeoff roll and reduce the ground speed needed to attain flying speed. Larger airports usually have several runways in different directions, so that one can be selected that is most nearly aligned with the wind. Airports with one runway are often constructed to be aligned with the prevailing wind.

Runway dimensions vary from as small as 245 m long and 8 m wide in smaller general aviation airports, to 5,500 m long and 80 m wide at large international airports built to accommodate the largest jets, to the huge 11 917 m × 274 m lake bed runway 17/35 at Edwards Air Force Base in California, which is a landing site for the Space Shuttle.

2. Declared Distances of the Runway

1) Take off Run Available (TORA)

TORA is the length of runway declared available and suitable for the ground run of an airplane taking off.

2) Take off Distance Available (TODA)

TODA is the length of the takeoff run available plus the length of the clearway, if clearway is provided.

The clearway length allowed must lie within the aerodrome or airport boundary. According to the Federal Aviation Regulations and Joint Aviation Requirements (JAR), TODA is the lesser of TORA plus clearway or 1.5 times TORA.

3) Accelerate-Stop Distance Available (ASDA)

ASDA is the length of the takeoff run available plus the length of the stopway, if stopway is provided.

4) Landing Distance Available (LDA)

LDA the length of runway which is declared available and suitable for the ground run of an airplane landing.

5) Emergency Distance Available (EDA)

EDA is LDA (or TORA) plus a stopway.

3. Sections of the Runway

The runway safety area is the cleared, smoothed and graded area around the paved runway. It is kept free from any obstacles that might impede flight or ground roll of aircraft.

The runway is the surface from threshold to threshold, which typically features threshold markings, numbers, centerlines, but not overrun areas at both ends.

Blast pads, also known as overrun areas or stopways, are often constructed just before the start of a runway where jet blast produced by large planes during the takeoff roll could otherwise erode the ground and eventually damage the runway. Overrun areas are also constructed at the end of runways as emergency space to slowly stop planes that overrun the runway on a landing gone wrong, or to slowly stop a plane on a rejected takeoff or a take-off gone wrong. Blast pads are often not as strong as the main paved surface of the runway and are marked with yellow chevrons. Planes are not allowed to taxi, take-off or land on blast pads, except in an emergency.

Displaced thresholds may be used for taxiing, takeoff, and landing rollout, but not for touchdown. A displaced threshold often exists because obstacles just before the runway, runway strength, or noise restrictions may make the beginning section of runway unsuitable for landings. It is marked with white paint arrows that lead up to the beginning of the landing portion of the runway.

4. Runway Lighting

1) History

The first runway lighting appeared in 1930 at Cleveland Municipal Airport in Cleveland, Ohio. A line of lights on an airfield or elsewhere to guide aircraft in taking off or coming in to land or an illuminated runway is sometimes also known as a flare path.

2) Technical specifications

Runway lighting is used at airports which allow night landings. Seen from the air, runway lights form an outline of the runway. A particular runway may have some or all of the followings:

① Runway End Identification Lights (REIL), which are unidirectional (facing approach direction) or omnidirectional pair of synchronized flashing lights installed at the runway threshold, one on each side.

② Runway end lights, which are a pair of four lights on each side of the runway on precision instrument runways to extend along the full width of the runway. These lights show green when viewed by approaching aircraft and red when seen from the runway.

③ Runway edge lights, which are white elevated lights that run the length of the runway on either side. On precision instrument runways, the edge-lighting becomes yellow in the last 610 m of the runway. Taxiways are differentiated by being bordered by blue lights, or by having green centre lights, depending on the width of the taxiway, and the complexity of the taxi pattern.

④ Runway Centerline Lighting System (RCLS), which is lights embedded into the surface of the runway at 15 m intervals along the runway centerline on some precision instrument runways.

⑤ Touchdown Zone Lights (TDZL), which are rows of white light bars on either side of the centerline over the first 914 m of the runway.

⑥ Taxiway centerline lead-off lights, which are installed along lead-off markings, alternate green and yellow lights embedded into the runway pavement. It starts with green light about runway centerline to the position of first centerline light beyond holding position on taxiway.

⑦ Taxiway centerline lead-on lights, which are installed the same way as taxiway centerline lead-off lights.

⑧ Land and hold short lights, which are a row of white pulsating lights installed across the runway to indicate hold short position on some runways which are facilitating Land and Hold Short Operations (LAHSO).

⑨ Approach Lighting System (ALS), which is a lighting system installed on the approach end of an airport runway and consists of a series of light bars, strobe lights, or a combination of the two that extends outward from the runway end.

According to Transport Canada's regulations, the runway-edge lighting must be visible for at least 3 km. Additionally, a new system of advisory lighting, Runway Status Lights, is currently being tested in the United States.

The edge lights must be arranged such that: the minimum distance between lines is 23 m, and maximum is 61 m; the maximum distance between lights within each line is 61 m; the minimum length of parallel lines is 427 m; the minimum number of lights in the line is eight.

3) Control of Lighting System

Typically the lights are controlled by a control tower, a flight service station or another designated authority. Some airports are equipped with Pilot Controlled Lighting, so that pilots can temporarily turn on the lights when the relevant authority is not available. This avoids the need for automatic systems or staff to turn the lights on at night or in other low visibility situations. This also avoids the cost of having the lighting system on for extended periods. Smaller airports may not have lighted runways or runway markings. Particularly at private airfields for light planes, there may be nothing more than a windsock beside a landing strip.

5. Runway Markings

There are runway markings and signs on any runway. Larger runways have a distance remaining sign. This sign uses a single number to indicate the thousands of feet remaining, so 7 will indicate 7,000 ft (2,134 m) remaining. The runway threshold is marked by a line of green lights.

It should be noted that there are national variants for runway markings.

In Australia, Canada, Japan, the United Kingdom, as well as some other countries all 3-stripe and 2-stripe touchdown zones for precision runways are replaced with one-stripe touchdown zones.

In Australia, precision runways consist of only an aiming point and one 1-stripe touchdown zone. Furthermore, many non-precision and visual runways lack an aiming point.

In some Latin American countries like Colombia, Ecuador and Peru one 3-stripe is added and a 2-stripe is replaced with the aiming point.

Some European countries replace the aiming point with a 3-stripe touchdown zone.

Runways in Norway have yellow markings instead of the usual white ones. This also occurs on some airports in Japan. The yellow markings are used to ensure better contrast against snow.

6. Runway Types

There are three types of runways:

First, visual runways. They are used at small airstrips and are usually just a strip of grass, gravel, asphalt or concrete. Although there are usually no markings on a visual runway, they may have threshold markings, designators, and centerlines. Additionally, they do not provide an instrument-based landing procedure; pilots must be able to see the runway to use it. Also, radio communication may not be available and pilots must be self-reliant.

Second, non-precision instrument runways. They are often used at small-size to medium-size airports. These runways, depending on the surface, may be marked with threshold markings, designators, centerlines. They provide horizontal position guidance to planes on instrument approach via non-directional beacon (NDB), VHF omnidirectional range (VOR), Global Positioning System (GPS), etc.

Third, precision instrument runways. They are found at medium-size and large-size airports; and consist of blast pads, thresholds, designators, centerlines, aiming points and touchdown zone marks. Precision instrument runways provide both horizontal and vertical guidance for instrument approaches.

Runways may have different types on each end. To cut costs, many airports do not install precision guidance equipment on both ends. Runways with one precision end and any other type of end can install the full set of touchdown zones, even if some are past the midpoint. If a runway has precision markings on both ends, touchdown zones within 274 m of the midpoint are omitted, to avoid pilot confusion over which end the marking belongs to.

7. Runway Pavement

The choice of material used to construct the runway depends on the use and the local ground conditions. For a major airport, where the ground conditions permit, the most satisfactory type of pavement for long-term minimum maintenance is concrete. Although certain airports have used reinforcement in concrete pavements, this is generally found to be unnecessary, with the exception of expansion joints across the runway where a dowel assembly, which permits relative movement of the concrete slabs, is placed in the concrete. Where it can be anticipated that major settlements of the runway will occur over the years because of unstable ground conditions, it is preferable to install asphalt concrete surface, as it is easier to patch on a periodic basis. For fields with very low traffic of light planes, it is possible to use a sod surface.

For pavement designs borings are taken to determine the subgrade condition, and based on the relative bearing capacity of the subgrade, the specifications are established. For heavy-duty commercial aircraft, the pavement thickness, no matter what the top surface is, varies from 250 mm to 1,000 mm, including subgrade.

Airport pavements have been designed by two methods. The first, Westergaard, is based on the assumption that the pavement is an elastic plate supported on a heavy fluid base with a uniform reaction coefficient known as the K value. Experience has shown that K values on which the formula was developed are not applicable for newer aircraft with very large footprint pressures. The second method is called the California bearing ratio and was developed in the late 1940s. It is an extrapolation of the original test results, which are not applicable to modern aircraft pavements or to modern aircraft landing gear. Some designs were made by a mixture of these two design theories.

A more recent method is an analytical system based on the introduction of vehicle response as an important design parameter. Essentially it takes into account all factors, including the traffic conditions, service life, materials used in the construction, and, especially important, the dynamic response of the vehicles using the landing area.

Because airport pavement construction is so expensive, every effort is made to minimize the stresses imparted to the pavement by aircraft. Manufacturers of the larger planes design landing gear so that the weight of the plane is supported on larger and more numerous tires. Attention is also paid to the characteristics of the landing gear itself, so that adverse effects on the pavement are minimized. Sometimes it is possible to reinforce a pavement for higher loading by applying an overlay of asphaltic concrete or Portland cement concrete that is bonded to the original slab.

Post-tensioning concrete has been developed for the runway surface. This permits the use of thinner pavements and should result in longer concrete pavement life. Because of the susceptibility of thinner pavements to frost heave, this process is generally applicable only where there is no appreciable frost action.

Runway pavement surface is prepared and maintained to maximize friction for wheel braking. To minimize hydroplaning following heavy rain, the pavement surface is usually grooved so that the surface water film flows into the grooves and the peaks between grooves will still be in contact with the aircraft tires. To maintain the macrotexturing built into the runway by the grooves, maintenance crews engage in airfield rubber removal or hydrocleaning in order to meet required FAA friction levels.

8. Runway Length

A runway of at least 1,829 m in length is usually adequate for aircraft weights below approximately 90,718 kg. Larger aircraft including widebodies will usually require at least 2,438 m at sea level and somewhat more at higher altitude airports. International widebody flights, which carry substantial amounts of fuel and are therefore heavier, may also have landing requirements of 3,048 m or more and takeoff requirements of 3,962 m.

At sea level, 3,048 m can be considered an adequate length to land virtually any aircraft. For example, at O'Hare International, when landing simultaneously on 22R and 27L or parallel 27R, it is routine for arrivals from the Far East which would normally be vectored for 22R (2,286 m) or 27R (2,438 m) to request 27L (3,048 m). It is always accommodated, although occasionally with a delay.

An aircraft will need a longer runway at a higher altitude due to decreased density of air at higher altitudes, which reduces lift and engine power. An aircraft will also require a longer runway in hotter or more humid conditions.

Text C Best Practices for Airport Portland Cement
Concrete Pavement Construction

1. Introduction

The first concrete pavement for airport use was constructed during 1927 and 1928 at the Ford Terminal in Dearborn, Michigan. Since then, concrete pavements have been widely used for constructing runways, taxiways, and apron areas at airports. The design and construction procedures used for airport pavements evolved through experience, practice, field trials, and application of theoretical considerations. Concrete pavements have a long and successful history of use at civilian airports and at military airfields in the United States.

Air transportation is one of the key industries in the United States. The high cost of shutdowns for pavement maintenance and rehabilitation at airports results in significant impact on local and regional economies, in addition to unnecessary delays to the traveling public. A similar concern exists at military airfields where operational readiness can be impacted by poor pavements. For airport pavements to perform well, it is essential that these pavements are designed and constructed to a high degree of quality. A well-designed and constructed concrete pavement will withstand the anticipated aircraft loadings under the local climatic conditions over the desired period of time with minimum maintenance and repair.

Desirable concrete pavement performance can be obtained by ensuring that the occurrences of various distresses that can develop are minimized. Distresses that may develop in airport concrete pavements include the followings: Cracking (corner, longitudinal, transverse, durability/materials related); Joint related (spalling, pumping, joint seal damage); Surface defects (scaling, popouts, map cracking).

The development of concrete pavement distresses can be minimized by a few operations as follows selecting the proper pavement thickness; providing adequate foundation support including a free draining non-erodible base; performing proper joint layout and installation; designing and installing adequate load transfer at joints; selecting proper constituents for the concrete; ensuring adequate concrete consolidation; providing proper finishing to the concrete surface; maintaining joint sealant in good condition.

Another important concern for concrete pavement construction is minimizing the probability of early-age distress, typically in the form of cracking and spalling. This is accomplished by the use of sound design principles and by implementing good construction techniques.

1) Purpose

The information presented here is a compendium of good construction and inspection practices that lead to long-term pavement performance. In addition to highlighting good construction practices, this manual also includes a discussion of practices that are known to result in poor pavement performance. Simply stated, good construction practices mean the systematic application of the collective know-how derived through years of field experience and application of technical knowledge.

This manual does not directly address concrete pavement design issues. However, it is emphasized that both good structural and good geometric design elements are critical to successful early age and long-term performance of airport concrete pavements. Pavements that will perform to expectations will only occur when good designs are implemented through good construction practices.

2) Scope

This manual presents construction practices that are accepted by the industry as practices that produce quality concrete pavements. Specifically, the scope of this manual includes the followings:

① Documentation of good construction techniques and practices.

② Discussion of advantages and disadvantages of techniques or practices where more than one method is available.

③ Identification of practices that result in long-term performance of airport PCC pavements.

④ Identification of practices that result in early age or premature failures and poor long-term performance and discussion on how to mitigate problems when they do occur.

⑤ Discussion of commonly encountered problems in meeting project specifications.

3) Quality in Constructed Projects

A fundamental assumption made during the preparation of this manual is that a quality pavement performs well. Quality is an inherent property of a well-constructed pavement. Quality is not a hit or miss proposition. As defined by the American Society of Civil Engineers (ASCE), "Quality is never an accident. It is always the result of high intentions, intelligent direction, and skilled execution. It represents a wise choice amongst many alternatives."

Quality construction requires dedication from the project management down to the execution by labor. The management, as well as the field crews, needs to buy into the concept of quality in construction, not necessarily because it is mandated, but because it is the right approach. The contractor needs to emphasize teamwork and collective accountability for constructing long lasting pavements.

Good materials and construction practices are vital for producing high quality and long lasting airfield concrete pavements. Even if a pavement is designed to the highest standards, it will not perform well if it is not constructed well. A pavement that is constructed well will require less maintenance and repairs over the years. As such, construction requirements and specifications need to be well defined, able to be measured, and not arbitrary. The project specifications need sufficient flexibility to allow for innovations by the contractor.

2. Consideration of Design Issues

1) Introduction

Factors affecting long-term airport pavement performance can be broadly divided into the following categories: adequate design of pavement structure; use of quality materials; use of proper construction procedures; timely maintenance and repairs.

Airports in the United States are either civil airports or military airports. The overall process of designing a concrete pavement at an airport involves the following steps:

① Soil Investigation: Soil borings are performed to determine the properties of the subsurface strata and to obtain depth to groundwater. Soil samples are obtained for soil classification and laboratory testing.

② Evaluate Subgrade Support at Design Grade: The information obtained from the soil investigation is used to evaluate the subgrade conditions at and below the design grade.

③ Design Pavement Section: An appropriate base type and thickness are determined. Then the appropriate design procedure is used to obtain the thickness of the PCC pavement.

④ Select Jointing Plan: A slab size has to be selected and a jointing plan has to be developed. Appropriate longitudinal and transverse joint details have to be developed. Also, proper details are required for joints and transition slabs at tie-ins to existing pavements.

⑤ Develop Plans and Specifications: The design details are translated into plans and specifications.

The critical design features that influence the long-term performance of concrete pavements:

① Subgrade support uniformity and stability.

② Base and subbase uniformity (type and thickness), including drainage provisions.

③ Pavement thickness.

④ Concrete properties, as specified, including: uniformity (ability of concrete to produce consistent properties), workability (ability of concrete to be placed, consolidated and finished), strength (ability of concrete to support traffic and environmental conditions), and durability (ability of concrete to provide long-term service).

⑤ Jointing details, including slab dimensions, load transfer at joints, and joint sealing provisions.

For each project, the design engineer establishes the acceptable parameters for each of the design variables. It is then expected that during construction, the quality of the design will be provided as expected (in terms of specifications) or better. It is a common experience that when several marginal features are built into a pavement, either because of design deficiency or because of poor construction or a combination of both, then the pavement will exhibit premature failures or provide less than expected performance over the long term.

Several examples are given to illustrate the criticality of various construction operations:

① Grading: Proper grading is an important construction item. Proper grading facilitates drainage and placement of successive layers.

② Jointing: Jointing is provided to control slab cracking. This minimizes the potential for random cracking. Random cracking is a maintenance concern and may affect the load capacity of the pavement. Shallow joint sawing and late sawing are some of the causes of random cracking. If dowel bars are misaligned or bonded to the concrete, joints will not function and random cracking can develop in adjacent slab panels.

③ Subgrade and subbase/base quality: If the compaction of the subgrade, subbase and base is compromised, the pavement may deflect too much under aircraft loading and corner cracking may develop.

④ Concrete strength: Low strength concrete will result in early fatigue cracking of the pavement. Concrete flexural strength at 28 days for airport paving is typically 4,100 to 5,200 kPa. For fast track construction, these strength levels may be required at an earlier age.

⑤ Concrete durability: Concrete that is not durable (a result of poor or reactive materials, a poor air-void system, or due to over-finishing) may deteriorate prematurely.

⑥ Concrete curing: Concrete that has not cured adequately can deteriorate prematurely. Poorly cured concrete can also result in early agespalling.

⑦ Concrete finishability: Concrete that is over-finished or requires excessive manipulation to provide finishability will deteriorate prematurely. Poorly finished concrete may also result in poor surface condition.

⑧ Paver Operation: The paver operation has a significant impact on pavement smoothness and in-place quality of concrete.

Airport concrete pavements are typically jointed plain concrete pavements. Very few agencies specify jointed concrete pavements incorporating steel or continuous reinforcement for production paving. This manual focuses on jointed plain concrete pavements.

Airport concrete pavements are typically designed on the basis of mixed aircraft loadings to provide low maintenance service for 20~30 years. The pavements, typically plain jointed concrete pavements, are designed on the basis of expected aircraft repetitions over the design period. For larger commercial airports that receive wide-body aircraft, pavement thickness may range from 400 mm to 500 mm and transverse joint spacing may range from 4.6 m to 7.6 m. Longitudinal joint spacing may range from 3.8 m to 7.6 m. Also, most designers now specify dowel bars for longitudinal construction joints. For general aviation airports, slab thickness may range from 125 mm to 300 mm and transverse and longitudinal joint spacing may range from 2.4 m to 4.6 m.

2) Consideration Variability

Pavement performance is significantly affected by the variability in the properties of key design features. While a certain amount of variability is unavoidable, excessive variability in the construction process can lead to random performance of pavements, as well as higher cost to the contractor. Construction variability can be controlled by making effective use of quality management plans.

3) Summary

A successful airport concrete pavement project depends on ensuring that the design process has been optimized and quality in construction has been implemented.

The design engineer needs to ensure that pavement designs and associated construction specifications are practical and the quality requirements are achievable and necessary. Also, methods to measure specific requirements need to be well defined. Finally, it is advisable that on larger or time-sensitive projects, the design engineer or his representative be available on site on a regular basis to resolve design-related issues that develop during layout and construction.

Integrated Skills: Making a Powerpoint File

The on-the-sea airport construction is one of the trends for the development of current and future airports. By means of Internet search and the review of special literatures, you should collect engineering pictures and technological information about on-the-sea airports, and make a relevant PowerPoint file. Then make a conference whose background is *On-The-Sea Airports Construction* in class, and report your ideas in this conference. When preparing your report, you can make a reference to the erected on-the-sea airports such as Hong Kong International Airport (Figure 11-1), Macau International Airport (Figure 11-2), Kansai International Airport (Figure 11-3), etc.

Figure 11-1　Hong Kong International Airport

Figure 11-2　Macau International Airport

Figure 11-3　Kansai International Airport, Japan

Unit 12   Civil Engineering Materials

Lead-in

1. Which groups may be the building materials classified into?
2. Which materials are much stronger in tension than in compression?
3. What are the structural materials?

Vocabulary Warm-up: Matching

_____1. adhesive	a. 硬化剂
_____2. accelerator	b. 混合物；附加剂
_____3. setting agent	c. 黏合剂
_____4. admixture	d. 超增塑剂
_____5. superplasticizer	e. 水合，水合作用
_____6. hydration	f. 催化剂

Text A   Building Materials

The materials are the basic elements of any building. Building materials may be classified into three groups, according to the purposes they are used for. The first group is structural materials. Structural materials are those that hold the building up, keep it rigid, form its outer covering of walls and roof, and divide its interior into rooms. The second group is materials for the equipment inside the building, such as the plumbing, heating and lighting systems. Finally, there are materials that are used to protect or decorate the structural materials.

The basic structural materials are wood, *clay*, bricks, stone, *steel*, *plastic*, and *concrete*.

1. Wood

Wood is the oldest of the structural materials. It has been used since prehistoric times. Wood can be shaped with simple tools, and it can be easily fastened with nails, screws, or *adhesive*s. It is available either as *lumber*, which is a natural wood, or as glued laminated timber. This latter form is much stronger and consists of specially selected and prepared wood laminations securely bonded together with glue or adhesives. The laminations do not exceed in thickness. Glued laminated beams are available in a variety of shapes. Both lumber and glued timber are widely used in building construction.

Wood contains certain natural defects, such as knots, shakes, etc., which require that it be stress-traded to establish design or allowable stress values for particular members. Another important factor is the grain of the wood. The allowable stress in compression parallel to the grain, for instance, is higher than the allowable stress perpendicular to the grain.

2. Clay

The essential feature of most clay is that it contains a variety of intimately mixed minerals. The balance of the various types depends on the origin of the clay and widely in differing properties results accordingly. The nature of clay varies greatly with depth and from region to region, and the manufacturing method and properties of the resultant unit vary accordingly. Clay is used in the form of bricks and *tile*. Most bricks today are made from clay which is pressed into *mould*s either by hand or machine. Since there are many types of clay, there are many types of brick of varying strengths. The strongest bricks, the type frequently used to support the ends of bridges, are called engineering bricks. A well-known engineering brick is dark blue in color. There are also red engineering bricks, which, like the blue bricks, are very hard, dense and strong.

3. Bricks

All bricks are much stronger in compression than in tension, and their tensile strength is so low that they are unsuitable for resisting forces which would produce tension. It is not possible to obtain an accurate estimate of the strength of *brickwork* in compression by testing to *destruction* of individual bricks, since both the nature of the *mortar* and the quality of the *workmanship* have influences on strength.

The great advantage of brick construction is its low cost. Clay is cheap and plentiful, and bricks are small enough to be laid easily by hand.

4. Stone

Stone is strong and durable. In many respects it is an ideal building material. However, it is difficult to handle. For use in a large building, stones must be cut from the solid rock in a *quarry*. All this labor makes stone very expensive. Because it is expensive and difficult to work with, stone is used today mainly as a thin outside covering for large buildings and for the decorative walls and floors of entrance lobbies.

Properties of stone depend on what nature has provided. Therefore, the designer does not have the choice of properties and color available in some of the manufactured building units. The most the stone producer can do for purchasers is to avoid quarrying certain stone beds that have been proved by experience to have poor strength or poor durability.

5. Steel

Steel is one of the most useful structural building materials. Because of its high strength, it is particularly suitable for use with heavy loads and over long spans. But it must be painted to prevent *rusting*. It has some of wood's *elastic* quality. And it can be formed into almost any shape. In modern construction steel is used for the supporting framework of large buildings.

6. Plastic

Plastic is modified with plasticizers, *fillers*, or other *ingredient*s. Consequently, each base material forms the nucleus for a large number of products having a wide variety of properties. Plastics are quite different in their *composition* and structure from other materials, such as metals, their behavior under stress and under other conditions is likely to be different from other materials. Just as steel is markedly different and is used for different *application*s, so are the various plastics materials — some hard and *brittle*, others soft and extensible.

More than many other materials, plastics are sensitive to temperature and to the rate and time of application of load. This *viscoelastic* behavior, combining elastic and viscous or plastic reaction to stress, is unlike the behavior of materials which are traditionally considered to behave only elastically.

7. Concrete

Concrete is a man-made material created by the proper mixing of coarse aggregate, such as *gravel*, fine aggregate, sand and cement, with adequate and controlled amounts of water. Concrete is very strong in compression, and ultimate strength of 34.5 N/mm^2 to 41.4 N/mm^2 can be obtained. Concrete is very weak in tension, so has to be adequately reinforced with steel bars. The concrete then takes the compression, and the steel takes all the tension. Both materials work together as a composite material. This material is then called reinforced concrete.

Concrete solidifies and hardens after mixing and placement due to a chemical process known as *hydration*. The water reacts with the cement, which bonds the other components together, eventually creating a stone-like material. It is used to make pavements, architectural structures, foundations, *motorway*s (roads), *overpass*es, parking structures, brick (block) walls and footings for gates, fences and poles.

Concrete is used more than any other man-made material on the planet. As for 2005, about six billion cubic meters of concrete were made each year, which equaled one cubic meter for every person on Earth. Concrete powers a US $ 35 billion industry which employs more than two million workers in the United States alone. More than 55,000 miles of freeways and highways in America are made of this material. The People's Republic of China currently consumes 40% of the world's cement (concrete) production.

Many other kinds of materials besides the basic structural materials are needed for a building. There is window glass admitting light and providing a view. There is insulation keeping the building warm in winter and cool in summer. A big office or apartment building contains vast and complicated heating, air conditioning, plumbing, electrical, and lighting systems. These will use great quantities of metals and alloys, plastics, adhesives, coatings, such as paint for walls and tar to keep roofs from leaking. By the end of World War II the nonstructural materials used in a building added up to more than half the total cost of the building itself.

Glossary

Complete the glossary using words in the text.

1. _____	黏土	2. _____	钢
3. _____	塑料	4. _____	混凝土
5. _____	黏合剂	6. _____	木材,木料
7. _____	瓦片	8. _____	铸模,模型
9. _____	砌砖	10. _____	破坏
11. _____	砂浆	12. _____	工艺
13. _____	石场	14. _____	生锈
15. _____	有弹性的	16. _____	填充料
17. _____	成分	18. _____	构成,成分
19. _____	运用	20. _____	脆性的
21. _____	黏弹性的	22. _____	沙砾
23. _____	水合,水合作用	24. _____	高速公路
25. _____	立交桥		

Exercises

1. Decide whether the following statements are true (T) or false (F) according to the text.

1) All bricks are much stronger in tension than in compression, and their tensile strength is so low that they are unsuitable for resisting forces which would produce tension.

2) Concrete solidifies and hardens before mixing and placement due to a chemical process known as hydration.

2. Write down the words with the following prefix or suffix.

man- _____

hydr- _____

visco- _____

-work _____

-crete _____

3. Translate the following paragraph into Chinese.

1) Building materials may be classified into three groups, according to the purposes they are used for. The first group is structural materials. Structural materials are those that hold the building up, keep it rigid, form its outer covering of walls and roof, and divide its interior into rooms. The second group is materials for the equipment inside the building, such as the plumbing, heating and lighting systems. Finally, there are materials that are used to protect or decorate the structural materials.

2) Concrete is a man-made material created by the proper mixing of coarse

aggregate, such as gravel, fine aggregate, sand and cement, with adequate and controlled amounts of water. Concrete is very strong in compression, and ultimate strength of 34. 5 N/mm² to 41. 4 N/mm² can be obtained. Concrete is very weak in tension, so has to be adequately reinforced with steel bars. The concrete then takes the compression, and the steel takes all the tension. Both materials work together as a composite material. This material is then called reinforced concrete.

3) Concrete solidifies and hardens after mixing and placement due to a chemical process known as hydration. The water reacts with the cement, which bonds the other components together, eventually creating a stone-like material.

Text B Cementitious Materials

Cementitious materials include the many products that are mixed with either water or some other liquid or both to form a cementing paste that may be formed or molded while plastic will set into a rigid shape. When sand is added to the paste, mortar is formed. A combination of coarse and fine aggregate (sand) added to the paste forms concrete.

Types of Cementitious Materials

There are many varieties of cement and numerous ways of classification. One of the simplest classification is by the chemical constituent that is responsible for the setting or hardening of the cement. On this basis, the silicate and aluminate cement, wherein the setting agents are calcium silicates and aluminates, constitute the most important group of modern cement. Included in this group are the Portland, aluminous, and natural cement. Lime, wherein the hardening is due to the conversion of hydroxides to carbonates, was formerly widely used as the sole cementitious material, but its slow setting and hardening was not compatible with modern requirements. Hence, its principal function today is to plasticize the otherwise harsh cement and add resilience to mortar and stucco. Use of lime is beneficial in that its slow setting promotes healing, the recementing of hairline cracks.

Another class of cement is composed of calcined gypsum and its related products. The gypsum cement is widely used in interior plaster and for fabrication of boards and blocks; but the solubility of gypsum prevents its use in construction exposed to any but extremely dry climates. Oxychloride cement constitutes a class of specialty cement of unusual properties. Their cost prohibits their general use in competition with the cheaper cement; but for special uses, such as the production of sparkproof floors, they cannot be equaled.

Masonry cement or mortar cement is widely used because of their convenience. While they are, in general, mixtures of one or more of the above-mentioned cement with some admixtures, they deserve special consideration because of their economies.

Other cementitious materials, such as polymers, fly ash, and silica fume, may be used as a cement replacement in concrete. Polymers are plastics with long-chain molecules. Concrete made with them have many qualities much superior to those of ordinary concrete. Silica fume, also known as microsilica, is a waste product of electric-arc furnaces. The silica reacts with lime in concrete to form a cementitious material. A fume particle has a diameter only 1% of that of a cement particle.

Portland Cement

Portland cement, the most common of the modern cement, is made by carefully blending selected raw materials to produce a finished material meeting the requirements of ASTM C150/C150M-2012 for one of eight specific cement types. Four major compounds[lime (CaO), iron (Fe_2O_3), silica (SiO_2) and alumina (Al_2O_3)] and two minor compounds (gypsum ($CaSO_4 \cdot 2H_2O$) and magnesia (MgO)) constitute the raw materials. The calcareous (CaO) materials typically come from limestone, calcite, marl, or shale. The argillaceous (SiO_2 and Al_2O_3) materials are derived from clay, shale, and sand. The materials used for the manufacture of any specific cement are dependent on the manufacturing plant's location and availability of raw materials. Portland cement can be made of a wide variety of industrial by-products.

In the manufacture of cement, the raw materials are first mined and then ground to a powder before blending in predetermined proportions. The blend is fed into the upper end of a rotary kiln heated to $2,600\sim3,000$ °F by burning oil, gas, or powdered coal. Because cement production is an energy-intensive process, reheaters and the use of alternative fuel sources, such as old tires, are used to reduce the fuel cost. (Burning tires provide heat to produce the clinker and the steel belts provide the iron constituent.) Exposure to the elevated temperature chemically fuses the raw materials together into hard nodules called cement clinker. After cooling, the clinker is passed through a ball mill and ground to a fineness where essentially all of it will pass a No. 200 sieve (75 um). During the grinding, gypsum is added in small amounts to control the temperature and regulate the cement setting time. The material that exits the ball mill is Portland cement. It is normally sold in bags containing 94 lb of cement.

Concrete, the most common use for Portland cement, is a complex material consisting of Portland cement, aggregates, water, and possibly chemical and mineral admixtures. Only rarely is Portland cement used alone, such as for a cement slurry for filling well holes or for a fine grout. Therefore, it is important to examine the relationship between the various Portland cement properties and their potential effect upon the finished concrete. Portland cement concrete is generally selected for structural use because of its strength and durability. Strength is easily measured and can be used as a general directly proportional indicator of overall durability. Durability cannot be easily measured but can be specified by controlling the cement chemistry and aggregate properties.

Aluminous Cement

This is prepared by fusing a mixture of aluminous and calcareous materials (usually bauxite and limestone) and grinding the resultant product to a fine powder. This cement is characterized by its rapid-hardening properties and the high strength developed at early ages.

Since a large amount of heat is liberated with rapidly by aluminous cement during hydration, care must be taken not to use the cement in places where this heat cannot be dissipated. It is usually not desirable to place aluminous-cement concrete in lifts of over 12 in; otherwise the temperature rise may cause serious weakening of the concrete.

Aluminous cement is much more resistant to the action of sulfate water than is Portland cement. It also appears to be much more resistant to attack by water containing aggressive carbon dioxide or weak mineral acids than the silicate cement. Its principal use is in concrete where advantage may be taken of its very high early strength or of its sulfate resistance, and where the extra cost of the cement is not an important factor.

Another use of aluminous cement is in combination with firebrick to make refractory concrete. As temperatures increase, dehydration of the hydration products occurs. Ultimately, these compounds create a ceramic bond with the aggregates.

Natural Cement

Natural cement is formed by calcining a naturally occurring mixture of calcareous and argillaceous substances at a temperature below that at which sintering takes place. Since natural cement is derived from naturally occurring materials and no particular effort is made to adjust the composition, both the composition and properties vary rather widely. Some natural cement may be almost the equivalent of Portland cement in properties; others are much weaker. Natural cement is principally used in masonry mortars and as an admixture in Portland cement concrete.

Text C Types of Concrete

Various types of concrete have been developed for specialist application and have become known by these names.

1. Regular Concrete

Regular concrete is the lay term describing concrete that is produced by following the mixing instructions that are commonly published on packets of cement, typically using sand or other common material as the aggregate, and often mixed in improvised containers. This concrete can be produced to yield a varying strength from about 10 MPa to about 40 MPa, depending on the purpose, ranging from blinding to structural concrete respectively. Many types of premixed concrete are available which include powdered cement mixed with aggregate, needing only water.

2. High-Strength Concrete

High-strength concrete has a compressive strength generally greater than 40 MPa.

High-strength concrete is made by lowering the water-cement (w/c) ratio to 0.35 or lower. Often silica fume is added to prevent the formation of free calcium hydroxide crystals in the cement matrix, which might reduce the strength at the cement-aggregate bond.

Low w/c ratios and the use of silica fume make concrete mixes significantly less workable, which is particularly likely to be a problem in high-strength concrete applications where dense rebar cages are likely to be used. To compensate for the reduced workability, superplasticizers are commonly added to high-strength mixtures. Aggregate must be selected carefully for high-strength mixes, as weaker aggregates may not be strong enough to resist the loads imposed on the concrete and cause a failure to start in the aggregate rather than in the matrix or at a void, as normally occurs in regular concrete.

In some applications of high-strength concrete the design criterion is the elastic modulus rather than the ultimate compressive strength.

3. High-Performance Concrete

High-performance concrete (HPC) is a relatively new term used to describe concrete that conforms to a set of standards above those of the most common applications, but not limited to strength. While all high-strength concrete is also high-performance, not all high-performance concrete is high-strength. Some examples of such standards currently used in relation to HPC are: ease of placement, compaction without segregation, early age strength, long-term mechanical properties, permeability, density, heat of hydration, toughness, volume stability, and long life in severe environments.

4. Self-Compacting Concrete

During the 1980s, a number of countries including Japan, Sweden and France developed concrete that is self-compacting, known as self-consolidating concrete in the United States. This self-compacting concrete (SCC) is characterized by: extreme fluidity as measured by flow, typically between 650~750 mm on a flow table, rather than slump (height); no need for vibrators to compact the concrete; placement being easier; and no bleed water, or aggregate segregation.

SCC can save up to 50% in labor costs due to 80% faster pouring and reduced wear and tear on formwork. As for 2005, self-compacting concrete accounts for 10%~15% of concrete sales in some European countries. In the US precast concrete industry, SCC represents over 75% of concrete production. 38 departments of transportation in the US accept the use of SCC for road and bridge projects.

5. Shotcrete

Shotcrete uses compressed air to shoot (cast) concrete onto (or into) a frame or structure. Shotcrete is frequently used against vertical soil or rock surfaces, as it eliminates the need for formwork. It is sometimes used for rock support, especially in tunneling. Today there are two application methods for shotcrete: the dry-mix and the wet-mix procedure. In dry-mix the dry mixture of cement and aggregates is filled into the machine and conveyed with compressed air through the hoses. The water needed for the hydration is added at the nozzle. In wet-mix, the mixtures are prepared with all necessary water for hydration. The mixtures are pumped through the hoses. At the nozzle compressed air is added for spraying. For both methods additives such as accelerators and fiber reinforcement may be used.

The term Gunite is occasionally used for shotcrete, but properly refers only to dry-mix shotcrete, and once was a proprietary name.

6. Pervious Concrete

Pervious concrete is sometimes specified by engineers and architects when porosity is required to allow some air movement or to facilitate the drainage and flow of water through structures. Pervious concrete is referred to as "no fines" concrete because it is manufactured by leaving out the sand or "fine aggregate". A pervious concrete mixture contains little or no sand (fines), creating a substantial void content. Using sufficient paste to coat and bind the aggregate particles together create a system of highly permeable, interconnected voids that drains quickly. Typically, between 15% and 25% voids are achieved in the hardened concrete, and flow rates for water through pervious concrete are typically around 0.34 cm/s, although they can be much higher. Both the low mortar content and high porosity also reduce strength compared to conventional concrete mixtures, but sufficient strength for many applications is readily achieved.

Pervious concrete pavement is a unique and effective means to address important environmental issues and support sustainable growth. By capturing rainwater and allowing it to seep into the ground, porous concrete is instrumental in recharging groundwater, reducing stormwater runoff, and meeting US Environmental Protection Agency (EPA) stormwater regulations. The use of pervious concrete is among the Best Management Practices (BMP) recommended by the EPA, and by other agencies and geotechnical engineers across the country, for the management of stormwater runoff on a regional and local basis. This pavement technology creates more efficient land use by eliminating the need for retention ponds, swales, and other stormwater management devices. In doing so, pervious concrete has the ability to lower overall project costs on a first-cost basis.

7. Aerated Concrete

Aerated concrete produced by the addition of an air entraining agent to the concrete (or a lightweight aggregate like expanded clay pellets or cork granules and vermiculite) is sometimes called Cellular concrete. See also aerated autoclaved concrete. Cork granules are obtained during production of bottle stoppers from the treated bark of Cork oak or Quercus suber trees. These trees are mainly found in Portugal, Spain and North Africa. Portugal is the largest cork producing country, followed by Spain. The waste cork granules have a density of about 300 kg/m³, which is lower than that of most of the lightweight aggregates used for making lightweight concrete. It has been found that cork granules do not significantly influence cement hydration. However, cork dust can influence hydration. Cork cement composites have several advantages over standard concrete, such as lower thermal conductivities, lower densities and good energy absorption characteristics. These composites can be made of density from 400 to 1,500 kg/m³, compressive strength from 1 to 26 MPa, and flexural strength from 0.5 to 4.0 MPa.

8. Roller-Compacted Concrete

Roller-compacted concrete, sometimes called rollcrete, is a low-cement-content stiff concrete placed using techniques borrowed from earthmoving and paving work. The concrete is placed on the surface to be covered, and is compacted in place using large heavy rollers typically used in earthwork. The concrete mix achieves a high density and cures over time into a strong monolithic block. Roller-compacted concrete is typically used for concrete pavement, but has also been used to build concrete dams, as the low cement content causes less heat to be generated while curing than typical for conventionally placed massive concrete pours.

9. Others

The use of recycled glass as aggregate in concrete has become popular in modern times, with large scale research being carried out at Columbia University in New York. This greatly enhances the aesthetic appeal of the concrete. Recent research findings has shown that concrete made with recycled glass aggregates has shown better long term strength and better thermal insulation due to its better thermal properties of the glass aggregates.

Strictly speaking, asphalt is a form of concrete as well, with bituminous materials replacing cement as the binder. This type of concrete is able to develop high resistance within few hours after been manufactured. This feature has advantages such as removing the formwork early and to move forward in the building process at record time, repair road surfaces that become fully operational in just few hours.

Integrated Skills: Research and Presentation

By means of the field research, Internet search or the review of special literatures, you should research for some technological information on building materials that you are interested in. The information includes basic facts, structures materials, materials for the equipment, decorative materials and so on. Then you should write a scientific essay based on one of the above aspects, and finally present your essay in class and exchange your ideas with others.

Unit 13　Civil Engineering Equipments

Lead-in

1. What does HVAC mean?
2. What different purposes is ventilation utilized for?
3. Why is the use of axial fans steadily increasing?

Vocabulary Warm-up：Matching

_____1. heating	a. 制冷；冷藏，冷冻
_____2. ventilation	b. 过滤效率
_____3. air conditioning	c. 暖气之装置（设备）
_____4. refrigeration	d. （空气中的）湿度
_____5. humidity	e. 区域加热
_____6. air-change rate	f. 空气流通；通风设备
_____7. filtration efficiency	g. 非集中供热
_____8. thermal coefficients	h. 保温系数
_____9. non-central heating	i. 换气率
_____10. zoned heating	j. 空气调节装置

Text A　Heating，Ventilation and Air Conditioning

The necessity of *heating*，*ventilation* and air-conditioning（HVAC）control of *environmental* conditions within buildings has been well established over the years as being highly desirable for various types of occupancy and comfort conditions as well as for many industrial manufacturing processes. In fact，without HVAC systems，many manufactured products produced by industry that is literally taken for granted would not be available today.

1. Definition

1）Adiabatic Process：A *thermodynamic* process that takes place without any heat being added or subtracted and at constant total heat.

2）Air *Makeup*：New or fresh air brought into a building to replace losses due to exflitration and exhausts，such as those from ventilation and chemical *hood*s.

3）Air Return：Air that leaves a conditioned space and is returned to the air conditioning equipment for treatment.

4）Air Saturated：Air that is fully saturated with water *vapor*（100% *humidity*），with the air and water vapor at the same temperature.

5) Air Change: The complete replacement of room air volume with new supply air.

6) Air Conditioning: The process of altering air supply to control simultaneously its humidity, temperature, cleanliness, and distribute on to meet specific criteria for a space. Air conditioning may either increase or decrease the space temperature.

7) Air Conditioning Comfort: Use of air conditioning solely for human comfort, as compared with conditioning for industrial processes or manufacturing.

8) Air Conditioning Industrial: Use of air conditioning in industrial plants where the prime objective is *enhancement* of a manufacturing process rather than human comfort.

9) Baseboard Radiation: A heat-surface device, such as a finned tube with a decorative cover.

10) Blow: *Horizontal* distance from a supply-air discharge register to a point at which the supply-air *velocity* reduces to 50 ft/min.

11) Boiler: A cast-iron or steel container fired with solid, liquid, or gaseous fuels to generate hot water or *steam* for use in heating a building through an appropriate distribution system.

12) Boiler-Burner Unit: A boiler with a matching burner whose heat-release capacity equals the boiler heating capacity with less certain losses.

13) Central Heating or Cooling Plant: One large heating or cooling unit used to heat or cool many rooms, spaces, or zones or several buildings, as compared to individual room, zone, or building units.

14) Coefficient of Performance: For machinery and heat pumps, the ratio of the effect produced to the total power of electrical input consumed.

15) Comfort Zone: An area plotted on a psychometric chart to indicate a combination of temperatures and humidities at which, in controlled tests, more than 50% of the persons were comfortable.

16) *Condensate*: Liquid formed by the condensation of steam or water vapor.

17) Condensers: Special equipment used in air conditioning to liquefy a gas.

18) Condensing Unit: A complete refrigerating system in one assembly, including the refrigerant compressor, motor, condenser, receiver, and other necessary accessories.

2. Heat and Humidity

People have always struggled with the problem of being comfortable in their environment. First attempts were to use fire directly to provide heat through cold winters. It was only in recent times that interest and technology permitted development of greater understanding of heat and heating, and substantial improvements in comfort were made. Comfort heating now is a highly developed science and, in conjunction with air conditioning, provides comfort conditions in all seasons in all parts of the world.

As more was learned about humidity and the capacity of the air to contain various amounts of water vapor, greater achievements in environmental control were made.

Control of humidity in buildings now is a very important part of heating, ventilation, and air conditioning, and in many cases is extremely important in meeting manufacturing requirements. Today, it is possible to alter the atmosphere or environment in buildings in any manner, to suit any particular need, with great *precision* and control.

3. Ventilation

Ventilation is utilized for many different purposes, the most common being control of humidity and condensation. Other well-known uses include exhaust hoods in restaurants, heat removal in industrial plants, fresh air in buildings, odor removal, and chemical and fume hood exhausts. In commercial buildings, ventilation air is used for replacement of stale, vitiated air, odor control, and smoke removal. Ventilation air contributes greatly to the comfort of the building's occupants. It is considered to be of such importance that many building codes contain specific requirements for minimum quantities of fresh, or outside, air that must be supplied to occupied areas.

Ventilation is also the prime method for reducing employees' exposure to excessive airborne contaminants that result from industrial operations. Ventilation is used to dilute contaminants to safe levels or to capture them at their point of origin before they pollute the employees' working environment. The *Occupational Safety and Health Act* (OSHA) standards set the legal limits for employees' exposure to many types of *toxic substance*s.

4. Movement of Air with Fans

Inasmuch as most ventilation systems are designed as mechanical ventilation systems that utilize various kinds of fans, knowledge of the types of fans in use will be of value in selection of ventilation fans. Fans are used to create a pressure differential that causes air to flow in a system. They generally incorporate one of several types of *impeller*s mounted in an appropriate housing or enclosure. An electric motor usually drives the impeller to move the air.

Two types of fans are commonly used in air-handling and air-moving systems: axial and centrifugal. They differ in the direction of airflow through the impeller.

Centrifugal fans are enclosed in a scroll-shaped housing, which is designed for efficient *airstream* energy transfer. This type of fan has the most versatility and low first cost and is the workhorse of the industry. Impeller *blade*s may be *radial*, forward-*curve*d, backward-inclined, or *airfoil*. When large volumes of air are moved, airfoil or backward-inclined blades are preferable because of higher efficiencies. For smaller volumes of air, forward, curved blades are used with satisfactory results. Centrifugal fans are manufactured with capacities of up to 500,000 ft³/min and can operate against pressures up to 30 in water gage.

Axial-flow fans are versatile and sometimes less costly than centrifugal fans. The use of axial fans is steadily increasing, because of the availability of controllable-pitch units, with increased emphasis on energy savings. Substantial energy savings can be realized by varying the blade pitch to meet specific duty loads. Axial fans develop *static* pressure by changing the velocity of the air through the impeller and converting it into static pressure. Axial fans are quite noisy and are generally used by industry where the noise level can be tolerated. When used for HVAC installations, sound *attenuator*s are almost always used in series with the fan for noise *abatement*.

Glossary

Complete the glossary using words in the text.

1. _____	暖气装置	2. _____	通风设备
3. _____	环境的	4. _____	热力学的
5. _____	补充,补足	6. _____	防护罩,罩
7. _____	水汽,水蒸气	8. _____	湿度
9. _____	提高	10. _____	水平的
11. _____	速度	12. _____	蒸汽
13. _____	冷凝物	14. _____	精确度
15. _____	有毒的	16. _____	物质
17. _____	推进者,叶轮	18. _____	气流
19. _____	叶片,桨叶	20. _____	辐射状的
21. _____	弧形的	22. _____	机翼,螺旋桨
23. _____	静止的,静力的	24. _____	衰减器
25. _____	减少,减轻,减退		

Exercises

1. Decide whether the following statements are true (T) or false (F) according to the text.

_____ 1) Ventilation is also the prime method for reducing employees' exposure to excessive airborne contaminants that result from industrial operations.

_____ 2) Axial and centrifugal are commonly used in air-handling and air-moving systems.

2. Write down the words with the following prefix or suffix.

therm(o)- _____

sim- _____

liqu- _____

-press- _____

-um _____

3. Translate the following paragraph into Chinese.

1) The necessity of heating, ventilation and air-conditioning (HVAC) control of environmental conditions within buildings has been well established over the years as being highly desirable for various types of occupancy and comfort conditions as well as for many industrial manufacturing processes. In fact, without HVAC systems, many

manufactured products produced by industry that is literally taken for granted would not be available today.

2) Fans are used to create a pressure differential that causes air to flow in a system. They generally incorporate one of several types of impellers mounted in an appropriate housing or enclosure. An electric motor usually drives the impeller to move the air.

Text B　HVAC

HVAC is an initialism or acronym that stands for "heating, ventilating, and air conditioning". HVAC is sometimes referred to as climate control and is particularly important in the design of medium to large industrial and office buildings such as skyscrapers and in marine environments such as aquariums, where humidity and temperature must all be closely regulated whilst maintaining safe and healthy conditions within. In certain regions (e. g. , UK) the term "Building Services" is also used, but may also include plumbing and electrical systems. Refrigeration is sometimes added to the field's abbreviation as HVAC & R or HVACR, or ventilating is dropped as HACR (such as the designation of HACR-rated circuit breakers).

Heating, ventilating and air conditioning is based on the basic principles of thermodynamics, fluid mechanics, heat transfer, also two inventions and discoveries made by Michael Faraday, Willis Carrier, Reuben Trane, James Joule, William Rankine, Sadi Carnot, and many others. The invention of the components of HVAC systems goes hand-in-hand with the industrial revolution, and new methods of modernization, higher efficiency, and system control are constantly introduced by companies and inventors all over the world.

The three functions of heating, ventilating, and air-conditioning are closely interrelated. All seek to provide thermal comfort, acceptable indoor air quality, and reasonable installation, operation, and maintenance costs. HVAC systems can provide ventilation, reduce air infiltration, and maintain pressure relationships between spaces. How air is delivered to, and removed from spaces is known as room air distribution.

In modern buildings the design, installation, and control systems of these functions are integrated into one or more HVAC systems. For very small buildings, contractors normally "size" and select HVAC systems and equipment. For larger buildings where required by law, "building services" designers and engineers, such as mechanical, architectural, or building services engineers analyze, design, and specify the HVAC systems, and specialty mechanical contractors build and commission them. In all buildings, building permits and code-compliance inspections of the installations are the norm.

The HVAC industry is a worldwide enterprise, with career opportunities including operation and maintenance, system design and construction, equipment manufacturing and sales, and in education and research. The HVAC industry had been historically regulated by the manufacturers of HVAC equipment, but Regulating and Standards organizations such as ASHRAE, SMACNA, ACCA, Uniform Mechanical Code, International Mechanical Code, and AMCA have been established to support the industry and encourage high standards and achievement.

Heating systems may be classified as central or local. Central heating is often used in cold climates to heat private houses and public buildings. Such a system contains a boiler, furnace, or heat pump to heat water, steam, or air, all in a central location such as a furnace room in a home or a mechanical room in a large building. The system also contains either ductwork, for forced air systems, or piping to distribute a heated fluid and radiators to transfer this heat to the air. The term "radiator" in this context is misleading since most heat transferred from the heat exchanger is by convection, not radiation. The radiators may be mounted on walls or buried in the floor to give under-floor heat.

"Mechanical" or "forced" ventilation is used to control indoor air quality. Excess humidity, odors, and contaminants can often be controlled via dilution or replacement with outside air. However, in humid climates much energy is required to remove excess moisture from ventilation air.

Natural ventilation is the ventilation of a building with outside air without the use of a fan or other mechanical system. It can be achieved with operable windows when the spaces to ventilate are small and the architecture permits. In more complex systems warm air in the building can be allowed to rise and flow out upper openings to the outside (stack effect) thus forcing cool outside air to be drawn into the building naturally through openings in the lower areas. These systems use very little energy but care must be taken to ensure the occupants' comfort. In warm or humid months, in many climates, maintaining thermal comfort via solely natural ventilation may not be possible so conventional air conditioning systems are used as backups. Air-side economizers perform the same function as natural ventilation, but use mechanical systems' fans, ducts, dampers, and control systems to introduce and distribute cool outdoor air when appropriate.

Air conditioning and refrigeration are provided through the removal of heat. The definition of cold is the absence of heat and all air conditioning systems work on this basic principle. Heat can be removed through the process of radiation, convection, and conduction using mediums such as water, air, ice, and chemicals referred to as refrigerants. In order to remove heat from something, you simply need to provide a medium that is colder — this is how all air conditioning and refrigeration systems work.

Air-conditioned buildings often have sealed windows, because open windows would disrupt the attempts of the HVAC system to maintain constant indoor air conditions.

For the last 20~30 years, manufacturers of HVAC equipment have been making an effort to make the systems they manufacture more efficient. This was originally driven by rising energy costs, and has more recently been driven by increased awareness over environmental issues. There are several methods for making HVAC systems more efficient.

1. Heating Energy

Water heating is more efficient for heating buildings and was the standard many years ago. Today forced air systems can double for air conditioning and are more popular. The most efficient central heating method is geothermal heating.

Energy efficiency can be improved even more in central heating systems by introducing zoned heating. This allows a more granular application of heat, similar to non-central heating systems. Zones are controlled by multiple thermostats. In water heating systems the thermostats control zone valves, and in forced air systems they control zone dampers inside the vents which selectively block the flow of air.

Ventilating is the process of "changing" or replacing air in any space to control temperature or remove moisture, odors, smoke, heat, dust and airborne bacteria. Ventilation includes both the exchange of air to the outside as well as circulation of air within the building. It is one of the most important factors for maintaining acceptable indoor air quality in buildings. Methods for ventilating a building may be divided into mechanical/forced and natural types. Ventilation is used to remove unpleasant smells and excessive moisture, introduce outside air, and to keep interior building air circulating, to prevent stagnation of the interior air.

2. Ventilation Energy Recovery

Energy recovery systems sometimes utilize heat recovery ventilation or energy recovery ventilation systems that employ heat exchangers or enthalpy wheels to recover sensible or latent heat from exhausted air. This is done by transfer of energy to the incoming outside fresh air.

3. Air Conditioning Energy

The performance of vapor compression refrigeration cycles is limited by thermodynamics. These AC and heat pump devices move heat rather than convert it from one form to another, so thermal efficiencies do not appropriately describe the performance of these devices. The Coefficient-of-Performance (COP) measures performance, but this dimensionless measure has not been adopted, rather the Energy Efficiency Ratio (EER). To more accurately describe the performance of air conditioning equipment over a typical cooling season a modified version of the EER is used, and is the Seasonal Energy Efficiency Ratio (SEER). The SEER article describes it further, and presents some economic comparisons using this useful performance measure.

Text C   Major Factors in HVAC Design

This article presents the necessary concepts for management of heat energy and aims at development of a better understanding of its effects on human comfort. The concepts must be well understood if they are to be applied successfully to modification of the environment in building interiors, computer facilities, and manufacturing processes.

1. Significance of Design Criteria

Achievement of the desired performance of any HVAC system, whether designed for human comfort, industrial production or industrial process requirements, is significantly related to the development of appropriate and accurate design criteria.

Some of the more common items that are generally considered are as follows: outside design temperatures[winter and summer; dry bulb (DB); wet bulb (WB)]; design temperatures (winter: heating F DB and relative humidity; summer: cooling F DB and relative humidity); filtration efficiency of supply air; ventilation requirements; exhaust requirements; humidification; dehumidification; air-change rates; positive-pressure areas; negative-pressure areas; balanced-pressure areas; contaminated exhausts; chemical exhausts and fume hoods; energy conservation devices; economizer system; enthalpy control system; infiltration; exfiltration; controls.

2. Design Criteria Accuracy

Some engineers apply much effort to determination of design conditions with great accuracy. This is usually not necessary, because of the great number of variables involved in the design process. Strict design criteria will increase the cost of the necessary machinery for such optimum conditions and may be unnecessary. It is generally recognized that it is impossible to provide a specific indoor condition that will satisfy every occupant at all times. Hence, HVAC engineers tend to be practical in their designs and accept the fact that the occupants will adapt to minor variations from ideal conditions. Engineers also know that human comfort depends on the type and quantity of clothing worn by the occupants, the types of activities performed, environmental conditions, duration of occupancy, ventilation air, and closeness of and number of people within the conditioned space and recognize that these conditions are usually unpredictable.

3. Outline of Design Procedure

Design of an HVAC system is not a simple task. The procedure varies considerably from one application or project to another, and important considerations for one project may have little impact on another. But for all projects, to some extent, the following major steps have to be taken:

1) Determine all applicable design conditions, such as inside and outside temperature and humidity conditions for winter and summer conditions, including prevailing winds and speeds.

2) Determine all particular and peculiar interior space conditions that will be maintained.

3) Estimate, for every space, heating or cooling loads from adjacent unheated or uncooled spaces.

4) Carefully check architectural drawings for all building materials used for walls, roofs, floors, ceilings, doors, etc., and determine the necessary thermal coefficients for each.

5）Establish values for air infiltration and exfiltration quantities, for use in determining heat losses and heat gains.

6）Determine ventilation quantities and corresponding loads for heat losses and heat gains.

7）Determine heat or cooling loads due to internal machinery, equipment, lights, motors, etc.

8）Include allowance for effects of solar load.

9）Total the heat losses requiring heating of spaces and heat gains requiring cooling of spaces, to determine equipment capacities.

10）Determine system type and control method to be applied.

4. Temperatures Determined by Heat Balances

In cold weather, comfortable indoor temperatures may have to be maintained by a heating device. It should provide heat to the space at the same rate as the space is losing heat. Similarly, when cooling is required, heat should be removed from the space at the same rate that it is gaining heat. In each case, there must be a heat-balance between heat in and heat out when heating and the reverse in cooling. Comfortable inside conditions can only be maintained if this heat balance can be controlled or maintained.

The rate at which heat is gained or lost is a function of the difference between the inside air temperature to be maintained and the outside air temperature. Such temperatures must be established for design purposes in order to properly size and select HVAC equipment that will maintain the desired design conditions. Many other conditions that also affect the flow of heat in and out of buildings, however, should also be considered in selection of equipment.

5. Methods of Heat Transfer

Heat always flows from a hot to a cold object, in strict compliance with the second law of thermodynamics. This direction of heat flow occurs by conduction, convection, or radiation and in any combination of these forms.

Thermal conduction is a process in which heat energy is transferred through matter by the transmission of kinetic energy from molecule to molecule or atom to atom.

Thermal convection is a means of transferring heat in air by natural or forced movements of air or a gas. Natural convection is usually a rotary or circular motion caused by warm air rising and cooler air falling. Convection can be mechanically produced (forced convection), usually by use of a fan or blower.

Thermal radiation transfers energy in wave form from a hot body to a relatively cold body. The transfer occurs independently of any material between the two bodies. Radiation energy is converted from one source to a very long wave form of electromagnetic energy. Interception of this long wave by solid matter will convert the radiant energy back to heat.

Integrated Skills: Writing a Scientific Essay

By means of the field research, Internet search or the review of special literatures, you should search for some technological information on HVAC that you are interested in. The information includes basic facts, design criterion and methods of heating and cooling, air-conditioning control. Then you should write a scientific essay based on one of the above aspects, and finally present your essay in class and exchange your ideas with others.

Unit 14 Civil Engineering Construction

Lead-in

1. Which are the types of project contract?
2. Which are the goals of project management?
3. Which are the basic functions of construction engineering?

Vocabulary Warm-up: Matching

_____1. 合同	a. working drawings
_____2. 施工图	b. bidding
_____3. 供应商	c. supervision
_____4. 诉讼	d. litigation
_____5. 融资	e. contract
_____6. 投标	f. dispute
_____7. 谈判	g. jurisdiction
_____8. 纠纷	h. financing
_____9. 监理	i. vendor
_____10. 裁判权	j. negotiate

Text A   Construction Engineering

The construction industry is typically divided into specialty areas, with each area requiring different skills, resources, and knowledge to participate effectively in it. The area classifications typically used are residential (single and multifamily housing), building (all buildings other than housing), heavy/highway (dams, bridges, ports, *sewage-treatment plant*s, highways), utility (sanitary and storm drainage, water lines, electrical and telephone lines, pumping stations), and industrial (*refineries*, mills, power plants, chemical plants, heavy manufacturing facilities). Civil engineers can be heavily involved in all of these areas of construction, although fewer are involved in residential. Due to the differences in each of these market areas, most engineers specialize in only one or two of the areas during their careers.

Construction projects are complex and *time-consuming undertaking*s that require the interaction and cooperation of many different persons to accomplish. All projects must be completed in accordance with specific project plans and specifications, along with other *contract* restrictions that may be imposed on the production operations. Essentially, all civil engineering construction projects are unique. Regardless of the similarity to other projects, there are always distinguishing elements of each project that make it unique, such as the type of soil, the exposure to weather, the human resources assigned to the project, the social and political climate, and so on. In manufacturing, raw resources are brought to a factory with a fairly controlled environment; in construction, the "factory" is set up on site, and production is accomplished in an uncertain environment.

It is this diversity among projects that makes the preparation for a civil engineering project interesting and challenging. Although it is often difficult to control the environment of the project, it is the duty of the contractor to predict the possible situations that may be encountered and to develop *contingency* strategies accordingly. The dilemma of this situation is that the contractor who allows for contingencies in project cost estimates will have a difficult time competing against other less competent or less cautious contractors. The failure rate in the construction industry is the highest in the U.S.; one of the leading causes for failure is the inability to manage in such a highly competitive market and to realize a fair return on investment.

1. Participants in the Construction Process

There are several *participant*s in the construction process, all with important roles in developing a successful project. The owner, either private or public, is the party that initiates the demand for the project and ultimately pays for its completion. The owner's role in the process varies considerably; however, the primary role of the owner is to effectively communicate the scope of work desired to the other parties. The designer is responsible for developing adequate *working drawing*s and specifications, in accordance with current design practices and codes, to communicate the product desired by the owner upon completion of the project. The prime contractor is responsible for managing the resources needed to carry out the construction process in a manner that ensures the project will be conducted safely, within budget, and on schedule, and that it meets or exceeds the quality requirements of the plans and specifications. Subcontractors are specialty contractors who contract with the prime contractor to conduct a specific portion of the project within the overall project schedule. Suppliers are the *vendor*s who contract to supply required materials for the project within the project specifications and schedule. The success of any project depends on the coordination of the efforts of all parties involved, hopefully to the financial advantage of all. In recent years, these relationships have become more adversarial, with much conflict and *litigation*, often to the *detriment* of the projects.

2. Construction Contracts

Construction projects are done under a variety of contract arrangements for each of the parties involved.

They range from a single contract for a single element of the project to a single contract for the whole project, including the *financing*, design, construction, and operation of the facility. Typical contract types include lump sum, unit price, cost plus, and construction management.

These contract systems can be used with either the competitive **bidding** process or with **negotiate**d processes. A contract system becoming more popular with owners is design-build, in which all of the responsibilities can be placed with one party for the owner to deal with. Each type of contract impacts the roles and responsibilities of each of the parties on a project. It also impacts the management functions to be carried out by the contractor on the project, especially the cost engineering function.

A major development in business relationships in the construction industry is partnering. Partnering is an approach to conduct business that confronts the economic and technological challenges in industry in the 21st century. This new approach focuses on making **long-term** commitments with mutual goals for all parties involved to achieve mutual success. It requires changing traditional relationships to a shared culture without regard to normal organizational boundaries. Participants seek to avoid the adversarial problems typical for many business ventures. Most of all, a relationship must be based upon trust. Although partnering in its pure form relates to a long-term business relationship for multiple projects, many single-project partnering relationships have been developed, primarily for public owner projects. Partnering is an excellent vehicle to attain improved quality on construction projects and to avoid serious conflicts.

3. Goals of Project Management

Most construction teams have the same performance goals:

Cost: Complete the project within the cost budget, including the budgeted costs of all change orders.

Time: Complete the project by the scheduled completion date or within the allowance for work days.

Quality: Perform all work on the project, meeting or exceeding the project plans and specifications.

Safety: Complete the project with zero lost-time accidents.

Conflict: Resolve **dispute**s at the lowest practical level and have zero disputes.

Project startup: Successfully start up the completed project (by the owner) with zero rework.

4. Basic Functions of Construction Engineering

The activities involved in the construction engineering for projects include the following basic functions:

Cost engineering: The cost estimating, cost accounting, and cost-control activities related to a project, plus the development of cost databases.

Project planning and scheduling: The development of initial project plans and schedules, project monitoring and updating, and the development of **as-built** project schedules.

Equipment planning and management: The selection of needed equipment for projects, productivity planning to accomplish the project with the selected equipment in the required project schedule and estimate, and the management of the equipment fleet.

Design of temporary structures: The design of temporary structures required for the construction of the project, such as concrete formwork, scaffolding, shoring, and bracing.

Contract management: The management of the activities of the project to comply with contract provisions and document contract changes and to minimize contract disputes.

Human resource management: The selection, training, and supervision of the personnel needed to complete the project work within schedule.

Project safety: The establishment of safe working practices and conditions for the project, the communication of these safety requirements to all project personnel, the maintenance of safety records, and the enforcement of these requirements.

5. *Innovations* in Construction

There are several innovative developments in technological tools that have been implemented or are being considered for implementation for construction projects. New tools such as CAD systems, expert systems, bar coding, and automated equipment offer excellent potential for improved productivity and cost effectiveness in industry. Companies who ignore these new technologies will have difficulty competing in the future.

Glossary

Complete the glossary using words in the text.

1. _____	污水处理厂	2. _____	炼油厂
3. _____	费时的	4. _____	任务
5. _____	合同	6. _____	偶然发生的事故
7. _____	参与者	8. _____	施工图
9. _____	供应商	10. _____	诉讼
11. _____	损害	12. _____	融资
13. _____	投标	14. _____	谈判
15. _____	长期的	16. _____	纠纷
17. _____	刚竣工的	18. _____	改革,创新

Exercises

1. Decide whether the following statements are true (T) or false (F) according to the text.

_____ 1) Construction projects are complex and time-consuming undertakings that require the interaction and cooperation of many different persons to accomplish.

_____2）Partnering focuses on making short-term commitments with mutual goals for all parties involved to achieve mutual success.

2. Write down the words with the following prefix or suffix.

co- _____

pro- _____

-tract _____

-age _____

-ation _____

3. Translate the following paragraph into Chinese.

1）The area classifications typically used are residential（single and multifamily housing）, building（all buildings other than housing）, heavy/highway（dams, bridges, ports, sewage-treatment plants, highways）, utility（sanitary and storm drainage, water lines, electrical and telephone lines, pumping stations）, and industrial（refineries, mills, power plants, chemical plants, heavy manufacturing facilities）.

2）Partnering is an approach to conducting business that confronts the economic and technological challenges in industry in the 21st century. This new approach focuses on making long-term commitments with mutual goals for all parties involved to achieve mutual success. It requires changing traditional relationships to a shared culture without regard to normal organizational boundaries. Participants seek to avoid the adversarial problems typical for many business ventures.

Text B   Traditional Construction Procedure

Construction under the traditional construction procedure is performed by contractors. While they would like to satisfy the owner and the building designers, contractors have the main objective of making a profit. Hence, their initial task is to prepare a bid price based on an accurate estimate of construction costs. This requires development of a concept for performance of the work and a construction time schedule. After a contract has been awarded, contractors must furnish and pay for all materials, equipment, power, labor, and supervision required for construction. The owner compensates the contractors for construction costs and services.

A general contractor assumes overall responsibility for construction of a building. The contractor engages subcontractors who take responsibility for the work of the various trades required for construction. For example, a plumbing contractor installs the plumbing, an electrical contractor installs the electrical system, a steel erector installs structural steel, and an elevator contractor installs elevators. Their contracts are with the general contractor, and they are paid by the general contractor.

Sometimes, in addition to a general contractor, the owner contracts separately with specialty contractors, such as electrical and mechanical contractors, who perform a substantial amount of the work required for a building. Such contractors are called prime contractors. Their work is scheduled and coordinated by the general contractor, but they are paid directly by the owner.

Sometimes also, the owner may use the design-build method and award a contract to an organization for both the design and construction of a building. Such organizations are called design-build contractors. One variation of this type of contract is employed by developers from groups of one-family homes or low-rise apartment buildings. The homebuilder designs and constructs the dwellings, but the design is substantially completed before owners purchase the homes.

Administration of the construction procedure is often difficult. Consequently, some owners seek assistance from an expert, called a professional construction manager, with extensive construction experience, who receives a fee. The construction manager negotiates with general contractors and helps select one to construct the building. Managers usually supervise selection of subcontractors too. During construction, they help control costs, expedite equipment and material deliveries, and keep the work on schedule. In some cases, instead, the owner may prefer to engage a construction manager, to assist in administrating both design and construction.

Construction contractors employ labor that may or may not be unionized. Unionized craftspeople are members of unions that are organized by construction trades, such as carpenter, plumber, and electrician unions. Union members will perform only the work assigned to their trade. On the job, groups of workers are supervised by crew supervisors, all of whom report to a superintendent.

During construction, all work should be inspected. For this purpose, the owner, often through the architect and consultants, engages inspectors. The field inspectors may be placed under the control of an owner's representative, who may be titled clerk of the works, architect's superintendent, engineer's superintendent, or resident engineer. The inspectors have the responsibility of ensuring that construction meets the requirements of the contract documents and is performed under safe conditions. Such inspections may be made at frequent intervals.

In addition, inspections also are made by representatives of one or more governmental agencies. They have the responsibility of ensuring that construction meets legal requirements and have little or no concern with detailed conformance with the contract documents. Such legal inspections are made periodically or at the end of certain stages of construction. One agency that will make frequent inspections is the local or state building department, which ever has jurisdiction. The purpose of these inspections is to ensure conformance with the local or state building code.

During construction, standards, regulations, and procedures of the Occupational Safety and Health Administration should be observed.

Following is a description of the basic traditional construction procedure for a multistory building: After the award of a construction contract to a general contractor, the owner may ask the contractor to start a portion of the work before signing of the contract by giving the contractor a letter of intent or after signing of the contract by issuing a written notice to proceed. The contractor then obtains construction permits, as required, from governmental agencies, such as the local building, water, sewer, and highway departments.

The general contractor plans and schedules construction operations in detail and mobilizes equipment and personnel for the project. Subcontractors are notified of the contract award and issued letters of intent or awarded subcontracts, then are given, at appropriate times, notices to proceed.

Before construction starts, the general contractor orders a survey to be made of adjacent structures and terrain, both for the record and to become knowledgeable of local conditions. A survey is then made to lay out construction.

Field offices for the contractor are erected on or near the site. If desirable for safety reasons to protect passersby, the contractor erects a fence around the site and an overhead protective cover, called a bridge. Structures required to be removed from the site are demolished and the debris is carted away.

Next, the site is prepared to receive the building. This work may involve grading the top surface to bring it to the proper elevations, excavating to required depths for basement and foundations, and shifting of utility piping. For deep excavations, earth sides are braced and the bottom is drained.

Major construction starts with the placement of foundations, on which the building rests. This is followed by the erection of load-bearing walls and structural framing. Depending on the height of the building, ladders, stairs, or elevators may be installed to enable construction personnel to travel from floor to floor and eventually to the roof. Also, hoists may be installed to lift materials to upper levels. If needed, temporary flooring may be placed for use of personnel.

As the building rises, pipes, ducts, and electric conduit and wiring are installed. Then, permanent floors, exterior walls, and windows are constructed. At the appropriate time, permanent elevators are installed. If required, fireproofing is placed for steel framing. Next, fixed partitions are built and the roof and its covering, or roofing, are put in place.

Finishing operations follow — These include installation of the followings: ceilings; tile; wallboard; wall paneling; plumbing fixtures; heating furnaces; air-conditioning equipment; heating and cooling devices for rooms; escalators; floor coverings; window glass; movable partitions; doors; hardware; electrical equipment and apparatus, including lighting fixtures, switches, outlets, transformers, and controls; and other items called for in the drawings and specifications. Field offices, fences, bridges, and other temporary construction must be removed from the site. Utilities, such as gas, electricity, and water, are hooked up to the building. The site is landscaped and paved.

Finally, the building interior is painted and cleaned.

The owner's representatives then give the building a final inspection. If they find that the structure conforms with the contract documents, the owner accepts the project and gives the general contractor final payment on issuance by the building department of a certificate of occupancy, which indicates that the completed building meets building-code requirements.

Text C　Concrete Placement

Good Practice

The principles governing proper placement of concrete are: Segregation must be avoided during all operations between the mixer and the point of placement, including final consolidation and finishing.

The concrete must be thoroughly consolidated, worked solidly around all embedded items, and should fill all angles and corners of the forms. Where fresh concrete is placed against or on hardened concrete, a good bond must be developed. Unconfined concrete must not be placed under water.

The temperature of fresh concrete must be controlled from the time of mixing through final placement, and protected after placement.

Methods of Placing

Concrete may be conveyed from a mixer to point of placement by a variety of methods and equipment, if properly transported to avoid segregation. Selection of the most appropriate technique for economy depends on jobsite conditions, especially project size, equipment, and the contractor's experience. In building construction, concrete usually is placed with hand or power-operated buggies; drop-bottom buckets with a crane; inclined chutes; flexible and rigid pipe by pumping; shotcrete, in which either dry materials and water are sprayed separately or mixed concrete is shot against the forms; and for underwater placing, tremie chutes (closed flexible tubes). For mass-concrete construction, side-dump cars on narrow-gage track or belt conveyers may be used. For pavement, concrete may be placed by bucket from the swinging boom of a paving mixer, directly by dump truck or mixer truck, or indirectly by trucks into a spreader.

A special method of placing concrete suitable for a number of unusual conditions consists of grout-filling preplaced coarse aggregate. This method is particularly useful for underwater concreting, because grout, introduced into the aggregate through a vertical pipe gradually lifted, displaces the water, which is lighter than the grout. Because of bearing contact of the aggregate, less than usual overall shrinkage is also achieved.

Excess Water

Even within the specified limits on slump and water-cementitious materials ratio, excess water must be avoided. In this context, excess water is present for the conditions of placing if evidence of water rise (vertical segregation) or water flow (horizontal segregation) occurs. Excess water also tends to aggravate surface defects by increased leakage through form openings. The result may be honeycomb, sandstreaks, variations in color, or soft spots at the surface.

In vertical formwork, water rise causes weak planes between each layer deposited. In addition to the deleterious structural effect, such planes, when hardened, contain voids through which water may pass.

In horizontal elements, such as floor slabs, excess water rises and causes a weak laitance layer at the top. This layer suffers from low strength, low abrasion resistance, high shrinkage, and generally poor quality.

Consolid Action

The purpose of consolidation is to eliminate voids of entrapped air and to ensure intimate complete contact of the concrete with the surfaces of the forms and the reinforcement. Intense vibration, however, may also reduce the volume of desirable entrained air; but this reduction can be compensated by adjustment of the mix proportions.

Powered internal vibrators are usually used to achieve consolidation. For thin slabs, however, high-quality, low-slump concrete can be effectively consolidated, without excess water, by mechanical surface vibrators. For precast elements in rigid, watertight forms, external vibration (of the form itself) is highly effective. External vibration is also effective with in-place forms, but should not be used unless the formwork is specially designed for the temporary increase in internal pressures to full fluid head plus the impact of the vibrator.

Except in certain paving operations, vibration of the reinforcement should be avoided. Although it is effective, the necessary control to prevent over-vibration is difficult. Also, when concrete is placed in several lifts of layers, vibration of vertical rebars passing into partly set concrete below may be harmful. Note, however, that revibration of concrete before the final set, under controlled conditions, can improve concrete strength markedly and reduce surface voids (bugholes). This technique is too difficult to control for general use on field-cast vertical elements, but it is very effective in finishing slabs with powered vibrating equipment.

Manual spading is most efficient for removal of entrapped air at form surfaces. This method is particularly effective where smooth impermeable form material is used and the surface is upward sloping.

On the usual building project, different conditions of placement are usually encountered that make it desirable to provide for various combinations of the techniques described. One precaution generally applicable is that the vibrators not be used to move the concrete laterally.

Concreting Vertical Elements

The interior of columns is usually congested; it contains a large volume of reinforcing steel compared with the volume of concrete, and has a large height compared with its cross-sectional dimensions. Therefore, though columns should be continuously cast, the concrete should be placed in 2 to 4 ft deep increments and consolidated with internal vibrators. These should be lifted after each increment has been vibrated. If delay occurs in concrete supply before a column has been completed, every effort should be made to avoid a cold joint. When the remainder of the column is cast, the first increment should be small, and should be vibrated to penetrate the previous portion slightly.

In all columns and reinforced narrow walls, concrete placing should begin with 2 to 4 in of grout. Otherwise, loose stone will collect at the bottom, resulting in the formation of honeycomb. This grout should be proportioned for about the same slump as the concrete or slightly more, but at the same or lower water-cementitious material ratio. (Some engineers prefer to start vertical placement with a mix having the same proportions of water, cement, and fine aggregate, but with one-half the quantity of coarse aggregate, as in the design mix, and to place a starting layer 6 to 12 in deep.)

When concrete is placed for walls, the only practicable means to avoid segregation is to place no more than a 24 in layer in one pass. Each layer should be vibrated separately and kept nearly level.

For walls deeper than 4 ft, concrete should be placed through vertical, flexible trunks or chutes located about 8 ft apart. The trunks may be flexible or rigid, and come in sections so that they can be lifted as the level of concrete in place rises. The concrete should not fall free, from the end of the trunk, more than 4 ft or segregation will occur, with the coarse aggregate ricocheting off the forms to lodge on one side. Successive layers after the initial layer should be penetrated by internal vibrators for a depth of about 4 to 6 in to ensure complete integration at the surface of each layer. Deeper penetration can be beneficial (revibration), but control under variable jobsite conditions is too uncertain for recommendation of this practice for general use.

The results of poor placement in walls are frequently observed: sloping layer lines; honeycombs, leaking, if water is present; and, if cores are taken at successive heights, up to a 50% reduction in strength from bottom to top. Some precautions necessary to avoid these ill effects are: Place concrete in level layers through closely spaced trunks or chutes; Do not place concrete full depth at each placing point; Do not move concrete laterally with vibrators; For deep, long walls, reduce the slump for upper layers 2 to 3 in below the slump for the starting layer; on any delay between placing of layers, vibrate the concrete thoroughly at the interface.

If concreting must be suspended between planned horizontal construction joints, level off the layer cast, remove any laitance and excess water, and make a straight, level construction joint, if possible, with a small cleat attached to the form on the exposed face.

Concreting Horizontal Elements

For concrete slabs, careless placing methods result in horizontal segregation, with desired properties in the wrong location, the top consisting of excess water and fines with low abrasion and weather resistance, and high shrinkage. For a good surface in a one-course slab, low-slump concrete and a minimum of vibration and finishing are desirable. Immediate screeding with a power-vibrated screed is helpful in distributing low-slump, high-quality concrete. No further finishing should be undertaken until free water, if any, disappears. A powered, rotary tamping float can smooth very-low-slump concrete at this stage. Final troweling should be delayed, if necessary, until the surface can support the weight of the finisher.

When concrete is placed for deep beams that are monolithic with a slab, the beam should be filled first. Then, a short delay for settlement should ensue before slab concrete is cast. Vibration through the top slab should penetrate the beam concrete sufficiently to ensure thorough consolidation.

When a slab is cast, successive batches of concrete should be placed on the edge of previous batches, to maintain progressive filling without segregation. For slabs with sloping surfaces, concrete placing should usually begin at the lower edge.

For thin shells in steeply sloping areas, placing should precede downslope. Slump should be adjusted and finishing coordinated to prevent restraint by horizontal reinforcing bars from causing plastic cracking in the fresh concrete.

Integrated Skills: Presenting an Essay

By means of the field research, Internet search or the review of special literatures, you should search for some technological information on construction engineering that you are interested in. The information includes procedure, state codes, participants, contracts, safety and administration of construction engineering. Then you should write a scientific essay based on one of the above aspects, and finally present your essay in class and exchange your ideas with others.

Unit 15　Civil Engineering Management Ⅰ: Cost Estimation

Lead-in

1. What does the cost of a bridge include?
2. Talk about the approaches to cost estimation.
3. What are the types of construction cost estimate?

Vocabulary Warm-up: Matching

_____1. parameter   a. 常数

_____2. constant   b. 汇总

_____3. regression   c. 应急费用

_____4. bid   d. 外包,外部采购

_____5. aggregation   e. 参数

_____6. appropriation   f. 花费,支出,开销

_____7. contingency   g. 回归

_____8. inflation   h. 投标,招标

_____9. lump sum   i. 管理费

_____10. outlays   j. 总价

_____11. outsourcing   k. 通货膨胀

_____12. overhead   l. 拨款

Text A　Costs Associated with Constructed Facilities

1. Introduction

The costs of a constructed *facility* to the *owner* include both the initial capital cost and the subsequent operation and maintenance costs. Each of these major cost categories consists of a number of cost components.

The capital cost for a construction project includes the expenses related to the initial establishment of the facility: land acquisition, including assembly, holding and improvement; planning and feasibility studies; architectural and engineering design; construction, including materials, equipment and labor; field *supervision* of construction; construction financing; insurance and taxes during construction; owner's general office overhead; equipment and furnishings not included in construction; inspection and testing.

The operation and maintenance cost in subsequent years over the project life cycle includes the following expenses: land rent, if applicable; operating staff; labor and material for maintenance and repairs; periodic *renovation*s; insurance and taxes; financing costs; utilities; owner's other expenses.

The *magnitude* of each of these cost components depends on the nature, size and location of the project as well as the management organization, among many *consideration*s. The owner is interested in achieving the lowest possible overall project cost that is consistent with its investment objectives.

It is important for design professionals and construction managers to realize that while the construction cost may be the single largest component of the capital cost, other cost components are not insignificant. For example, land acquisition costs are a major expenditure for building construction in high-density urban areas, and construction financing costs can reach the same order of magnitude as the construction cost in large projects such as the construction of nuclear power plants.

From the owner's *perspective*, it is equally important to estimate the corresponding operation and maintenance cost of each alternative for a proposed facility in order to analyze the life cycle costs. The large expenditures needed for facility maintenance, especially for publicly owned infrastructure, are reminders of the neglect in the past to consider fully the implications of operation and maintenance cost in the design stage.

In most construction budgets, there is an allowance for *contingencies* or unexpected costs occurring during construction. This contingency amount may be included within each cost item or be included in a single category of construction contingency. The amount of contingency is based on historical experience and the expected difficulty of a particular construction project. For example, one construction firm makes estimates of the expected cost in five different areas: design development changes; schedule adjustments; general administration changes (such as wage rates); differing site conditions for those expected, and third-party requirements imposed during construction, such as new permits.

Contingent amounts not spent for construction can be released near the end of construction to the owner or to add additional project elements.

2. Approaches to Cost Estimation

Cost estimating is one of the most important steps in project management. A cost estimate establishes the base line of the project cost at different stages of development in the project. A cost estimate at a given stage of project development represents a prediction provided by the cost engineer or estimator on the basis of available data.

Virtually, all cost estimation is performed according to one or some combination of the following basic approaches:

1) Production Function

A production function relates the amount or volume of output to the various inputs of labor, material and equipment. For example, the amount of output Q may be derived as a function of various input factors x_1, x_2, ..., x_n by means of mathematical and/or *statistical* methods. Thus, for a specified level of output, we may attempt to find a set of values for the input factors so as to minimize the production cost. The relationship between the sizes of a building project (expressed in square feet) to the input labor (expressed in labor hours per square foot) is an example of a production function for construction.

2) Empirical Cost Inference

Empirical estimation of cost functions requires statistical techniques which relate the cost of constructing or operating a facility to a few important characteristics or attributes of the system. The role of statistical inference is to estimate the best *parameter* values or *constants* in an assumed cost function. Usually, this is accomplished by means of *regression* analysis techniques.

3) Unit Costs for Bill of Quantities

A unit cost is assigned to each of the facility components or tasks as represented by the bill of quantities. The total cost is the summation of the products of the quantities multiplied by the corresponding unit costs. The unit cost method is straightforward in principle but quite laborious in application. The initial step is to break down a process into a number of tasks. Collectively, these tasks must be completed for the construction of a facility. Once these tasks are defined and quantities representing these tasks are assessed, a unit cost is assigned to each and then the total cost is determined by summing the costs incurred in each task. The level of detail in decomposing into tasks will vary considerably from one estimate to another.

4) Allocation of Joint Costs

Allocations of cost from existing accounts may be used to develop a cost function of an operation. The basic idea in this method is that each expenditure item can be assigned to particular characteristics of the operation. Ideally, the allocation of joint costs should be causally related to the category of basic costs in an allocation process. In many instances, however, a causal relationship between the allocation factor and the cost item cannot be identified or may not exist. For example, in construction projects, the accounts for basic costs may be classified according to labor, material, construction equipment, construction supervision, and general office overhead. These basic costs may then be allocated proportionally to various tasks which are subdivisions of a project.

3. Types of Construction Cost Estimates

The required levels of accuracy on construction cost estimates vary at different stages of project development, ranging from *ballpark figures* in the early stage to fairly reliable figures for budget control prior to construction. Generally, the accuracy of a cost estimate will reflect the information available at the time of estimation.

Construction cost estimates may be viewed from different perspectives because of different institutional requirements. In spite of the many types of cost estimates used at different stages of a project, cost estimates can best be classified into three major categories according to their functions. A construction cost estimate serves one of the three basic functions: design, *bid* and control.

1) Design Estimates

In the planning and design stages of a project, various design estimates reflect the progress of the design. At the very early stage, the order of magnitude estimate is usually made before the facility is designed, and must therefore rely on the cost data of similar facilities built in the past. A conceptual estimate is based on the conceptual design of the facility at the state that the basic technologies for the design are known. The detailed estimate is made when the scope of work is clearly defined and the detailed design is in progress so that the essential features of the facility are identifiable. The engineer's estimate is based on the completed plans and specifications when they are ready for the owner to *solicit* bids from construction *contractor*s. In preparing these estimates, the design professional will include expected amounts for contractors' overhead and profits.

The costs associated with a facility may be decomposed into a *hierarchy* of levels that are *appropriate* for the purpose of cost estimation. The level of detail in decomposing the facility into tasks depends on the type of cost estimate to be prepared.

As an example, consider the cost estimates for a proposed bridge across a river. An order of magnitude estimate is made for each of the potential alternatives, such as a tied arch bridge or a cantilever truss bridge. As the bridge type is selected, e.g. the technology is chosen to be a tied arch bridge instead of some new bridge form, a conceptual estimate is made on the basis of the layout of the selected bridge form on the basis of the conceptual design. When the detailed design has progressed to a point when the essential details are known, a detailed estimate is made on the basis of the well defined scope of the project. When the detailed plans and specifications are completed, an engineer's estimate can be made on the basis of items and quantities of work.

2) Bid Estimates

The contractor's bid estimates often reflect the desire of the contractor to secure the job as well as the estimating tools at its disposal. Some contractors have well established cost estimating procedures while others do not. Since only the lowest *bidder* will be the winner of the contract in most bidding contests, any effort devoted to cost estimating is a loss to the contractor who is not a successful bidder. Consequently, the contractor may put in the least amount of possible effort for making a cost estimate if he believes that his chance of success is not high.

If a general contractor intends to use subcontractors in the construction of a facility, it may solicit price quotations for various tasks to be subcontracted to specialty subcontractors. Thus, the general subcontractor will shift the burden of cost estimating to subcontractors. If all or part of the construction is to be undertaken by the general contractor, a bid estimate may be prepared on the basis of the quantity takeoffs from the plans provided by the owner or on the basis of the construction procedures devised by the contractor for implementing the project. For example, the cost of a footing of a certain type and size may be found in commercial publications on cost data which can be used to facilitate cost estimates from quantity takeoffs. However, the contractor may want to assess the actual cost of construction by considering the actual construction procedures to be used and the associated costs if the project is deemed to be different from typical designs. Hence, items such as labor, material and equipment needed to perform various tasks may be used as parameters for the cost estimates.

3) Control Estimates

Both the owner and the contractor must adopt some base line for cost control during the construction. For the owner, a budget estimate must be adopted early enough for planning long term financing of the facility. Consequently, the detailed estimate is often used as the budget estimate since it is sufficient definitive to reflect the project scope and is available long before the engineer's estimate. As the work progresses, the budgeted cost must be revised periodically to reflect the estimated cost to completion. A revised estimated cost is necessary either because of change orders initiated by the owner or due to unexpected cost overruns or savings.

For the contractor, the bid estimate is usually regarded as the budget estimate, which will be used for control purposes as well as for planning construction financing. The budgeted cost should also be updated periodically to reflect the estimated cost to completion as well as to insure adequate cash flows for the completion of the project.

Glossary

Complete the glossary using words in the text.

1. _____	设施,建筑物,场所	2. _____	业主,建设单位
3. _____	监督,管理	4. _____	维修,翻新
5. _____	大小,数量,量值		
6. _____	必须考虑的因素,要考虑的事		
7. _____	观点,角度	8. _____	应急,意外 偶然
9. _____	事实上,实际上		
10. _____	统计的,统计上的,统计学的		
11. _____	经验的,实证的	12. _____	参数
13. _____	常数	14. _____	回归
15. _____	大概数字	16. _____	投标,招标
17. _____	寻求		
18. _____	承包商,施工单位		
19. _____	层次,层级		
20. _____	合适的,适宜的		
21. _____	投标人		

Exercises

1. Write down the words with the following prefix or suffix.

pre-   _____

re-   _____

nov-   _____

-gress _____

-meter _____

2. Translate the following paragraph into Chinese.

1) The capital cost for a construction project includes the expenses related to the initial establishment of the facility: land acquisition, including assembly, holding and improvement; planning and feasibility studies; architectural and engineering design; construction, including materials, equipment and labor; field supervision of construction; construction financing; insurance and taxes during construction; owner's general office overhead; equipment and furnishings not included in construction; inspection and testing.

2) At the very early stage, the order of magnitude estimate is usually made before the facility is designed, and must therefore rely on the cost data of similar facilities built in the past. A conceptual estimate is based on the conceptual design of the facility at the state when the basic technologies for the design are known. The detailed estimate is made when the scope of work is clearly defined and the detailed design is in progress so that the essential features of the facility are identifiable. The engineer's estimate is based on the completed plans and specifications when they are ready for the owner to solicit bids from construction contractors.

Text B   Effects of Scale on Construction Cost

Screening cost estimates are often based on a single variable representing the capacity or some physical measure of the design such as floor area in buildings, length of highways, and volume of storage bins and production volumes of processing plants. Costs do not always vary linearly with respect to different facility sizes. Typically, scale economies or diseconomies exist. If the average cost per unit of capacity is declining, then scale economies exist. Conversely, scale diseconomies exist if average costs increase with greater size. Empirical data are sought to establish the economies of scale for various types of facility, if they exist, in order to take advantage of lower costs per unit of capacity.

Let x be a variable representing the facility capacity, and y be the resulting construction cost. Then, a linear cost relationship can be expressed in the form:

$$y = a + bx \tag{15-1}$$

where a and b are positive constants to be determined on the basis of historical data. Note that in Equation (15-1), a fixed cost of $y = a$ at $x = 0$ is implied as shown in

Figure 15-1. In general, this relationship is applicable only in a certain range of the variable x, such as between $x = c$ and $x = d$. If the values of y corresponding to $x = c$ and $x = d$ are known, then the cost of a facility corresponding to any x within the specified range may be obtained by linear interpolation. For example, the construction cost of a school building can be estimated on the basis of a linear relationship between cost and floor area if the unit cost per square foot of floor area is known for school buildings within certain limits of size.

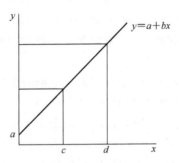

Figure 15-1　Linear cost relationship with economies of scale

A nonlinear cost relationship between the facility capacity x and construction cost y can often be represented in the form:

$$y = ax^b \tag{15-2}$$

where a and b are positive constants to be determined on the basis of historical data. For $0 < b < 1$, Equation (15-2) represents the case of increasing returns to scale, and for $b > 1$, the relationship becomes the case of decreasing returns to scale, as shown in Figure 15-2. Taking the logarithm of both sides of this equation, a linear relationship can be obtained as follows:

$$ln\ y = ln\ a + b\ ln\ x \tag{15-3}$$

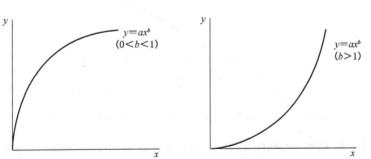

(a) **Increasing return to scale**　　　(b) **Decreasing return to scale**

Figure 15-2　Nonlinear cost relationship with increasing or decreasing economies of scale

Although no fixed cost is implied in Equation (15-2), the equation is usually applicable only for a certain range of x. The same limitation applies to Equation (15-3). A nonlinear cost relationship often used in estimating the cost of a new industrial processing plant from the known cost of an existing facility with a different size is known as the exponential rule. Let y_n be the known cost of an existing facility with capacity Q_n, and y be the estimated cost of the new facility which has a capacity Q. Then, from the empirical data, it can be assumed that:

$$y = y_n \left(\frac{Q}{Q_n} \right)^m \qquad (15\text{-}4)$$

where m usually varies from 0.5 to 0.9, depending on a specific type of facility. A value of $m = 0.6$ is often used for chemical processing plants. The exponential rule can be applied to estimate the total cost of a complete facility or the cost of some particular component of a facility.

Example 15-1: Determination of m for the exponential rule

The empirical cost data from a number of sewage treatment plants are plotted on a log-log scale for $\ln(Q/Q_n)$ and $\ln(y/y_n)$ and a linear relationship between these logarithmic ratios is shown in Figure 15-3. For $(Q/Q_n) = 1$ or $\ln(Q/Q_n) = 0$, $\ln(y/y_n) = 0$; and for $Q/Q_n = 2$ or $\ln(Q/Q_n) = 0.301$, $\ln(y/y_n) = 0.176,5$. Since m is the slope of the line in the figure, it can be determined from the geometric relation as follows:

$$m = \frac{0.176,5}{0.301} = 0.585$$

For $\ln(y/y_n) = 0.176,5$, $y/y_n = 1.5$, while the corresponding value of Q/Q_n is 2. In words, for $m = 0.585$, the cost of a plant increases only 1.5 times when the capacity is doubled.

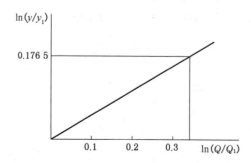

Figure 15-3   Log-log scale graph of exponential rule example

Example 15-2: Cost exponents for water and wastewater treatment plants

The magnitude of the cost exponent m in the exponential rule provides a simple measure of the economy of scale associated with building extra capacity for future growth and system reliability for the present in the design of treatment plants. When m is small, there is considerable incentive to provide extra capacity since scale economies exist. When m is close to 1, the cost is directly proportional to the design capacity. The value of m tends to increase as the number of duplicate units in a system increases. The values of m for several types of treatment plants with different plant components derived from statistical correlation of actual construction costs are shown in Table 15-1.

Table 15-1 Estimated values of cost exponents for water treatment plants

Treatment plant type	Exponent m	Capacity range (millions of gallons per day)
1. Water treatment	0.67	1~100
2. Waste treatment		
Primary with digestion (small)	0.55	0.1~10
Primary with digestion (large)	0.75	0.7~100
Trickling filter	0.60	0.1~20
Activated sludge	0.77	0.1~100
Stabilization ponds	0.57	0.1~100

Source: Data are collected from various sources by P.M. Berthouex.

Text C   Method of Estimation

If the design technology for a facility has been specified, the project can be decomposed into elements at various levels of detail for the purpose of cost estimation. The unit cost for each element in the bill of quantities must be assessed in order to compute the total construction cost. This concept is applicable to both design estimates and bid estimates, although different elements may be selected in the decomposition.

For design estimates, the unit cost method is commonly used when the project is decomposed into elements at various levels of a hierarchy as follows:

Preliminary Estimates. The project is decomposed into major structural systems or production equipment items, e.g. the entire floor of a building or a cooling system for a processing plant.

Detailed Estimates. The project is decomposed into components of various major systems, i.e., a single floor panel for a building or a heat exchanger for a cooling system.

Engineer's Estimates. The project is decomposed into detailed items of various components as warranted by the available cost data. Examples of detailed items are slabs and beams in a floor panel, or the piping and connections for a heat exchanger.

For bid estimates, the unit cost method can also be applied even though the contractor may choose to decompose the project into different levels in a hierarchy as follows:

Subcontractor Quotations. The decomposition of a project into subcontractor items for quotation involves a minimum amount of work for the general contractor. However, the accuracy of the resulting estimate depends on the reliability of the subcontractors since the general contractor selects one among several contractor quotations submitted for each item of subcontracted work.

Quantity Takeoffs. The decomposition of a project into items of quantities that are measured (or taken off) from the engineer's plan will result in a procedure similar to that adopted for a detailed estimate or an engineer's estimate by the design professional. The levels of detail may vary according to the desire of the general contractor and the availability of cost data.

Construction Procedures. If the construction procedure of a proposed project is used as the basis of a cost estimate, the project may be decomposed into items such as labor, material and equipment needed to perform various tasks in the projects.

1. Simple Unit Cost Formula

Suppose that a project is decomposed into n elements for cost estimation. Let Q_i be the quantity of the ith element and u_i be the corresponding unit cost. Then, the total cost of the project is given by:

$$y = \sum_{i=1}^{n} u_i Q_i \qquad (15\text{-}5)$$

where n is the number of units. Based on characteristics of the construction site, the technology employed, or the management of the construction process, the estimated unit cost, and u_i for each element may be adjusted.

2. Factored Estimate Formulas

A special application of the unit cost method is the "factored estimate" commonly used in process industries. Usually, an industrial process requires several major equipment components such as furnaces, towers, drums and pumps in a chemical processing plant, plus ancillary items such as piping, valves and electrical elements. The total cost of a project is dominated by the costs of purchasing and installing the major equipment components and their ancillary items. Let C_i be the purchase cost of a major equipment component i and f_i be a factor accounting for the cost of ancillary items needed for the installation of this equipment component i. Then, the total cost of a project is estimated by:

$$y = \sum_{i=1}^{n} C_i + \sum_{i=1}^{n} C_i f_i = \sum_{i=1}^{n} C_i (1 + f_i) \qquad (15\text{-}6)$$

where n is the number of major equipment components included in the project. The factored method is essentially based on the principle of computing the cost of ancillary items such as piping and valves as a fraction or a multiple of the costs of the major equipment items. The value of C_i may be obtained by applying the exponential rule so the use of Equation (15-6) may involve a combination of cost estimation methods.

3. Formula Based on Labor, Material and Equipment

Consider the simple case for which costs of labor, material and equipment are assigned to all tasks. Suppose that a project is decomposed into n tasks. Let Q_i be the quantity of work for task i, M_i be the unit material cost of task i, E_i be the unit equipment rate for task i, L_i be the units of labor required per unit of Q_i, and W_i be the wage rate associated with L_i. In this case, the total cost y is:

$$y = \sum_{i=1}^{n} y_i = \sum_{i=1}^{n} Q_i (M_i + E_i + W_i L_i) \qquad (15-7)$$

Note that $W_i L_i$ yields the labor cost per unit of Q_i, or the labor unit cost of task i. Consequently, the units for all terms in Equation (15-7) are consistent.

Example 15-3: Decomposition of a building foundation into design and construction elements.

The concept of decomposition is illustrated by the example of estimating the costs of a building foundation excluding excavation as shown in Table 15-2 in which the decomposed design elements are shown on horizontal lines and the decomposed contract elements are shown in vertical columns. For a design estimate, the decomposition of the project into footings, foundation walls and elevator pit is preferred since the designer can easily keep track of these design elements; however, for a bid estimate, the decomposition of the project into formwork, reinforcing bars and concrete may be preferred since the contractor can get quotations of such contract items more conveniently from specialty subcontractors.

Table 15-2   Illustrative decomposition of building foundation costs

Design elements	Contract elements			
	Formwork	Rebars	Concrete	Total cost
Footings	$ 5,000	$ 10,000	$ 13,000	$ 28,000
Foundation walls	$ 15,000	$ 18,000	$ 28,000	$ 61,000
Elevator pit	$ 9,000	$ 15,000	$ 16,000	$ 40,000
Total cost	$ 29,000	$ 43,000	$ 57,000	$ 129,000

Example 15-4: Cost estimate using labor, material and equipment rates.

For the given quantities of work Q_i for the concrete foundation of a building and the labor, material and equipment rates in Table 15-3, the cost estimate is computed on the basis of Equation (15-7). The result is tabulated in the last column of the same table.

Table 15-3   Illustrative cost estimate using labor, material and equipment rates

Description	Quantity Q_i	Material unit cost M_i	Equipment unit cost E_i	Wage rate W_i	Labor input L_i	Labor unit cost $W_i L_i$	Direct cost Y_i
Formwork	12,000 ft²	$ 0.4/ft²	$ 0.8/ft²	$ 15/hr	0.2 hr/ft²	$ 3.0/ft²	$ 50,400
Rebars	4,000 lb	0.2/lb	0.3/lb	$ 15/hr	0.04 hr/lb	0.6/lb	$ 4,400
Concrete	500 yd³	5.0/yd³	50/yd³	$ 15/hr	0.8 hr/yd³	12.0/yd³	$ 33,500
Total							$ 88,300

The principle of allocating joint costs to various elements in a project is often used in cost estimating. Because of the difficulty in establishing casual relationship between each element and its associated cost, the joint costs are often prorated in proportion to the basic costs for various elements.

One common application is found in the allocation of field supervision cost among the basic costs of various elements based on labor, material and equipment costs, and the allocation of the general overhead cost to various elements according to the basic and field supervision cost. Suppose that a project is decomposed into n tasks. Let y be the total basic cost for the project and y_i be the total basic cost for task i. If F is the total field supervision cost and F_i is the proration of that cost to task i, then a typical proportional allocation is:

$$F_i = F\frac{y_i}{y} \tag{15-8}$$

Similarly, let z be the total direct field cost which includes the total basic cost and the field supervision cost of the project, and z_i be the direct field cost for task i. If G is the general office overhead for proration to all tasks, and G_i is the share for task i, then

$$G_i = G\frac{z_i}{z} \tag{15-9}$$

Finally, let w be the grand total cost of the project which includes the direct field cost and the general office overhead cost charged to the project and w_i be that attributable task i. Then,

$$z = F + y = F + \sum_{i=1}^{n} y_i \tag{15-10}$$

and

$$w = G + z = G + \sum_{i=1}^{n} z_i \tag{15-11}$$

Example 15-5: Prorated costs for field supervision and office overhead

If the field supervision cost is \$13,245 for the project in Table 15-3 with a total direct cost of \$88,300, find the prorated field supervision costs for various elements of the project. Furthermore, if the general office overhead charged to the project is 4% of the direct field cost which is the sum of basic costs and field supervision cost, find the prorated general office overhead costs for various elements of the project.

For the project, $y = $ \$88,300 and $F = $ \$13,245. Hence: $z = 13,245 + 88,300 = $ \$101,545, $G = (0.04)(101,545) = $ \$4,062, $w = 101,545 + 4,062 = $ \$105,607.

The results of the proration of costs to various elements are shown in Table 15-4.

Table 15-4   Proration of field supervision and office overhead costs

Description	Basic cost y_i	Allocated field supervision cost F_i	Total field cost z_i	Allocated overhead cost G_i	Total cost
Formwork	$ 50,400	$ 7,560	$ 57,960	$ 2,319	$ 60,279
Rebars	$ 4,400	$ 660	$ 5,060	$ 202	$ 5,262
Concrete	$ 33,500	$ 5,025	$ 38,525	$ 1,541	$ 40,066
Total	$ 88,300	$ 13,245	$ 101,545	$ 4,062	$ 105,607

Example 15-6: A standard cost report for allocating overhead

The reliance on labor expenses as a means of allocating overhead burdens in typical management accounting systems can be illustrated by the example of a particular product's standard cost sheet. Table 15-5 is an actual product's standard cost sheet of a company following the procedure of using overhead burden rates assessed per direct labor hour. The material and labor costs for manufacturing a type of valve were estimated from engineering studies and from current material and labor prices. These amounts are summarized in Columns 2 and 3 of Table 15-5. The overhead costs shown in Column 4 of Table 15-5 were obtained by allocating the expenses of several departments to the various products manufactured in these departments in proportion to the labor cost. As shown in the last line of the table, the material cost represents 29% of the total cost, while labor costs are 11% of the total cost. The allocated overhead cost constitutes 60% of the total cost. Even though material costs exceed labor costs, only the labor costs are used in allocating overhead. Although this type of allocation method is common in industry, the arbitrary allocation of joint costs introduces unintended cross subsidies among products and may produce adverse consequences on sales and profits. For example, a particular type of part may incur few overhead expenses in practice, but this phenomenon would not be reflected in the standard cost report.

Table 15-5   Standard cost report for a type of valve

	Material cost	Labor cost	Overhead cost	Total cost
Purchased part	$ 1.198,0			$ 1.198,0
Operation				
Drill, face, tap		$ 0.043,8	$ 0.240,4	$ 0.284,2
Degrease		0.003,1	0.033,7	0.036,8
Remove burs		0.057,7	0.324,1	0.381,8
Total cost, this item	1.198,0	0.104,6	0.598,2	1.900,8
Other subassemblies	0.325,3	0.299,4	1.851,9	2.476,6
Total cost, subassemblies	1.523,3	0.404,0	2.450,1	4.377,4

Continued 15-5

	Material cost	Labor cost	Overhead cost	Total cost
Assmble and test		0.146,9	0.498,7	0.645,6
Pack without paper		0.023,4	0.134,9	0.158,3
Total cost, this item	$ 1.523,3	$ 0.574,3	$ 3.083,7	$ 5.181,3
Cost component, %	29%	11%	60%	100%

Source: H. T. johnson and R. S. Kaphan, Relevance lost: The Rise and Fall of Management Accounting, Aharvord Business School Press, boston

Preparing cost estimates normally requires the use of historical data on construction costs. Historical cost data will be useful for cost estimation only if they are collected and organized in a way that is compatible with future applications. Organizations which are engaged in cost estimation continually should keep a file for their own use. The information must be updated with respect to changes that will inevitably occur. The format of cost data, such as unit costs for various items, should be organized according to the current standard of usage in the organization.

Construction cost data are published in various forms by a number of organizations. These publications are useful as references for comparison. Basically, the following types of information are available:

1) Catalogs of vendors' data on important features and specifications relating to their products for which cost quotations are either published or can be obtained. A major source of vendors' information for building products is *Sweets' Catalog* published by McGraw-Hill Information Systems Company.

2) Periodicals containing construction cost data and indices. One source of such information is ENR, the *McGraw-Hill Construction Weekly*, which contains extensive cost data including quarterly cost reports. *Cost Engineering*, a journal of the American Society of Cost Engineers, also publishes useful cost data periodically.

3) Commercial cost reference manuals for estimating guides. An example is the *Building Construction Cost Data* published annually by R. S. Means Company, Inc., which contains unit prices on building construction items. *Dodge Manual for Building Construction*, published by McGraw-Hill, provides similar information.

(4) Digests of actual project costs. *The Dodge Digest of Building Costs and Specifications* provides descriptions of design features and costs of actual projects by building type. Once a week, ENR publishes the bid prices of a project chosen from all types of construction projects.

Historical cost data must be used cautiously. Changes in relative prices may have substantial impacts on construction costs which have increased in relative price. Unfortunately, systematic changes over a long period of time for such factors are difficult to predict. Errors in analysis also serve to introduce uncertainty into cost estimates. It is difficult, of course, to foresee all the problems which may occur in construction and operation of facilities. There is some evidence that estimates of construction and operating costs have tended to persistently understate the actual costs. This is due to the effects of greater than anticipated increases in costs, changes in design during the construction process, or over-optimism.

Integrated Skills: Cost Estimation

1) In making a screening estimate of an industrial plant for the production of batteries, an empirical formula based on data of a similar building completed before 1987 was proposed:

$$C = 16,000(Q + 50,000)^{\frac{1}{2}}$$

where Q is the daily production capacity of batteries and C is the cost of the building in 1987 dollars. If a similar plant is planned for a daily production capacity of 200,000 batteries, find the screening estimate of the building in 1987 dollars.

2) In making a preliminary estimate of a chemical processing plant, several major types of equipment are the most significant components in affecting the installation cost. The cost of piping and other ancillary items for each type of equipment can often be expressed as a percentage of that type of equipment for a given capacity. The standard costs for the major equipment types for two plants with different daily production capacities are as shown in the following table. It has been established that the installation cost of all equipment for a plant with daily production capacity between 150,000bbl and 600,000 bbl can best be estimated by using liner interpolation of the standard data. A new chemical processing plant with a daily production capacity of 400,000 bbl is being planned. Assuming that all other factors remain the same, estimate the cost of the new plant.

Table 15-6

Equipment type	Equipment cost ($ 1,000)		Factor for ancillary items	
	150,000 bbl	600,000 bbl	150,000 bbl	600,000 bbl
Furnaces	$ 3,000	$ 10,000	0.32	0.24
Towers	$ 2,000	$ 6,000	0.42	0.36
Drums	$ 1,500	$ 5,000	0.42	0.32
Pumps, etc.	$ 1,000	$ 4,000	0.54	0.42

3) The total construction cost of a refinery with a production capacity of 100,000 bbl/day in Caracas, Venezuela, completed in 1977 was $ 40 million. It was proposed that a similar refinery with a production capacity of $ 160,000 bbl/day be built in New Orleans, LA for completion in 1980. For the additional information given below, make

a screening estimate of the cost of the proposed plant.

In the total construction cost for the Caracas, Venezuela plant, there was an item of $ 2 million for site preparation and travel which is not typical for similar plants.

The variation of sizes of the refineries can be approximated by the exponential law with $m = 0.6$.

The inflation rate in U. S. dollars was approximately 9% per year from 1977 to 1980.

An adjustment factor of 1. 40 was suggested for the project to account for the increase of labor cost from Caracas, Venezuela to New Orleans, LA.

New air pollution equipment for the New Orleans, LA plant cost $ 4 million in 1980 dollars (not required for the Caracas plant).

The site condition at New Orleans required special piling foundation which cost $ 2 million in 1980 dollars.

Unit 16　Civil Engineering Management II: Construction Planning

Lead-in

1. Why is construction planning a fundamental and challenging activity?
2. What is the most demanding for you during the construction planning?

Vocabulary Warm-up: Matching

_____ 1. arrow diagram method（ADM）	a. 单代号网络图
_____ 2. activity on arrow network	b. 网络分支图
_____ 3. activity on node network	c. 双代号网络图
_____ 4. critical path method（CPM）	d. 横道图
_____ 5. critical path network（CPN）	e. 网络图,前导网络图
_____ 6. bar chart	f. 关键事件进度表
_____ 7. float trend charts	g. 关键路径网络图
_____ 8. fixed-duration scheduling	h. 箭线图方法
_____ 9. key event schedule	i. 确定的网络图
_____ 10. network branching	j. 关键路径法
_____ 11. deterministic network	k. 时差趋势图
_____ 12. dependency diagram	l. 固定工期进度安排

Text A　Construction Planning

1. Basic Concept in the Development of Construction Plans

Construction planning is a fundamental and challenging activity in the management and execution of construction projects. It involves the choice of technology, the definition of work tasks, the estimation of the required resources and *duration*s for individual tasks, and the identification of any *interaction*s among the different work tasks. A good construction plan is the basis for developing the budget and the *schedule* for work. Developing the construction plan is a critical task in the management of construction, even if the plan is not written or otherwise formally recorded. In addition to these technical aspects of construction planning, it may also be necessary to make organizational decisions about the relationships between project participants and even which organizations to include in a project. For example, the extent to which sub-contractors will be used on a project is often determined during construction planning.

In developing a construction plan, it is common to adopt a primary emphasis on either cost control or on schedule control. Some projects are primarily divided into expense categories with associated costs. In these cases, construction planning is cost or expense *orient*ed. Within the categories of expenditure, a distinction is made between costs *incur*red directly in the performance of an activity and indirectly for the accomplishment of the project. For example, borrowing expenses for project financing and overhead items are commonly treated as indirect costs. For other projects, scheduling of work activities over time is critical and is emphasized in the planning process. In this case, the planner insures that the proper *precedence* among activities is maintained and that efficient scheduling of the available resources *prevail*s. Traditional scheduling procedures emphasize the maintenance of task precedence (resulting in critical path scheduling procedures) or efficient use of resources over time (resulting in job shop scheduling procedures). Finally, most complex projects require consideration of both cost and scheduling over time, so that planning, monitoring and record keeping must consider both *dimension*s. In these cases, the integration of schedule and budget information is a major concern.

2. Choices of Technology and Construction Method

As in the development of *appropriate* alternatives for facility design, choices of appropriate technology and methods for construction are often ill-structured yet critical ingredients in the success of the project. For example, a decision whether to pump or to transport concrete in buckets will directly affect the cost and duration of tasks involved in building construction. A decision between these two alternatives should consider the relative costs, reliabilities, and availability of equipment for the two transport methods. Unfortunately, the exact implications of different methods depend upon numerous considerations for which information may be sketchy during the planning phase, such as the experience and expertise of workers or the particular underground condition at a site.

In selecting among alternative methods and technologies, it may be necessary to formulate a number of construction plans based on alternative methods or assumptions. Once the full plan is available, then the cost, time and reliability impacts of the alternative approaches can be reviewed. This examination of several alternatives is often made explicit in bidding competitions in which several alternative designs may be proposed or value engineering for alternative construction methods may be permitted. In this case, potential constructors may wish to prepare plans for each alternative design using the suggested construction method as well as to prepare plans for alternative construction methods which would be proposed as part of the value engineering process.

In forming a construction plan, a useful approach is to *simulate* the construction process either in the imagination of the planner or with a formal computer based simulation technique. By observing the result, comparisons among different plans or problems with the existing plan can be identified. For example, a decision to use a particular piece of equipment for an operation immediately leads to the question of whether or not there is sufficient access space for the equipment. Three dimensional *geometric models in a computer aided design* (**CAD**) *system may be helpful in simulating* space requirements for operations and for identifying any *interference*.

3. Defining Work Tasks

At the same time that the choice of technology and general method are considered, a *parallel* step in the planning process is to define the various work tasks that must be accomplished. These work tasks represent the necessary framework to permit scheduling of construction activities, along with estimating the resources required by the individual work tasks, and any necessary precedence or required sequence among the tasks. The scheduling problem is to determine an appropriate set of activity start time, resource allocations and completion times that will result in completion of the project in a timely and efficient fashion. Construction planning is the necessary fore-runner to scheduling. In this planning, defining work tasks, technology and construction method is typically done either simultaneously or in a series of *iterations*.

More formally, an activity is any subdivision of project tasks. The set of activities defined for a project should be comprehensive or completely *exhaustive* so that all necessary work tasks are included in one or more activities. Typically, each design element in the planned facility will have one or more associated project activities. Execution of an activity requires time and resources, including manpower and equipment. The time required to perform an activity is called the duration of the activity. The beginning and the end of activities are signposts or milestones, indicating the progress of the project.

It is generally advantageous to introduce an explicit *hierarchy* of work activities for the purpose of simplifying the *presentation* and development of a schedule. For example, the initial plan might define a single activity associated with "site clearance". Later, this single activity might be sub-divided into "re-locating utilities", "removing *vegetation*", "*grading*", etc. However, these activities could continue to be identified as sub-activities under the general activity of "site clearance". This hierarchical structure also facilitates the preparation of summary charts and reports in which detailed operations are combined into *aggregate* or "super"-activities.

In practice, the proper level of detail will depend upon the size, importance and difficulty of the project as well as the specific scheduling and accounting procedures which are adopted. However, it is generally the case that most schedules are prepared with too little detail than too much. It is important to keep in mind that task definition will serve as the basis for scheduling, for communicating the construction plan and for construction monitoring. Completion of tasks will also often serve as a basis for progress payments from the owner. Thus, more detailed task definitions can be quite useful. But more detailed task breakdowns are only valuable to the extent that the resources required, durations and activity relationships are realistically estimated for each activity.

Providing detailed work task breakdowns is not helpful without a **commensurate** effort to provide realistic resource requirement estimates. As more powerful, computer-based scheduling and monitoring procedures are introduced, the ease of defining and manipulating tasks will increase, and the number of work tasks can reasonably be expected to expand.

4. Defining Precedence Relationships among Activities

Once work activities have been defined, the relationships among the activities can be specified. Precedence relations among activities signify that the activities must take place in a particular sequence. Numerous natural sequences exist for construction activities due to requirements for structural *integrity*, regulations, and other technical requirements.

Some activities have a necessary technical or physical relationship that cannot be superseded. For example, concrete pours cannot proceed before formwork and *reinforcement* are in place. Some activities have a necessary precedence relationship over a continuous space rather than as discrete work task relationships. Some "precedence relationships" are not technically necessary but are imposed due to implicit decisions within the construction plan.

5. Estimating Activity Durations

In most scheduling procedures, each work activity has associated time duration. These durations are used extensively in preparing a schedule.

All formal scheduling procedures rely upon estimates of the durations of the various project activities as well as the definitions of the predecessor relationships among tasks. A straightforward approach to the estimation of activity durations is to keep historical records of particular activities and rely on the average durations from this experience in making new duration estimates.

6. Estimating Resource Requirements for Work Activities

In addition to precedence relationships and time durations, resource requirements are usually estimated for each activity. Since the work activities defined for a project are comprehensive, the total resources required for the project are the sum of the resources required for the various activities. By making resource requirement estimates for each activity, the requirements for particular resources during the course of the project can be identified. Potential *bottleneck*s can thus be identified, and schedule, resource allocation or technology changes made to avoid problems.

Many formal scheduling procedures can incorporate constraints imposed by the availability of particular resources. Another type of resource is space.

The initial problem in estimating resource requirements is to decide the extent and number of resources that might be defined. At a very aggregate level, resources categories might be limited to the amount of labor (measured in man-hours or in dollars), the amount of materials required for an activity, and the total cost of the activity. At this aggregate level, the resource estimates may be useful for purposes of project monitoring and cash flow planning. However, this aggregate definition of resource use would not reveal bottlenecks associated with particular types of equipment or workers.

More detailed definitions of required resources would include the number and type of both workers and equipment required by an activity as well as the amount and types of materials. Standard resource requirements for particular activities can be recorded and adjusted for the special conditions of particular projects. As a result, the resource types required for particular activities may already be defined. Reliance on historical or standard activity definitions of this type requires a standard coding system for activities.

From the planning perspective, the important decisions in estimating resource requirements are to determine the type of technology and equipment to employ and the number of *crew*s to *allocate* to each task. Clearly, assigning additional crews might result in faster completion of a particular activity. However, additional crews might result in *congestion* and coordination problems, so that work productivity might decline. Further, completing a particular activity earlier might not result in earlier completion of the entire project.

Glossary

Complete the glossary using words in the text.

1. _____	持续时间,期间	2. _____	交互,相互作用
3. _____	进度,日程安排		
4. _____	以……为中心,面向		
5. _____	招致,发生	6. _____	紧前任务
7. _____	优先	8. _____	维度
9. _____	合适的,适当的	10. _____	模拟,仿真
11. _____	几何的	12. _____	干涉,干扰
13. _____	平行的	14. _____	迭代,重复
15. _____	全面的,详尽的	16. _____	层级
17. _____	表达	18. _____	植被
19. _____	坡度	20. _____	总体
21. _____	相等的	22. _____	完整
23. _____	钢筋	24. _____	瓶颈
25. _____	班组	26. _____	分配,分派
27. _____	拥挤		

Exercises

1. Write down the words with the following prefix or suffix.

　　ill- _____

-fore-	_____
-form-	_____
-cur-	_____
-pose	_____

2. Translate the following paragraph into Chinese.

1) Consider a cold weather structure built by inflating a special rubber tent, spraying water on the tent, letting the water freeze, and then deflating and removing the tent. Develop a work breakdown for this structure, precedence relationships, and estimate the required resources. Assume that the tent is twenty feet by fifteen feet long by eight feet wide tall.

2) Develop a work breakdown and activity network for the project of designing a tower to support a radio transmission antenna.

3) Select a vacant site in your vicinity and define the various activities and precedence among these activities that would be required to prepare the site for the placement of pre-fabricated residences.

Text B Fundamental Scheduling Procedures（ I ）

1. Relevance of Construction Schedules

In addition to assigning dates to project activities, project scheduling is intended to match the resources of equipment, materials and labor with project work tasks over time. Good scheduling can eliminate problems due to production bottlenecks, facilitate the timely procurement of necessary materials, and otherwise insure the completion of a project as soon as possible. In contrast, poor scheduling can result in considerable waste as laborers and equipment wait for the availability of needed resources or the completion of preceding tasks. Delays in the completion of an entire project due to poor scheduling can also create havoc for owners who are eager to start using the constructed facilities.

Attitudes toward the formal scheduling of projects are often extreme. Many owners require detailed construction schedules to be submitted by contractors as a means of monitoring the work progress. The actual work performed is commonly compared to the schedule to determine if construction is proceeding satisfactorily. After the completion of construction, similar comparisons between the planned schedule and the actual accomplishments may be performed to allocate the liability for project delays due to changes requested by the owner, worker strikes or other unforeseen circumstances.

In contrast to these instances of reliance upon formal schedules, many field supervisors disdain and dislike formal scheduling procedures. In particular, the critical path method of scheduling is commonly required by owners and has been taught in universities for over two decades, but is often regarded in the field as irrelevant to actual operations and a time consuming distraction.

2. The Critical Path Method

The most widely used scheduling technique is the critical path method (CPM) for scheduling, often referred to as critical path scheduling. This method calculates the minimum completion time for a project along with the possible start and finish times for the project activities. Indeed, many texts and managers regard critical path scheduling as the only usable and practical scheduling procedure. Computer programs and algorithms for critical path scheduling are widely available and can efficiently handle projects with thousands of activities.

The critical path itself represents the set or sequence of predecessor/successor activities which will take the longest time to complete. The duration of the critical path is the sum of the activities' durations along the path. Thus, the critical path can be defined as the longest possible path through the "network" of project activities. The duration of the critical path represents the minimum time required to complete a project. Any delays along the critical path would imply that additional time would be required to complete the project.

Formally, critical path scheduling assumes that a project has been divided into activities of fixed duration and well defined predecessor relationships. A predecessor relationship implies that one activity must come before another in the schedule.

To use critical path scheduling in practice, construction planners often represent a resource constraint by a precedence relation. A constraint is simply a restriction on the options available to a manager, and a resource constraint is a constraint deriving from the limited availability of some resource of equipment, material, space or labor. Also, most critical path scheduling algorithms impose restrictions on the generality of the activity relationships or network geometries which are used. In essence, these restrictions imply that the construction plan can be represented by a network plan in which activities appear as nodes in a network. Nodes are numbered, and no two nodes can have the same number. Two nodes are introduced to represent the start and completion of the project itself.

The actual computer representation of the project schedule generally consists of a list of activities along with their associated durations, required resources and predecessor activities. Graphical network representations rather than a list are helpful for visualization of the plan and to insure that mathematical requirements are met. The actual input of the data to a computer program may be accomplished by filling in blanks on a screen menu, reading an existing data file, or typing data directly to the program with identifiers for the type of information being provided.

With an activity-on-branch network, dummy activities may be introduced for the purposes of providing unique activity designations and maintaining the correct sequence of activities. A dummy activity is assumed to have no time duration and can be graphically represented by a dashed line in a network. In general, dummy activities may be necessary to meet the requirements of specific computer scheduling algorithms, but it is important to limit the number of such dummy link insertions to the extent possible.

Example 16-1: Formulating a network diagram

Suppose that we wish to form an activity network for a seven-activity network with the following precedence:

Table 16-1

Activity	Predecessors	Activity	Predecessors
A	—	E	C
B	—	F	D
C	A,B	G	D,E

Forming an activity-on-branch network for this set of activities might begin to draw activities A, B and C. At this point, we note that two activities (A and B) lay between the same two events nodes; for clarity, we insert a dummy activity X and continue to place other activities. Placing activity G in the figure presents a problem, however, since we wish both activity D and activity E to be predecessors. Inserting an additional dummy activity Y along with activity G completes the activity network, as shown in Figure 16-1. A comparable activity-on-node representation is shown in Figure 16-2, including project start and finish nodes. Note that dummy activities are not required for expressing precedence relationships in activity-on-node networks.

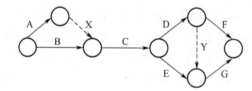

Figure 16-1   An activity-on-branch network for critical path scheduling

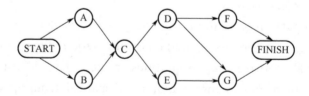

Figure 16-2   An activity-on-node network for critical path scheduling

3. Calculations for Critical Path Scheduling

With the background provided by the previous sections, we can formulate the critical path scheduling mathematically. We shall present an algorithm or set of instructions for critical path scheduling assuming an activity-on-branch project network. We also assume that all precedence is of a finish-to-start nature, so that a succeeding activity cannot start until the completion of a preceding activity. In a later section, we present a comparable algorithm for activity-on-node representations with multiple precedence types.

Suppose that our project network has $n+1$ nodes, the initial event being 0 and the last event being n. Let the time at which node events occur be x_1, x_2, ..., x_n, respectively. The start of the project at x_0 will be defined as time 0. Nodal event times must be consistent with activity durations, so that an activity's successor node event time must be larger than an activity's predecessor node event time plus its duration. For an activity defined as starting from event i and ending at event j, this relationship can be expressed as the inequality constraint, $x_j \geqslant x_i + D_{ij}$, where D_{ij} is the duration of activity (i, j). This same expression can be written for every activity and must hold true in any feasible schedule. Mathematically, the critical path scheduling problem is to minimize the time of project completion (x_n) subject to the constraints that each node completion event cannot occur until each of the predecessor activities have been completed.

The earliest event time algorithm computes the earliest possible time, $E(i)$, at which each event, i, in the network can occur. Earliest event times are computed as the maximum of the earliest start times plus activity durations for each of the activities immediately preceding an event. The earliest start time for each activity (i, j) is equal to the earliest possible time for the preceding event $E(i)$: $ES(i, j) = E(i)$. The earliest finish time of each activity (i, j) can be calculated by: $EF(i, j) = E(i) + D_{ij}$.

Activities are identified in this algorithm by the predecessor node (or event) i and the successor node j. The algorithm simply requires that each event in the network should be examined in turn beginning with the project start (node 0).

The latest event time algorithm computes the latest possible time, $L(j)$, at which each event j in the network can occur, given the desired completion time of the project, $L(n)$ for the last event n. Usually, the desired completion time will be equal to the earliest possible completion time, so that $E(n) = L(n)$ for the final node n. The procedure for finding the latest event time is analogous to that for the earliest event time except that the procedure begins with the final event and works backwards through the project activities. Thus, the earliest event time algorithm is often called a forward pass through the network, whereas the latest event time algorithm is the backward pass through the network. The latest finish time consistent with completion of the project in the desired time frame of $L(n)$ for each activity (i, j) is equal to the latest possible time $L(j)$ for the succeeding event: $LF(i, j) = L(j)$. The latest start time of each activity (i, j) can be calculated by: $LS(i, j) = L(j) - D_{ij}$.

The earliest start and latest finish times for each event are useful pieces of information in developing a project schedule. Events which have equal earliest and latest times, $E(i) = L(i)$, lie on the critical path or paths. An activity (i, j) is a critical activity if it satisfies all of the following conditions: $E(i) = L(i)$; $E(j) = L(j)$; $E(i) + Dij = L(j)$.

Hence, activities between critical events are also on a critical path as long as the activity's earliest start time equals its latest start time, $ES(i, j) = LS(i, j)$. To avoid delaying the project, all the activities on a critical path should begin as soon as possible, so each critical activity (i, j) must be scheduled to begin at the earliest possible start time, $E(i)$.

Text C   Fundamental Scheduling Procedures (Ⅱ)

Example 16-2: Critical path scheduling calculations

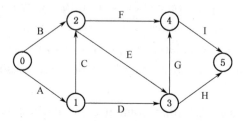

Figure 16-3   A nine-activity project network

Consider the network shown in Figure 16 - 3. Once the node numbers are established, a good aid for manual scheduling is to draw a small rectangle near each node with two possible entries. The left hand side would contain the earliest time the event could occur, whereas the right hand side would contain the latest time the event could occur without delaying the entire project.

Table 16-2   Precedence relations and durations for a nine-activity project example

Activity	Description	Predecessors	Duration
A	Site clearing	—	4
B	Removal of trees	—	3
C	General excavation	A	8
D	Grading general area	A	7
E	Excavation for trenches	B,C	9
F	Placing formwork and reinforcement for concrete	B,C	12
G	Installing sewer lines	D,E	2
H	Installing other utilities	D,E	5
I	Pouring conerete	F,G	6

For the network in Figure 16-3 with activity durations in Table 16-2, the earliest event time calculations proceed as follows:

Step 1 → $E(0) = 0$

Step 2: $j = 1 \rightarrow E(1) = \mathrm{Max}\{E(0) + D_{01}\} = \mathrm{Max}\{0 + 4\} = 4$

$j = 2 \rightarrow E(2) = \mathrm{Max}\{E(0) + D_{02}; E(1) + D_{12}\} = \mathrm{Max}\{0 + 3; 4 + 8\} = 12$

$j = 3 \rightarrow E(3) = \mathrm{Max}\{E(1) + D_{13}; E(2) + D_{23}\} = \mathrm{Max}\{4 + 7; 12 + 9\} = 21$

$j = 4 \rightarrow E(4) = \mathrm{Max}\{E(2) + D_{24}; E(3) + D_{34}\} = \mathrm{Max}\{12 + 12; 21 + 2\} = 24$

$j = 5 \rightarrow E(5) = \mathrm{Max}\{E(3) + D_{35}; E(4) + D_{45}\} = \mathrm{Max}\{21 + 5; 24 + 6\} = 30$

Thus, the minimum time required to complete the project is 30 since $E(5) = 30$. In this case, each event had at most two predecessors.

For the "backward pass", the latest event time calculations are:

Step 1 $\rightarrow L(5) = E(5) = 30$

Step 2: $j = 4 \rightarrow L(4) = \mathrm{Min}\{L(5) - D_{45}\} = \mathrm{Min}\{30 - 6\} = 24$

$j = 3 \rightarrow L(3) = \mathrm{Min}\{L(5) - D_{35}; L(4) - D_{34}\} = \mathrm{Min}\{30 - 5; 24 - 2\} = 22$

$j = 2 \rightarrow L(2) = \mathrm{Min}\{L(4) - D_{24}; L(3) - D_{23}\} = \mathrm{Min}\{24 - 12; 22 - 9\} = 12$

$j = 1 \rightarrow L(1) = \mathrm{Min}\{L(3) - D_{13}; L(2) - D_{12}\} = \mathrm{Min}\{22 - 7; 12 - 8\} = 4$

$j = 0 \rightarrow L(0) = \mathrm{Min}\{L(2) - D_{02}; L(1) - D_{01}\} = \mathrm{Min}\{12 - 3; 4 - 4\} = 0$

In this example, $E(0) = L(0)$, $E(1) = L(1)$, $E(2) = L(2)$, $E(4) = L(4)$, and $E(5) = L(5)$. As a result, all nodes but node 3 are in the critical path. Activities on the critical path include A(0, 1), C(1, 2), F(2, 4) and I(4, 5) as shown in Table 16-3.

Table 16-3　Identification of activities on the critical path for a nine-activity project

Activity	Duration D_{ij}	Earliest start time $E(i) = ES(i,j)$	Latest finish time $L(j) = LF(i,j)$	Latest start time $LS(i,j)$
A(0,1)	4	0*	4*	0
B(0,2)	3	0	12	9
C(1,2)	8	4*	12*	4
D(1,3)	7	4	22	15
E(2,3)	9	12	22	13
F(2,4)	12	12*	24*	12
G(3,4)	2	21	24	22
H(3,5)	5	21	30	25
I(4,5)	6	24	30*	24

* Activity on a critical path since $E(i) + D_{ij} = L(j)$

4. Activity Float and Schedules

A number of different activity schedules can be developed from the critical path scheduling procedure described in the previous section. An earliest time schedule would be developed by starting each activity as soon as possible, at $ES(i, j)$. Similarly, a latest time schedule would delay the start of each activity as long as possible but still finish the project in the minimum possible time. This late schedule can be developed by setting each activity's start time to $LS(i, j)$.

Activities that have different early and late start times [i. e., $ES(i, j) < LS(i, j)$] can be scheduled to start anytime between $ES(i, j)$ and $LS(i, j)$ as shown in Figure 16-4.

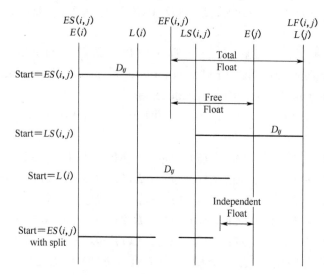

Figure 16-4　Illustration of activity float

The concept of float is to use part or all of this allowable range to schedule an activity without delaying the completion of the project. An activity that has the earliest time for its predecessor and successor nodes differing by more than its duration possesses a window in which it can be scheduled. That is, if $E(i) + D_{ij} < L(j)$, then some float is available in which to schedule this activity.

Float is a very valuable concept since it represents the scheduling flexibility or "maneuvering room" available to complete particular tasks. Activities on the critical path do not provide any flexibility for scheduling nor leeway in case of problems. For activities with some float, the actual starting time might be chosen to balance work loads over time, to correspond with material deliveries, or to improve the project's cash flow.

Of course, if one activity is allowed to float or change in the schedule, then the amount of float available for other activities may decrease. Three separate categories of float are defined in critical path scheduling.

Total float is the maximum amount of delay which can be assigned to any activity without delaying the entire project. The total float, $TF(i, j)$, for any activity (i, j) is calculated as: $TF(i, j) = L(j) - E(i) - D_{ij}$.

Free float is the amount of delay which can be assigned to any one activity without delaying subsequent activities. The free float, $FF(i, j)$, associated with activity (i, j) is: $FF(i, j) = E(j) - E(i) - D_{ij}$.

Independent float is the amount of delay which can be assigned to any one activity without delaying subsequent activities or restricting the scheduling of preceding activities. Independent float, $IF(i, j)$, for activity (i, j) is calculated as:

$$IF(i, j) = \begin{cases} 0 \\ E(j) - L(i) - D_{ij} \end{cases}$$

Each of these "floats" indicates an amount of flexibility associated with an activity. In all cases, total float equals or exceeds free float, while independent float is always less than or equal to free float. Also, any activity on a critical path has all three values of float equal to zero. The converse of this statement is also true, so any activity which has zero total float can be recognized as being on a critical path.

The various categories of activity float are illustrated in Figure 16-4 in which the activity is represented by a bar which can move back and forth in time depending upon its scheduling start. Three possible scheduled starts are shown, corresponding to the cases of starting each activity at the earliest event time, $E(i)$, the latest activity start time $LS(i, j)$, and at the latest event time $L(i)$. The three categories of float can be found directly from this figure. Finally, a fourth bar is included in the figure to illustrate the possibility that an activity might start, be temporarily halted, and then re-start. In this case, the temporary halt was sufficiently short that it was less than the independent float time and thus would not interfere with other activities. Whether or not such work splitting is possible or economical depends upon the nature of the activity.

As shown in Table 16-3, activity $D(1, 3)$ has free and independent floats of 10 for the project shown in Figure 16-4. Thus, the start of this activity could be scheduled anytime between time 4 and 14 after the project began without interfering with the schedule of other activities or with the earliest completion time of the project. As the total float of 11 units indicates, the start of activity D could also be delayed until time 15, but this would require that the schedule of other activities be restricted. For example, starting activity D at time 15 would require that activity G would begin as soon as activity D was completed. However, if this schedule was maintained, the overall completion date of the project would not be changed.

5. Presenting Project Schedules

Communicating the project schedule is a vital ingredient in successful project management. A good presentation will greatly ease the manager's problem of understanding the multitude of activities and their inter-relationships. Moreover, numerous individuals and parties are involved in any project, and they have to understand their assignments. Graphical presentations of project schedules are particularly useful since it is much easier to comprehend a graphical display of numerous pieces of information than to sift through a large table of numbers. Indeed, it has been common to use computer programs to perform critical path scheduling and then to produce bar charts of detailed activity schedules and resource assignments manually. With the availability of computer graphics, the cost and effort of producing graphical presentations has been significantly reduced and the production of presentation aids can be automated.

Network diagrams for projects have already been introduced. These diagrams provide a powerful visualization of the precedences and relationships among the various project activities. They are a basic means of communicating a project plan among the participating planners and project monitors. Project planning is often conducted by producing network representations of greater and greater refinement until the plan is satisfactory.

A useful variation on project network diagrams is to draw a time-scaled network. The activity diagrams shown in the previous section were topological networks in that only the relationship between nodes and branches were of interest. The actual diagram could be distorted in any way desired as long as the connections between nodes were not changed. In time-scaled network diagrams, activities on the network are plotted on a horizontal axis measuring the time since project commencement. In this time-scaled diagram, each node is shown at its earliest possible time. By looking over the horizontal axis, the time at which activity can begin can be observed. Obviously, this time-scaled diagram is produced as a display after activities are initially scheduled by the critical path method.

Another useful graphical representation tool is a bar or Gantt chart illustrating the scheduled time for each activity. The bar chart lists activities and shows their scheduled start, finish and duration. Activities are listed in the vertical axis of this figure, while time since project commencement is shown along the horizontal axis. During the course of monitoring a project, useful additions to the basic bar chart include a vertical line to indicate the current time plus small marks to indicate the current state of work on each activity.

Bar charts are particularly helpful for communicating the current state and schedule of activities on a project. As such, they have found wide acceptance as a project representation tool in the field. For planning purposes, bar charts are not as useful since they do not indicate the precedence relationships among activities. Thus, a planner must remember or record separately that a change in one activity's schedule may require changes to successor activities. There have been various schemes for mechanically linking activity bars to represent precedences, but it is now easier to use computer based tools to represent such relationships.

Integrated Skills: Constructing a Project Network

1) Construct an activity-on-branch network from the precedence relationships of activities in the project given in the table for the problem and determine the critical path and all slacks for the projects.

Table 16-4

Activity	A	B	C	D	E	F	G	H	I	J	K	L	M	N	O
Predecessors	—	A	A	—	B	C,D	C,D	D	H	F	E,J	F	G,J	G,I	L,N
Duration	6	7	1	14	5	8	9	3	5	3	4	12	6	2	7

2) Suppose that the precedence relationships in the table are all direct finish-to-start relationships with no lags except for the following:

C to D: S-S with a lag of 1

D to E: F-F with a lag of 3

A to F: S-S with a lag of 2

H to I: F-F with a lag of 4

L to M: S-S with a lag of 1

Table 16-5

Activity	A	B	C	D	E	F	G	H	I	J	K	L	M	N
Predecessors		A	B	C	D,G	A	E,J	—	H	I	F,J	H	L	K,M
Duration	5	6	3	4	5	8	3	3	2	7	2	7	4	3

Formulate an activity-on-node network representation and recompute the critical path with these precedence relationships.

3) Develop an example of a project network with three critical paths.

Unit 17　Civil Engineering Management Ⅲ: Quality Control

Lead-in

1. Should project managers always insure the conformance to the original designs and planning decisions? Why or why not?

2. What are the generally accepted specifications to be used during construction?

Vocabulary Warm-up: Matching

_____1. OSHA:　　a. 美国建筑规范学会

_____2. ASTM:　　b. 职业安全与健康管理局

_____3. ANSI:　　c. 可接受质量水平

_____4. CSI:　　d. 美国材料实验协会

_____5. AQL:　　e. 美国国家标准学会

Text A　Quality Control and Safety During Construction

1. Quality and Safety Concerns in Construction

Quality control and safety represent increasingly important concerns for project managers. *Defect*s or *failure*s in constructed facilities can result in very large costs. Even with minor defects, re-construction may be required and facility operations *impaire*d. Increased costs and *delay*s are the result. In the worst case, failures may cause personal injuries or fatalities. Accidents during the construction process can similarly result in personal injuries and large costs. Indirect costs of insurance, inspection and regulation are increasing rapidly due to these increased direct costs. Good project managers try to ensure that the job is done right the first time and that no major accidents occur on the project.

As with cost control, the most important decisions regarding the quality of a completed facility are made during the design and planning stages rather than during construction. It is during these *preliminary* stages that component configurations, material *specification*s and functional performance are decided. Quality control during construction consists largely of insuring *conformance* to these original designs and planning decisions.

While conformance to existing design decisions is the primary focus of quality control, there are *exception*s to this rule. First, *unforeseen* circumstances, incorrect design decisions or changes desired by an owner in the facility function may require re-evaluation of design decisions during the course of construction. While these changes may be motivated by the concern for quality, they represent occasions for re-design with all the *attendant* objectives and *constraints*. As a second case, some designs rely upon informed and appropriate decision making during the construction process itself. For example, some tunneling methods make decisions about the amount of shorings required at different locations based upon observation of soil conditions during the tunneling process. Since such decisions are based on better information concerning actual site conditions, the facility design may be more cost-effective as a result. Any special case of re-design during construction requires the various considerations.

With the attention to conformance as the measure of quality during the construction process, the specification of quality requirements in the design and contract documentation becomes *extremely* important. Quality requirements should be clear and verifiable, so that all parties in the project can understand the requirements for conformance. Much of the discussion in this chapter relates to the development and the implications of different quality requirements for construction as well as the issues associated with insuring conformance.

Safety during the construction project is also influenced in large part by decisions made during the planning and design process. Some designs or construction plans are *inherently* difficult and dangerous to implement, whereas other comparable plans may considerably reduce the *possibility* of accidents. For example, clear separation of traffic from construction zones during roadway rehabilitation can greatly reduce the possibility of accidental collisions. Beyond these design decisions, safety largely depends upon education, *vigilance* and cooperation during the construction process. Workers should be constantly alert to the possibilities of accidents and avoid taking unnecessary risks.

2. Organizing for Quality and Safety

A variety of different organizations are possible for quality and safety control during construction. One common model is to have a group responsible for quality *assurance* and another group primarily responsible for safety within an organization. In large organizations, departments dedicated to quality assurance and to safety might assign specific individuals to assume responsibility for these functions on particular projects. For smaller projects, the project manager or an assistant might assume these and other responsibilities. In either case, insuring safe and quality construction is a concern of the project manager in *overall* charge of the project in addition to the concerns of personnel, cost, time and other management issues.

*Inspector*s and quality assurance personnel will be involved in a project to represent a variety of different organizations. Each of the parties directly concerned with the project may have their own quality and safety inspectors, including the owner, the engineer/architect, and the various constructor firms. These inspectors may be contractors from specialized quality assurance organizations. In addition to on-site inspections, *sample*s of materials will commonly be tested by specialized laboratories to insure compliance. Inspectors to insure compliance with regulatory requirements will also be involved. Common examples are inspectors for the local government's building department, for environmental agencies, and for *occupational* health and safety agencies.

The US Occupational Safety and Health Administration (OSHA) routinely conduct site visits of work places in conjunction with approved state inspection agencies. OSHA inspectors are required by law to issue citations for all standard violations observed. Safety standards *prescribe* a variety of mechanical safeguards and procedures; for example, ladder safety is covered by over 140 regulations. In cases of extreme non-compliance with standards, OSHA inspectors can stop work on a project. However, only a small fraction of construction sites are visited by OSHA inspectors and most construction site accidents are not caused by violations of existing standards. As a result, safety is largely the responsibility of the managers on site rather than that of public inspectors.

While the multitude of participants involved in the construction process require the services of inspectors, it cannot be emphasized too strongly that inspectors are only a formal check on quality control. Quality control should be a primary objective for all the members of a project team. Managers should take responsibility for maintaining and improving quality control. Employee participation in quality control should be sought and rewarded, including the introduction of new ideas. Most important of all, quality improvement can serve as a *catalyst* for improved productivity. By suggesting new work methods, by avoiding rework, and by avoiding long term problems, good quality control can pay for itself. Owners should promote good quality control and seek out contractors who maintain such standards.

In addition to the various organizational bodies involved in quality control, issues of quality control arise in virtually all the functional areas of construction activities. For example, insuring accurate and useful information is an important part of maintaining quality performance. Other aspects of quality control include document control (including changes during the construction process), procurement, field inspection and testing, and final checkout of the facility.

3. Work and Material Specifications

Specifications of work quality are an important feature of facility designs. Specifications of required quality and components represent part of the necessary documentation to describe a facility. Typically, this documentation includes any special provisions of the facility design as well as references to generally accepted specifications to be used during construction.

General specifications of work quality are available in numerous fields and are issued in publications of organizations such as the American Society for Testing and Materials (ASTM), the American National Standards Institute (ANSI), or the Construction

Specifications Institute (CSI). Distinct specifications are formalized for particular types of construction activities, such as *welding* standards issued by the American Welding Society, or for particular facility types, such as the Standard Specifications for Highway Bridges issued by the American Association of State Highway and Transportation Officials. These general specifications must be modified to reflect local conditions, policies, available materials, local regulations and other special circumstances.

Construction specifications normally consist of a series of instructions or *prohibition*s for specific operations. For example, the following passage illustrates a typical specification, in this case for excavation for structures:

Conform to elevations and dimensions shown on plan within a *tolerance* of plus or minus 0.10 foot, and extending a sufficient distance from footings and foundations to permit placing and removal of concrete formwork, installation of services, other construction, and for inspection. In excavating for footings and foundations, take care not to disturb bottom of excavation. Excavate by hand to final grade just before concrete reinforcement is placed. Trim bottoms to required lines and grades to leave solid base to receive concrete.

This set of specifications requires judgment in application since some items are not precisely specified. For example, excavation must extend a "sufficient" distance to permit inspection and other activities. Obviously, the term "sufficient" in this case may be subject to varying *interpretation*s. In contrast, a specification that tolerances are within plus or minus a tenth of a foot is subject to direct measurement. However, specific requirements of the facility or characteristics of the site may make the standard tolerance of a tenth of a foot inappropriate. Writing specifications typically requires a trade-off between assuming reasonable behavior on the part of all the parties concerned in interpreting words such as "sufficient" versus the effort and possible inaccuracy in pre-specifying all operations.

In recent years, performance specifications have been developed for many construction operations. Rather than specifying the required construction process, these specifications refer to the required performance or quality of the finished facility. The exact method by which this performance is obtained is left to the construction contractor. For example, traditional specifications for *asphalt* pavement specified the composition of the asphalt material, the asphalt temperature during paving, and *compacting* procedures. In contrast, a performance specification for asphalt would detail the desired performance of the pavement with respect to impermeability, strength, etc. How the desired performance level was attained would be up to the paving contractor. In some cases, the payment for asphalt paving might increase with better quality of asphalt beyond some minimum level of performance.

Example 17-1: Concrete Pavement Strength

Concrete pavements of superior strength result in cost savings by delaying the time at which repairs or re-construction is required. In contrast, concrete of lower quality will necessitate more frequent overlays or other repair procedures. Contract provisions with adjustments to the amount of a contractor's compensation based on pavement quality have become increasingly common in recognition of the cost savings associated with higher quality construction. Even if a pavement does not meet the "ultimate" design standard, it is still worth using the lower quality pavement and re-surfacing later rather than completely rejecting the pavement. Based on these life cycle cost considerations, a typical pay schedule might be:

Table 17-1

Load Ratio	Pay Factor	Load Ratio	Pay Factor
<0.50	Reject	1.10~1.29	1.05
0.50~0.69	0.90	1.30~1.49	1.10
0.70~0.89	0.95	>1.50	1.12
0.90~1.09	1.00		

In this table, the Load Ratio is the ratio of the actual pavement strength to the desired design strength and the Pay Factor is a fraction by which the total pavement contract amount is multiplied to obtain the appropriate compensation to the contractor. For example, if a contractor achieves concrete strength twenty percent greater than the design specification, then the load ratio is 1.20 and the appropriate pay factor is 1.05, so the contractor receives a five percent bonus. Pay factors are computed after tests on the concrete actually used in a pavement. Note that a 90% pay factor exists in this case with even pavement quality only 50% of that originally desired. This high pay factor even with weak concrete strength might exist since much of the costs of pavements are incurred in preparing the pavement foundation. Concrete strengths of less than 50% are cause for complete rejection in this case, however.

Glossary

Complete the glossary using words in the text.

1. _____ 缺陷
2. _____ 失效
3. _____ 损害
4. _____ 延误
5. _____ 初步的
6. _____ 规格;规范
7. _____ 符合
8. _____ 例外
9. _____ 不可预见的,意料之外的
10. _____ 伴随的
11. _____ 约束
12. _____ 极端地
13. _____ 内在地,固有地
14. _____ 可能性
15. _____ 警觉,警惕
16. _____ 保证,确保

17. _____	总体的,全面的,综合的		
18. _____	检查员,视察员,巡视员		
19. _____	样本	20. _____	职业的
21. _____	规定	22. _____	推进剂
23. _____	焊接	24. _____	禁令,禁律
25. _____	公差,容差	26. _____	解释
27. _____	沥青	28. _____	压实

Exercises

1. Write down the words with the following prefix or suffix.

micro-/macro-   _____

-pact   _____

-ject   _____

-gram   _____

-sis   _____

2. Suppose that a sampling plan calls for a sample of size n = 50. To be acceptable, only three or fewer samples can be defective. Estimate the probability of accepting the lot if the average defective percentage is 15%, 5% or 2%. Do not use an approximation in this calculation.

3. Suppose that a project manager tested the strength of one tile out of a batch of 3,000 to be used on a building. This one sample measurement was compared with the design specification and, in this case, the sampled tile's strength exceeded that of the specification. On this basis, the project manager accepted the tile shipment. If the sampled tile had defects (with a strength less than the specification), the lot would be rejected by the project manager.

1) What is the probability that ninety percent of the tiles are substandard, even though the project manager's sample gave a satisfactory result?

2) Sketch out the operating characteristic curve for this sampling plan as a function of the actual fraction of defective tiles.

Text B   Quality Control（Ⅰ）

1. Total Quality Control

Quality control in construction typically involves insuring compliance with minimum standards of material and workmanship in order to insure the performance of the facility according to the design. These minimum standards are contained in the specifications described in the previous section. For the purpose of insuring compliance, random samples and statistical methods are commonly used as the basis for accepting or rejecting work completed and batches of materials. Rejection of a batch is based on non-conformance or violation of the relevant design specifications. Procedures for this quality control practice are described in the following sections.

An implicit assumption in these traditional quality control practices is the notion of an acceptable quality level which is an allowable fraction of defective items. Materials obtained from suppliers or work performed by an organization is inspected and passed as acceptable if the estimated defective percentage is within the acceptable quality level. Problems with materials or goods are corrected after delivery of the product.

In contrast to this traditional approach of quality control is the goal of total quality control. In this system, no defective items are allowed anywhere in the construction process. While the zero defects goal can never be permanently obtained, it provides a goal so that an organization is never satisfied with its quality control program even if defects are reduced by substantial amounts year after year. This concept and approach to quality control was first developed in manufacturing firms in Japan and Europe, but has since spread to many construction companies. The best known formal certification for quality improvement is the International Organization for Standardization's ISO 9000 standard. ISO 9000 emphasizes good documentation, quality goals and a series of cycles of planning, implementation and review.

Total quality control is a commitment to quality expressed in all parts of an organization and typically involves many elements. Design reviews to insure safety and effective construction procedures are a major element. Other elements include extensive training for personnel, shifting the responsibility for detecting defects from quality control inspectors to workers, and continually maintaining equipment. Worker involvement in improved quality control is often formalized in quality circles in which groups of workers meet regularly to make suggestions for quality improvement. Material suppliers are also required to insure zero defects in delivered goods. Initially, all materials from a supplier are inspected and batches of goods with any defective items are returned. Suppliers with good records can be certified and not subject to complete inspection subsequently.

The traditional microeconomic view of quality control is that there are an "optimum" proportion of defective items. Trying to achieve greater quality than this optimum would substantially increase costs of inspection and reduce worker productivity. However, many companies have found that commitment to total quality control has substantial economic benefits that had been unappreciated in traditional approaches. Expenses associated with inventory, rework, scrap and warranties were reduced. Worker enthusiasm and commitment improved. Customers often appreciated higher quality work and would pay a premium for good quality. As a result, improved quality control became a competitive advantage.

Of course, total quality control is difficult to apply, particular in construction. The unique nature of each facility, the variability in the workforce, the multitude of subcontractors and the cost of making necessary investments in education and procedures make programs of total quality control in construction difficult. Nevertheless, a commitment to improved quality even without endorsing the goal of zero defects can pay real dividends to organizations.

Example 17-2: Experience with Quality Circles

Quality circles represent a group of five to fifteen workers who meet on a frequent basis to identify, discuss and solve productivity and quality problems. A circle leader acts as liaison between the workers in the group and upper levels of management. Appearing below are some examples of reported quality circle accomplishments in construction:

On a highway project under construction by Taisei Corporation, it was found that the loss rate of ready-mixed concrete was too high. A quality circle composed of cement masons found out that the most important reason for this was due to an inaccurate checking method. By applying the circle's recommendations, the loss rate was reduced by 11.4%.

In a building project by Shimizu Construction Company, many cases of faulty reinforced concrete work were reported. The iron workers quality circle examined their work thoroughly and soon the faulty workmanship disappeared. A 10% increase in productivity was also achieved.

2. Quality Control by Statistical Methods

An ideal quality control program might test all materials and work on a particular facility. For example, non-destructive techniques such as x-ray inspection of welds can be used throughout a facility. An on-site inspector can witness the appropriateness and adequacy of construction methods at all times. Even better, individual craftsmen can perform continuing inspection of materials and their own work. Exhaustive or 100% testing of all materials and work by inspectors can be exceedingly expensive, however. In many instances, testing requires the destruction of a material sample, so exhaustive testing is not even possible. As a result, small samples are used to establish the basis of accepting or rejecting a particular work item or shipment of materials. Statistical methods are used to interpret the results of test on a small sample to reach a conclusion concerning the acceptability of an entire lot or batch of materials or work products.

The use of statistics is essential in interpreting the results of testing on a small sample. Without adequate interpretation, small sample testing results can be quite misleading. As an example, suppose that there are ten defective pieces of material in a lot of one hundred. In taking a sample of five pieces, the inspector might not find any defective pieces or might have all sample pieces defective. Drawing a direct inference that none or all pieces in the population are defective on the basis of these samples would be incorrect. Due to this random nature of the sample selection process, testing results can vary substantially. It is only with statistical methods that issues such as the chance of different levels of defective items in the full lot can be fully analyzed from a small sample test.

There are two types of statistical sampling which are commonly used for the purpose of quality control in batches of work or materials:

The acceptance or rejection of a lot is based on the number of defective (bad) or non-defective (good) items in the sample. This is referred to as sampling by attributes.

Instead of using defective and non-defective classifications for an item, a quantitative quality measure or the value of a measured variable is used as a quality indicator. This testing procedure is referred to as sampling by variables.

Whatever sampling plan is used in testing, it is always assumed that the samples are representative of the entire population under consideration. Samples are expected to be chosen randomly so that each member of the population is equally likely to be chosen. Convenient sampling plans such as sampling every twentieth piece, choosing a sample every two hours, or picking the top piece on a delivery truck may be adequate to insure a random sample if pieces are randomly mixed in a stack or in use. However, some convenient sampling plans can be inappropriate. For example, checking only easily accessible joints in a building component is inappropriate since joints that are hard to reach may be more likely to have erection or fabrication problems.

Another assumption implicit in statistical quality control procedures is that the quality of materials or work is expected to vary from one piece to another. This is certainly true in the field of construction. While a designer may assume that all concrete is exactly the same in a building, the variations in material properties, manufacturing, handling, pouring, and temperature during setting insure that concrete is actually heterogeneous in quality. Reducing such variations to a minimum is one aspect of quality construction. Insuring that the materials actually placed achieve some minimum quality level with respect to average properties or fraction of defectives is the task of quality control.

3. Statistical Quality Control with Sampling by Attributes

Sampling by attributes is a widely applied quality control method. The procedure is intended to determine whether or not a particular group of materials or work products is acceptable. In the literature of statistical quality control, a group of materials or work items to be tested is called a lot or batch. An assumption in the procedure is that each item in a batch can be tested and classified as either acceptable or deficient based upon mutually acceptable testing procedures and acceptance criteria. Each lot is tested to determine if it satisfies a minimum acceptable quality level (AQL) expressed as the maximum percentage of defective items in a lot or process.

In its basic form, sampling by attributes is applied by testing a pre-defined number of sample items from a lot. If the number of defective items is greater than a trigger level, then the lot is rejected as being likely to be of unacceptable quality. Otherwise, the lot is accepted. Developing this type of sampling plan requires consideration of probability, statistics and acceptable risk levels on the part of the supplier and consumer of the lot. Refinements to this basic application procedure are also possible. For example, if the number of defectives is greater than some pre-defined number, then additional sampling may be started rather than immediate rejection of the lot. In many cases, the trigger level is a single defective item in the sample. In the remainder of this section, the mathematical basis for interpreting this type of sampling plan is developed.

More formally, a lot is defined as acceptable if it contains a fraction p_1 or less defective items. Similarly, a lot is defined as unacceptable if it contains a fraction p_2 or more defective units. Generally, the acceptance fraction is less than or equal to the rejection fraction, $p_1 \leqslant p_2$, and the two fractions are often equal so that there is no ambiguous range of lot acceptability between p_1 and p_2. Given a sample size and a trigger level for lot rejection or acceptance, we would like to determine the probabilities that acceptable lots might be incorrectly rejected (termed producer's risk) or that deficient lots might be incorrectly accepted (termed consumer's risk).

Consider a lot of finite number N, in which m items are defective (bad) and the remaining (N-m) items are non-defective (good). If a random sample of n items is taken from this lot, then we can determine the probability of having different numbers of defective items in the sample. With a pre-defined acceptable number of defective items, we can then develop the probability of accepting a lot as a function of the sample size, the allowable number of defective items, and the actual fraction of defective items. This derivation appears below.

The number of different samples of size n that can be selected from a finite population N is termed a mathematical combination and is computed as:

$$\binom{N}{n} = \frac{N(N-1)\cdots(N-n+1)}{n!} = \frac{N!}{n!(N-n)!} \tag{17-1}$$

where a factorial, ($n!$) is $n \times (n-1) \times (n-2)\cdots(1)$ and zero factorial ($0!$) is one by convention. The number of possible samples with exactly x defectives is the combination associated with obtaining x defectives from m possible defective items and ($n-x$) good items from ($N-m$) good items:

$$\binom{m}{x}\binom{N-m}{n-x} = \frac{m!}{x!(m-x)!} \times \frac{(N-m)!}{(n-x)!(N-m-n+x)!} \tag{17-2}$$

Given these possible numbers of samples, the probability of having exactly x defective items in the sample is given by the ratio as the hyper geometric series:

$$P(X = x) = \frac{\binom{m}{x}\binom{N-m}{n-x}}{\binom{N}{n}} \tag{17-3}$$

With this function, we can calculate the probability of obtaining different numbers of defectives in a sample of a given size.

Text C　Quality Control Ⅱ

Suppose that the actual fraction of defectives in the lot is p and the actual fraction of non-defectives is q, then p plus q is one, resulting in $m = N_p$, and $N - m = N_q$. Then, a function $g(p)$ representing the probability of having r or less defective items in a sample of size n is obtained by substituting m and N into Equation (17-3) and summing over the acceptable defective number of items:

$$g(p) = \sum_{x=0}^{r} P(X = x) = \sum_{x=0}^{r} \frac{\binom{N_p}{x}\binom{N_p}{n-x}}{\binom{N}{n}} \tag{17-4}$$

If the number of items in the lot, N, is large in comparison with the sample size n, then the function $g(p)$ can be approximated by the binomial distribution:

$$g(p) = \sum_{x=0}^{r} \binom{n}{x} p^x q^{n-x} \tag{17-5}$$

The function $g(p)$ indicates the probability of accepting a lot, given the sample size n and the number of allowable defective items in the sample r. The function $g(p)$ can be represented graphical for each combination of sample size n and number of allowable defective items r, as shown in Figure 17-1. Each curve is referred to as the operating characteristic curve (OC curve) in this graph. For the special case of a single sample ($n = 1$), the function $g(p)$ can be simplified, so that the probability of accepting a lot is equal to the fraction of acceptable items in the lot. For example, there is a probability of 0.5 that the lot may be accepted from a single sample test even if fifty percent of the lot is defective.

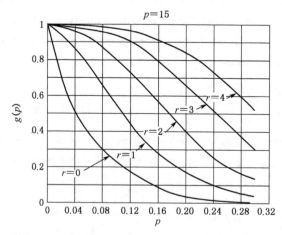

Figure 17-1　Operating characteristic curves indicating probability of lot acceptance

For any combination of n and r, we can read off the value of $g(p)$ for a given p from the corresponding OC curve. For example, $n = 15$ is specified in Figure 17-1.

Then, for various values of r, we find: $r = 0$, $p = 24\%$, $g(p) \approx 2\%$; $r = 0$, $p = 4\%$, $g(p) \approx 54\%$; $r = 1$, $p = 24\%$, $g(p) \approx 10\%$; $r = 1$, $p = 4\%$, $g(p) \approx 88\%$.

The producer's and consumer's risk can be related to various points on an operating characteristic curve. Producer's risk is the chance that otherwise acceptable lots fail the sampling plan (i. e. have more than the allowable number of defective items in the sample) solely due to random fluctuations in the selection of the sample. In contrast, consumer's risk is the chance that an unacceptable lot is acceptable (i. e. has less than the allowable number of defective items in the sample) due to a better than average quality in the sample. For example, suppose that a sample size of 15 is chosen with a trigger level for rejection of one item. With a four percent acceptable level and a greater than four percent defective fraction, the consumer's risk is at most eighty-eight percent. In contrast, with a four percent acceptable level and a four percent defective fraction, the producer's risk is at most $1 - 0.88 = 0.12$ or twelve percent.

In specifying the sampling plan implicit in the operating characteristic curve, the supplier and consumer of materials or work must agree on the levels of risk acceptable to themselves. If the lot is of acceptable quality, the supplier would like to minimize the chance or risk that a lot is rejected solely on the basis of a lower than average quality sample. Similarly, the consumer would like to minimize the risk of accepting under the sampling plan a deficient lot. In addition, both parties presumably would like to minimize the costs and delays associated with testing. Devising an acceptable sampling plan requires trade off the objectives of risk minimization among the parties involved and the cost of testing.

4. Statistical Quality Control with Sampling by Variables

As described in the previous section, sampling by attributes is based on a classification of items as good or defective. Many work and material attributes possess continuous properties, such as strength, density or length. With the sampling by attributes procedure, a particular level of a variable quantity must be defined as acceptable quality. More generally, two items classified as good might have quite different strengths or other attributes. Intuitively, it seems reasonable that some "credit" should be provided for exceptionally good items in a sample. Sampling by variables was developed for application to continuously measurable quantities of this type. The procedure uses measured values of an attribute in a sample to determine the overall acceptability of a batch or lot. Sampling by variables has the advantage of using more information from tests since it is based on actual measured values rather than a simple classification. As a result, acceptance sampling by variables can be more efficient than sampling by attributes in the sense that fewer samples are required to obtain a desired level of quality control.

In applying sampling by variables, an acceptable lot quality can be defined with respect to an upper limit U, a lower limit L, or both. With these boundary conditions, an acceptable quality level can be defined as a maximum allowable fraction of defective items, M. In Figure 17-2, the probability distribution of item attribute x is illustrated. With an upper limit U, the fraction of defective items is equal to the area under the distribution function to the right of U (so that $x \geqslant U$). This fraction of defective items would be compared to the allowable fraction M to determine the acceptability of a lot. With both a lower and an upper limit on acceptable quality, the fraction defective would be the fraction of items greater than the upper limit or less than the lower limit. Alternatively, the limits could be imposed upon the acceptable average level of the variable.

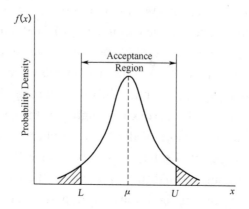

Figure 17-2   Variable probability distributions and acceptance regions

In sampling by variables, the fraction of defective items is estimated by using measured values from a sample of items. As with sampling by attributes, the procedure assumes a random sample of a given size is obtained from a lot or batch. In the application of sampling by variables plans, the measured characteristic is virtually always assumed to be normally distributed as illustrated in Figure 17-2. The normal distribution is likely to be a reasonably good assumption for many measured characteristics such as material density or degree of soil compaction. The Central Limit Theorem provides a general support for the assumption: if the source of variations is a large number of small and independent random effects, then the resulting distribution of values will approximate the normal distribution. If the distribution of measured values is not likely to be approximately normal, then sampling by attributes should be adopted. Deviations from normal distributions may appear as skewed or non-symmetric distributions, or as distributions with fixed upper and lower limits.

The fraction of defective items in a sample or the chance that the population average has different values is estimated from two statistics obtained from the sample: the sample mean and standard deviation. Mathematically, let n be the number of items in the sample and x_i, $i = 1, 2, 3, \cdots, n$, be the measured values of the variable characteristic x. Then an estimate of the overall population mean μ is the sample mean:

$$\mu = \overline{x} = \frac{1}{n}\sum_{i=1}^{n}x_i \tag{17-6}$$

An estimate of the population standard deviation is s, the square root of the sample variance statistic:

$$\sigma^2 \approx s^2 = \frac{1}{n-1}\sum_{i=1}^{n}(x_i - \overline{x})^2 = \frac{1}{n-1}\left(\sum_{i=1}^{n}x_i^2 - n\overline{x}^2\right) \tag{17-7}$$

Based on these two estimated parameters and the desired limits, the various fractions of interest for the population can be calculated.

The probability that the average value of a population is greater than a particular lower limit is calculated from the test statistic:

$$t_L = \frac{\overline{x}-L}{s/\sqrt{n}} = \frac{(\overline{x}-L)\sqrt{n}}{s} \tag{17-8}$$

which is t-distributed with $(n-1)$ degrees of freedom. If the population standard deviation is known in advance, then this known value is substituted for the estimates and the resulting test statistic would be normally distributed. The t-distribution is similar in appearance to a standard normal distribution, although the spread or variability in the function decreases as the degrees of freedom parameter increases. As the number of degrees of freedom becomes very large, the t-distribution coincides with the normal distribution.

With an upper limit, the calculations are similar, and the probability that the average value of a population is less than a particular upper limit can be calculated from the test statistic:

$$t_U = \frac{U-\overline{x}}{s/\sqrt{n}} = \frac{(U-\overline{x})\sqrt{n}}{s} \tag{17-9}$$

With both upper and lower limits, the sum of the probabilities of being above the upper limit or below the lower limit can be calculated.

The calculations to estimate the fraction of items above an upper limit or below a lower limit are very similar to those for the population average. The only difference is that the square root of the number of samples does not appear in the test statistic formulas:

$$t_{AL} = \frac{\overline{x}-L}{s} \text{ and } t_{AU} = \frac{U-\overline{x}}{s} \tag{17-10}$$

where t_{AL} is the test statistic for all items with a lower limit and t_{AU} is the test statistic for all items with a upper limit. For example, the test statistic for items above an upper limit of 5.5 with = 4.0, $s = 3.0$, and $n = 5$ is $t_{AU} = (8.5-5.0)/3.0 = 1.5$ with $n-1 = 4$ degrees of freedom.

Instead of using sampling plans that specify an allowable fraction of defective items, it saves computations to simply write specifications in terms of the allowable test statistic values themselves. This procedure is equivalent to requiring that the sample average be at least a pre-specified number of standard deviations away from an upper or lower limit. For example, with $\bar{x} = 4.0$, $U = 8.5$, $s = 3.0$ and $n = 41$, the sample mean is only about $(8.5 - 4.0)/3.0 = 1.5$ standard deviations away from the upper limit.

To summarize, the application of sampling by variables requires the specification of a sample size, the relevant upper or limits, and either the allowable fraction of items falling outside the designated limits or the allowable probability that the population average falls outside the designated limit. Random samples are drawn from a pre-defined population and tested to obtained measured values of a variable attribute. From these measurements, the sample mean, standard deviation, and quality control test statistic are calculated. Finally, the test statistic is compared to the allowable trigger level and the lot is either accepted or rejected. It is also possible to apply sequential sampling in this procedure, so that a batch may be subjected to additional sampling and testing to further refine the test statistic values.

With sampling by variables, it is notable that a producer of material or work can adopt two general strategies for meeting the required specifications. First, a producer may insure that the average quality level is quite high, even if the variability among items is high. This strategy is illustrated in Figure 17-3 as a "high quality average" strategy. Second, a producer may meet a desired quality target by reducing the variability within each batch. In Figure 17-3, this is labeled the "low variability" strategy. In either case, a producer should maintain high standards to avoid rejection of a batch.

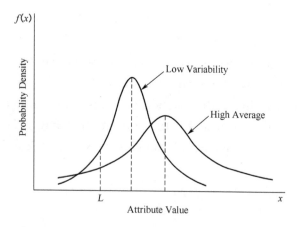

Figure 17-3　Testing for defective component strengths

Integrated Skills: Calculation

1. Suppose that a sampling-by-attributes plan is specified in which ten samples are taken at random from a large lot ($N = 100$) and at most one sample item is allowed to be defective for the lot to be acceptable.

1) If the actual percentage defective is five percent, what is the probability of lot acceptance? (Note: you may use relevant approximations in this calculation.)

2) What is the consumer's risk if an acceptable quality level is fifteen percent defective and the actual fraction defective is five percent?

3) What is the producer's risk with this sampling plan and an eight percent defective percentage?

2. The yield stress of a random sample of 25 pieces of steel was measured, yielding a mean of 52,800 psi. and an estimated standard deviation of $s = 4,600$ psi.

1) What is the probability that the population mean is less than 50,000 psi?

2) What is the estimated fraction of pieces with yield strength less than 50,000 psi?

3) Is this sampling procedure sampling-by-attributes or sampling-by-variable?

3. In a random sample of 40 blocks chosen from a production line, the mean length was 10.63 inches and the estimated standard deviation was 0.4 inch. Between what lengths can it be said that 98% of block lengths will lie?

Unit 18 Environment Engineering and Aesthetics in Civil Engineering

Lead-in

1. What factors can cause air pollution?
2. How can civil engineering improve our environment?

Vocabulary Warm-up: Matching

_____1. urbanization	a. 建筑声学
_____2. adverse effect	b. 园林建筑学
_____3. acute effect	c. 建筑气候学
_____4. greenhouse effect	d. 急性效应（作用）
_____5. architectural acoustics	e. 反作用
_____6. construction climatology	f. 热力学
_____7. landscape architecture	g. 都市化,文雅化
_____8. thermodynamics	h. 温室效应

Text A   Air Pollution and Effect of Noise

Quantitative discussions of air pollution are hampered by the lack of a clear definition for "clean air". Most scientists assume "air" to be a mixture of the gases such as Nitrogen, Oxygen, Argon, Carbon dioxide, Neon, Helium, Methane, Krypton, Nitrous oxide, Hydrogen, Xenon, Nitrogen dioxide and Ozone. If this is clean air, then any other *constituent* of air can be called a pollutant. However, one never finds such "clean air" in nature. It may thus be more appropriate to define air pollutants as those substances which exist in such concentrations as to cause an unwanted effect. These pollutants can be natural or man-made and can be in the form of gases or *particulates*.

Air pollution may arise from acts of nature. There have been many times when people have been forced to seek shelter indoors during a severe *sandstorm*, when the wind-borne ash from *erupting volcanoes* encompassed large portions of the surface of the earth. Early in the 1950's, forest fire in some of the southeastern states blanketed an area of about 300,000 square miles. The smoke from the forest fires was so intense that air flights had to be canceled in many cities.

But acts of nature are often beyond the control of man. Of chief concern is the second and more pressing source of air pollution — the man-made pollutants.

While air pollution is not a new phenomenon, it is now apparent that it is one of our most rapidly growing environmental problems. What are the factors contributing to this rather *recent trend* toward deterioration of the air environment? There are three major underlying factors which serve to explain this condition.

The first factor is population growth. The upward trends in population growth in the Unites States, since World War Ⅱ, have indeed been impressive. More people mean more manufactured goods and services. This, in turn, lends to the second factor.

The second one is *expansion* in industry and technology. The growth of industrial activity, in the same period, has likewise been remarkable in terms of expansion of existing plant capacity, and the increase in number of new manufacturing establishments. In addition, there has been the introduction of a great number of new processes, methods and products. The nature of the airborne wastes from some of these new technologies was completely unknown until *adverse effects* on man and the environment suddenly became manifest. New industries and processes were introduced on a large scale within recent decades. In most cases, the raw materials and by-products waste initially were of unknown *toxicity*, and knowledge of the methods and procedures for abatement of resulting pollution problems aged far behind the technology of manufacture. The combination of increasing quantities of atmospheric *emissions*, including material of undefined character, compounded the growth and complexity of atmospheric pollution.

The third one is social changes. Two important social changes occurred during the same period, and served to accelerate the trend of *burgeoning* air pollution:

1. *Urbanization*

The unrelenting movement of people from rural sections into urban centers has led to the rapid evolution of cities into large *metropolitan* complexes. The result of this development is an ever increasing density of population and of industrial and commercial activity. Thus, the producers of the airborne pollutants now, more than ever before, reside in close proximity to the potential receptors.

2. The Improvement of Living Standards

The other social factor which has indirectly contributed to the intensification of air pollution over relatively recent years has been raising standard of living which has prevailed during this period. Few families today are without a car, television set, refrigeration, automatic washing machines, etc. The vast majority of these conveniences require electric power.

Modern society produces greater per capita solid refuse than ever before. Great use of paper, plastic and similar materials for single service containers, and for packaging food and numerous domestic and commercial products of everyday life is placing enormous demands on solid waste disposal facilities. *Open burning* and incinerators of all types and sizes are emitting air-contaminating combustion products of increasing quantities and chemical complexity.

Some of these pollutants, such as automobile gases, are discharged into the air at street levels. Others, such as smoke from chimneys of apartment houses or power plants where electricity is generated, enter the atmosphere at higher levels.

The amount of pollution in the cities is affected by atmospheric conditions. Some conditions reduce the pollution and others increase it. If winds are strong enough, they blow pollutants up and away, also rain and snow wash the air. But these natural forces can be slow and also infrequent. Pollution is also lessened by action of currents in the air. Because the surface of the earth is normally warmer than the air above the surface, air current is set up that rises into the higher atmosphere, carrying the pollutants with it. In this way the amount of pollution on the surface where people live is reduced. But sometimes, due to natural causes, the air above the earth's surface is warmer than the air at the surface. When this happens, the warm air remains in a layer above the cold air at the surface, and stops the normal flow of rising warm air. This is known as a temperature inversion. The air pollution and smog-forming substances become trapped between the two layers and hang over the city, often with serious effects on people's comfort, health and even life.

The combined impact of population growth, expansion in industry and technology and social changes operating in our contemporary society can be regarded as the compounding factors which have resulted in serious degradation of the urban air environment in relatively recent years. In certain metropolitan areas, this trend has already reached alarming proportions. In those areas, the rate of pollution very frequently exceeds the capacity of the atmosphere to purify itself by natural processes of dilution and dispersion. During these periods, severe air pollution occurs and is clearly manifested by eye irritation, reduced visibility and other adverse effects.

We have learned about some knowledge of the air pollution. Below, let's learn some relevant knowledge on health effects of noise.

Noise is commonly defined as unwanted sound. In recent years, noise pollution has become increasingly serious with large-scale use of motor *vehicle*s and production equipment. It varies from the characteristics, quantity, distribution and protection of noise sources, as well as time and place. In the living environment, sound level is about 30 dB (*decibel*) in the relatively quiet environment at night; it is about 80 dB during the day with the frequent vehicle. It can be as high as 90 dB on both sides of the street near the factory or in some areas. Noise level in the work environment is relatively high, such as in the textile, machinery and printing industry, and noise level in some sites is more than 90 dB, sometimes as high as $100 \sim 105$ dB. On some special sites, such as the site using pneumatic tools, testing motors and vibration tables, noise level may be as high as 120 dB. Noise is higher near the airport *aviation*: it is up to 130 dB.

Speaking of the impact of noise, it is a normal quiet environment when the noise level is $30 \sim 40$ dB; when it is more than 50 dB, sleep and rest will be disturbed. Due to lack of rest, fatigue cannot be eliminated; to a certain extent, the normal physiological functions will be affected; the talk will be interfered when noise level is above 70 dB, thus causing upset, lack of concentration and low working efficiency, and even accidents may occur; if a man works or lives in environment with noise level more than 90 dB in a long term, it will seriously affect the hearing and lead to other diseases.

The most immediate and acute health effect of excessive noise is impairment of hearing. Hearing damage includes those that are acute and chronic. Exposure to strong noise can cause them *drumming* in the ears; as long as it is not long that the ears are in strong noise, they will return to normal after leaving the noisy environment: it is known as the *auditory* adaptation. If they're exposed to strong noise for a long time, hearing loss is more obvious. It takes a few hours, or even a dozen to twenty hours to return to normal after leaving the noisy environment: it is known as auditory fatigue. This is caused by damage to some parts of the auditory system.

Sound pressure waves caused by vibration set the ear drum (tympanic membrane) in motion. This activates the three bones in the middle ear. Acute damage can occur to the ear drum, but this occurs only with very loud sudden noise. More serious is the chronic damage to the tiny hair cells in the inner ear. Prolonged exposure to noise of a certain frequency pattern can cause either temporary hearing loss, which disappears in a few days, or permanent loss. Much of the hearing loss in industry occurs in the middle range of frequencies. Unfortunately, speech frequencies are in the same area, and *speech perception* is thus hindered. Many older people, while still able to hear jet planes and *rumbling* trains quite well, complain that "everyone is whispering". They have experienced damage to certain hair cells which hinders the reception of sounds of a specific frequency.

Hearing loss occurs with age advancing even without environmental damage. It is difficult, therefore, to develop epidemiological date to show the loss due to excessive noise. Research has, however, shown that hearing loss due to noise is real and not imagined.

Another problem with noise is its effect on other bodily functions such as the cardiovascular system. It has been discovered that noise alters the rhythm of the heartbeat, makes the blood thicker, dilates blood vessels and makes focusing difficult. It is no wonder that excessive noise has been blamed for headaches and irritability. Noise is especially annoying to people who do close work, like watchmakers.

All of the above reactions are those which our ancestral caveman also experienced. Noise to him meant danger and his senses and nerves were "up", and it is questionable how much of our physical ill is due to this.

We also know that man can't adapt to noise, in the sense that his body functions no longer react a certain way to excessive noise. People do not, therefore, get "used to" noise in the physiological sense.

In addition to the noise problem, it might be appropriate to mention the potential problems of very high or very low frequency sound, out of our usual $20 \sim 20,000$ Hz (*Hertz*) hearing range. The health effects of these, if any, remain to be studied. Numerous cases comparing patients in noisy and quiet hospitals point to increased convalescent time when the hospital was noisy. This can be translated directly to a dollar figure. Recent count cases have been won by workers seeking damages for hearing loss suffered in the job. The Veterans Administration spends huge millions of dollars every year for care of patients with hearing disorders.

Other costs, such as sleeping pills, lost time in industry, and apartment sound proofing are difficult to quantify. It is even more difficult to measure the effect that noise has on the quality of life. How much is noise to blame for irate husbands and grumpy wives, for grouchy taxi drivers and surly clerks? Children reared in a noisy neighborhood must be taught to listen. They can't focus their auditory senses on one sound, such as the voice of a teacher.

The harm of urban environment noise can be prevented by proper control of noise sources, rational planning of the factory city, reasonable layout of streets and residential areas, an additional effective noise protection facilities and developing noise reduction system of traffic management.

Depending on the purpose, environmental noise standards can be divided into three types: the noise should be controlled at $75 \sim 90$ dB in order to protect the hearing, the noise should be controlled at $55 \sim 70$ dB in order to ensure the work and learning, and the noise should be controlled at $35 \sim 50$ dB in order to ensure the rest and sleep. It is ideal for low-value; high-value is not allowed to exceed the limit.

Noise is a real and dangerous form of environmental pollution. Since people cannot adapt to it physiologically, we are perhaps adapting physiologically instead. Noise can keep our senses "on edge" and prevent us from relaxing. Our mental powers must therefore control this insult to our bodies. Since noise, in the context of human evolution, is a very recent development, we have not yet adapted to it, and must thus be living on our *buffer* capacity. One wonders how plentiful this is.

Glossary

Complete the glossary using words in the text.

1. _____ 成分,构成部分,要素　2. _____ 微粒
3. _____ 沙暴,沙漠地带的暴风沙
4. _____ 爆发,喷发　5. _____ 火山
6. _____ 近代的趋势　7. _____ 扩大,膨胀
8. _____ 反作用　9. _____ 毒性
10. _____ 排放物

11. _____　迅速成长的,迅速发展的

12. _____　都市化,文雅化

13. _____　大都会的,大城市的

14. _____　露天焚烧

15. _____　交通工具,车辆

16. _____　分贝　　　　17. _____　航空,航空学

18. _____　连续有节奏的声音

19. _____　听觉的,听觉器官的　20. _____　言语感受

21. _____　隆隆声,辘辘声　22. _____　赫兹

23. _____　起缓冲作用的人或物

Exercises

1. Fill in the blanks with the information given in the text.

1) It may thus be more _____ to define air pollutants as those substances which exist in such concentrations as to cause an unwanted effect.

2) The upward trends in population growth in the Unites States, since World War II, have indeed been _____ .

3) In most cases, the raw materials and by-products waste initially were of unknown _____ , and knowledge of the methods and procedures for _____ of resulting pollution problems aged far behind the technology of manufacture.

4) During these periods, severe air pollution occurs and is clearly manifested by eye _____ , reduced visibility and other adverse effects.

5) On some special sites, such as the site using pneumatic tools, testing motors and _____ tables, noise level may be as high as 120 dB. Noise is higher near the airport _____ : it is up to 130 dB.

6) Prolonged exposure to noise of a certain frequency pattern can cause either temporary hearing loss, which disappears in a few days, or _____ loss.

7) It has been discovered that noise alters the _____ of the heartbeat, makes the blood thicker, dilates blood vessels and makes focusing difficult.

8) How much is noise to blame for irate husbands and grumpy wives, for _____ taxi drivers and surly clerks?

2. Write down the words with the following prefix or suffix.

anti-　　_____

bi-/ di-　_____

bio-　　_____

-oxid(e)-　_____

-proof　_____

3. Translate the following passages from English into Chinese.

In the context of air pollution control, gaseous pollutants include substances that are gases at normal temperature and pressure as well as vapors of substance that are liquid or solid at normal temperature and pressure. Among the gaseous pollutants of great importance in terms of present knowledge are carbon monoxide, hydrocarbons, hydrogen sulfide, nitrogen oxides, ozone and other oxidants and sulfur oxides. Pollution emissions from industrial processes reflect the ingenuity of modern industrial technology. Thus, early imaginable form of pollutant is emitted in some quantity by some industrial operate.

The difference between the noise scale in dB and actual noise levels must be noted. Doubling the intensity by two identical sources of noise will increase the noise level by approximately 3 dB. In terms of hearing, about a 10 dB increase is necessary to make a sound seem twice as loud to a listener. If the noise level is too high when compared with a standard or criterion, noise abatement measures must be implemented. Such measures work best if they are aimed at the source of the noise. There are basically four different ways in which noise levels can be controlled or reducd: ①Protect the person exposed to the noise. ②Intercept the noise by blocking its path. ③Increase the distance from the source. ④Reduce the sound intensity at the source.

Text B   Aesthetics in Bridge Design

The appearance of the bridge is dominated by the shapes and sizes of the elements controlled by the engineer — the structural elements themselves, not by details, color, or surfaces. The Golden Gate Bridge owes its appeal to the graceful shape of its towers and cables, not to its reddish color. If the towers were ugly, painting them red would not make them attractive.

Periodically the suggestions is made that the engineers should delegate responsibility for the appearance of bridge to architects or other visually trained professionals such as urban designers, landscape architects, or sculptors, the notion being that this would result in better-looking structures. This is rarely successful. The usual result is an overly elaborate and overly costly structure, without a significant improvement in appearance. There are two reasons for this. The first is that the visual professional doesn't understand the special limitations and problems of bridge or, even more important, the *aesthetic* 于 opportunities of modern engineering materials for structural types. So, the designer falls back on the familiar conventions of architecture or sculpture and makes something that looks like a building or a piece of sculpture, not a bridge.

The second reason is that the delegation can never be complete. In the United States, engineers have the professional and legal responsibility for bridge. The organizations that commission, finance, and build bridges are usually run by engineers. None of these people are willing or legally able to delegate complete responsibility for a bridge to a nonengineer, no matter how talented. This usually means that the "important" (structural) aspects of the bridge are reserved for the engineers, while the "window dressing" (surface details, color) is left to the visual professional. Although the details are important parts of a structure's aesthetic impact, it is the structure itself — its spans, proportions, and major elements — that has the largest role in creating its effect. If the engineer does a poor job with these major elements, no amount of architectural add-ons will compensate.

Engineers must control the structural elements — that is their job. However, for the bridge to be a complete aesthetic success, the details, colors, and surfaces have to be in tune with the structural concept. Therefore, the engineers should control those as well. That said, visual professionals who are willing to take the time to understand the special nature of bridge can offer a positive contribution to engineers and be acting as aesthetic advisors or critics. Their role is comparable to that of other specialists, such as geologists or hydrologists, whom the engineer may call for advice. Engineers have accepted a responsibility to society for bridge design. For that reason, no engineer would knowingly build a bridge that is unsafe. For the same reason, no engineer should knowingly build a bridge that is ugly.

Aesthetic ability is not some mysterious quality bestowed by fate on a fortunate few. Though many engineers are not well prepared by their education or experience for the visual aspects of their responsibilities, they can learn what makes bridges attractive and they can develop their abilities to make their own bridges attractive.

Many tools are available to help engineers improve their abilities in the aesthetic aspect of engineering. Many books present a series of guidelines and the principles that underlie them. It is a useful base from which to consider the appearance of structures. But the guideline is only a beginning. Engineers should form their own opinions about what looks good, and work to develop their own aesthetic abilities.

Bridges must safely carry their loads for a long enough time to repay the investment. This basic fact imposes a hierarchy on design making:

1) Performance

Structural capacity, safety, durability, and maintainability

2) Cost

Construction and maintenance

3) Appearance

That says, this statement often leads to the assumption that three criteria are necessarily in conflict, that one must be sacrificed for the others. This is not true. For example, many people think improved appearance will automatically add cost. They believe good appearance derives from add-ons, like an unusual color, special materials such as stone or brick, or ornamental features. In fact, the greatest aesthetic impact is made by the structural members themselves — the cables, girders, and piers. These things have to be there anyway. If they are well shaped, the bridge will be attractive, without necessarily adding cost.

Many of the greatest bridge engineers were able to create that are beautiful, structurally sound, and less expensive than any alternative, all at the same. In decisions about appearance, as in decisions about strength and durability, the challenge is to achieve an improvement without an increase in cost. It can be done.

Now let's discuss the next item on the specific requirements for bridge aesthetics.

1) Conformity with Environment

A bridge must conform with its surroundings and environment. The setting or surroundings greatly affect its appearance. Those that are exposed to the river view are seen and appreciated more than others that are partly hidden by adjoining objects.

Generally, a structure must be in harmony with its environment and not appear as an intrusion. To secure harmony between the structure and its environment means the merging of its general outlines with those of the landscape. It should be remembered that the bridge will likely be seen from various angles, and that each viewpoint will cause its own individual impression.

The rule, generally, is to make the bridge more striking than its surroundings, so the eye will naturally be attracted to it. A good method is to make separate photographs of the site and the design to the same scale, and after placing the proposed bridge in the landscape view, to rephotograph the combination. Features of the design that fail to conform to the surroundings will then appear, and changes can be made until it is of satisfactory.

2) Economic Use of Material

Economic use of material is another standard of excellence. Beauty exists in every structure which is designed according to the principles of economy, with the greatest simplicity, the fewest members, and the most pleasing outline consistent with construction.

3) Exhibition of Purpose and Construction

The purpose of the bridge should be plainly evident, and generally the construction should be revealed. Expressiveness, to many people, is the main source of beauty. Strength and boldness should predominate.

4) Pleasing Outline and Proportions

A bridge is beautiful if its primary form or outline and its relative proportions are well and properly chosen. The proportions must satisfy the eye and the aesthetic feeling, and have optical harmony. Good general lines are necessary as a basis; a consistent scale or proportion of parts should follow. A form that admits of no explanation cannot be beautiful. It must have and show some purposes in its general relation. Each part of a bridge structure should be treated in such a way that its function therein is apparent and emphasized according to its importance. The different kinds of materials used in structures call for different treatment and varying aesthetic standards.

The underlying thought connecting these precepts is that the structure must fit the work it is to do, that it should express the truth, and that falsities are erroneous and outside the realm of rational aesthetics.

The achievement of good general lines is best attained by a study of the profile of the structure. The feature of a bridge is so pleasing to the eyes of all observers, either cultivated or ignorant, as perfect symmetry in the layout of spans; consequently is involved thereby. Unfortunately, the connections are not always favorable to the perfect symmetry of design. In such cases it becomes necessary to do the best one can with the unfavorable conditions, and to make the structure aesthetically pleasing, if not symmetrical.

5) Appropriate but Limited Use of Ornament

Mere ornamentation generally affronts the sense of harmony and fitness. In many bridges what would otherwise be a pleasing outline is spoiled by the introduction of massive ornamental portals at the ends. It is not advisable to correct the hard, rigid outline of a span by the use of additional parts that falsely proclaim a different function for the members or confuse their action in the structure. Ornament is not architecture, and a bridge of beautiful outline may easily be spoiled with an excessive amount of details. Superfluous decoration has a minifying effect.

6) Expressiveness

Expressiveness in a structure is, for many people, the greatest element of beauty, and the visible parts and lines show their purpose. A bridge must be a truthful creation and its appearance should show its purpose. As strength is a chief requisite, the element should be emphasized.

7) Symmetry and Simplicity

One of the most important factors of good design is symmetry. If conditions permit, the general outline on each side of the center should be the same, or nearly so. The absence of symmetry should be permitted only when the ground contour or other conditions are such as to make a symmetrical arrangement impossible.

Simplicity is important though not so essential as symmetry. Too many members are confusing, and fewer large pieces are preferable. Simplicity means primarily a truth-telling structure, having no subterfuges about the line of stress, no covering of concrete or steel with a stone facing.

Text C Solar Energy in Buildings and Ground-Source Heat Pump Air Condition System

The energy that the earth receives from the sun is called solar energy. The sun has provided, either directly or indirectly, almost all other sources of energy for the earth since its beginning. As we all know, solar energy is enormous. Each day the sun deposits an average of 1,400 Btu per square foot of area to the United States, therefore, a home of 1,000 square foot receives approximately 511 million Btu per year. However, solar energy can vary from season to season, from 2,000 Btu to as low as 500 Btu per square foot per day during a period from June to December. Factors such as cloud cover and geographical location affect the total amount of solar energy received. The solar energy is inexhaustible, clean and the renewable energy will be future primary energy. How to use the solar energy has already been the hot spot which in various countries scientific and technical workers are studying. The first application of developing solar energy in the construction lies in the field of solar energy heating, hot water supply, solar cookers, etc.

As known to all, the availability and cost of energy has become dominant factors in society today. Obviously, solving the "energy crisis" makes good sense. Many schemes have been proposed for conserving present energy resources and for developing new ones. It is always possible to use less energy in any process. Therefore, energy engineer emerged and developed. The first goal of energy engineer is to determine the methods by which energy utilization is reduced but the output remains the same or even increases. The second goal is to determine which methods of using less energy are cost-effective.

Meanwhile, looking for ideal energy sources is also very important to solve energy crisis. The recipe for an ideal energy source calls for one that is unlimited in supply, widely available, and inexpensive; it should not add to the earth's total heat burden or produce chemical air and water pollutants. Solar energy fulfills all of these criteria. Solar energy does not add excess heat to that which must be radiated from the earth. On a global basis, utilization of only a small fraction of solar energy reaching the earth could provide for all energy needs.

Solar energy is the most popular of the many alternate energy sources being discussed today. Volumes have been written and much said about solar energy. However, many people don't understand its possibilities and limitations.

Each year approximately 20 percent of the nation's energy use is for heating and cooling homes. Solar energy is an alternative that can reduce our dependence on scarce fossil fuels with their ever-increasing price.

The use of solar energy in construction is not only heating and hot substandard water supply, but also the solar energy refrigeration. Solar energy is unlimited in supply, but its exploitation and utilization are limited owing to the limitation of technology and conditions. Solar energy utilization needs an enormous amount of land, and there are economic and environmental problems related to the use of even a fraction of this amount land for solar energy collection. First, this energy from the sun is diffuse, i.e., it is spread out very thinly. It must therefore be collected by some means because only a small amount of it arrives in one place. Second, the energy received is intermittent because the sun shines only during the day and it is often obscured by clouds. Thus, the energy received must be stored until it is needed.

Solar thermal system can have energy storage, or operate without any storage. The storage is most useful when solar radiation availability differs in time from the heat demand. However, many systems can operate without thermal storage. In solar cooling systems, for instance, the availability of higher levels of solar radiation occurs usually at the same time when the cooling loads higher. Thus, the system may operate with a good efficiency without thermal storage. The same may happen in solar cogeneration systems, or in systems that provide an amount of useful energy which is much lower than the total load. The use of storage also leads to thermal losses. The combination of these effects is, in principle, positive, and solar fraction may be higher than without storage. In order to decide on the advantage of using thermal storage, both possibilities should be evaluated (storage and no storage). An ideal thermal storage would be able to receive heat and not increase its temperature.

Solar energy can be transformed either to electricity or to heat allowing, in theory, any refrigeration technology to be driven by it. Still, several constraints concerning both the quality and the quantity of solar energy limit the potential of solar driven or even solar assisted refrigeration technologies. Keeping in mind the characteristics of solar electricity and thermo-mechanical systems, and also for reasons of brevity, they will be discussed only to a limited extend.

The electrically driven systems are characterized by the limited useful power that can be achieved by solar means, and also by their fairly high initial cost. At the present time, solar refrigerant have mainly three kinds. The first is the solar sorption system, and the second is solar adsorption refrigeration system, and then finally is the ejector refrigeration system.

In a word, there is still a high research demand for the utilization of solar energy in air conditioning systems, and research mainly focused on solar collectors and, more intensely on the sorption cooling technologies. Solar assisted refrigeration appears to be a promising alternative to the conventional electrical driven air conditioning units also from an environmental point of view, since it results in decreased CO_2 emissions, and the elimination of CFCs and HCFCs in the case of the prevailing solar cooling technologies. The latter is expected to influence the developments in the air conditioning sector significantly.

We have introduced the application of solar energy in building. Now we come to understand the ground-source heat pump air condition system.

A heat pump is a refrigeration system whose purpose is to remove heat from one and supply another. In a conventional refrigerant, cooling is the only desired effect, while in a heat pump, either heating or cooling may be the desired effect. In most residential and commercial heat pump applications, heat is taken from the cooling outside air and "pumped" into the room for heating or heat is removed from the room and "pumped" to the warmer outside air for cooling. In the cooling mode the heat pump operates in heat pumps which use either groundwater or ice for either a source or sink, and there are many areas which could use these to advantage. At present, the vast majority of units are air to air heat pumps, river or lake to air heat pumps and soil to air heat pumps.

The working principle of a heat pump is similar to the refrigerant, the process of vapor compression heat pump is as follows. The working substance absorbs a quantity of heat from heat source (outdoor air or water) and gives it to the indoor one, which is the cold body (surrounding medium), and improves the door temperature.

In the 1950s, many heat pumps were installed in residences as the primary heating source. However, within a few years there are so many renewable energy sources used in heat pump, such as ground source heat pump, ground water source heat pump, surface water heat pump.

Ground source heat pump systems are "down to earth" heating, cooling and hot water systems designed to tap the earth's stored energy. The electrically powered unit pulls heat from the earth to warm your home during the winter. And, with the flick of a switch, the same system pulls heat from your home in the summer and transfers it to the earth. In addition, water heat from the system can be used to heat water at a very low cost. Ground source heat pump systems provide optimum performance, dependable service, high efficiency and much more.

The down to earth option is an electrically powered system that capitalizes in the earth's moderate temperature. Water or an antifreeze solution is circulated through plastic pipes buried beneath the earth's surface. In the winter, this solution collects heat from the earth, carries it through the system and into your home. The ground source heat pump system provides you with constant warmth and comfort during the cold winter months.

In summer, the ground source heat pump system reverses itself to cool your home. Operating like the refrigerator in your kitchen, this system pulls heat from your home. The heat is then carried by the fluid in the pipes through the system and transferred to earth. A ground source heat pump system guarantees you constant cool relief during hot summer months. And, as an added benefit, you can utilize the waste heat from your home in the summer to heat water at substantial savings.

The down to earth energy option is the smart, efficient alternative to fossil fuels because the system works with Mother Earth by moving heat rather than making heat. By operating on the simple premise that heat always moves from hot to cold, ground source heat pump systems can efficiently heat and cool your home by operating around the earth's moderate temperature.

Ground source heat pump systems are neatly-wrapped, energy efficient packages to help you with your household expenses. They offer you an economical method for managing your utility bills.

With a ground source heat pump system, you can escape the headache of balancing numerous utility bills. Each month you will receive one bill from your home heating, cooling and hot water cost.

No matter how hot or cold it is above ground, the temperature underground stays comfortable year round. But you don't have to live underground to be comfortable. In the winter, a machine called a ground source heat pump system can take heat from the ground and put it into your home. And in the summer it does the opposite. The ground source heat pump system pulls heat from your home and put it in the ground. The ground source heat pump system saves money and energy. It is also quiet and small enough to fit your house.

What can you expect from a ground source heat pump system?

1. Saving

Ground source heat pump system can put your home heating costs as much as 60 percent in the winter, reduce your home cooling costs up to 25 percent in the summer, and provide hot water for normal household use.

2. Conservation

Ground source heat pump systems work with the environment by utilizing the earth's moderate temperature to heat your home in winter and cool your home in summer.

3. Cleanliness

Ground source heat pump systems, a clean alternative for heating and cooling, help preserve nature. A ground source heat pump system minimizes the present environmental problems like acid rain, air pollution or the destruction of the ozone layer.

4. Durability

Ground source heat pump systems last longer than conventional systems because they are self-contained systems housed entirely within your home and underground. These systems must endure.

5. Low Maintenance

Ground source heat pump systems are not prone to breakdowns after frequent use like some conventional systems. Similar in concept to a refrigerator, a ground source heat pump system has few moving parts subject to breakdown. The heat exchanger in a ground source heat pump system, which transfers heat to and from the earth, is made of engineered plastic. It can operate efficiently fifty years after installation.

6. Low Noise

Aside from cool relief and warm comfort, a ground source heat pump system will offer no additional clues to its hard work. Ground source heat pump systems have no noisy, rattling units to disturb your family or neighbors. Without these loud reminders, you may even forget your ground source heat pump system is there.

Integrated Skills: Presenting a Scientific Essay

By means of the field research, Internet search or the review of special literatures, you should search for some information on environment engineering or aesthetics in civil engineering that you are interested in. Then you should write a scientific essay based on one of the above aspects, and finally present your essay in class and exchange your ideas with others.

第二部分　专业英语实践篇

Part Ⅱ　*Special English Practice*

Unit 19　Skills of Special Communication in English

19.1　概述

随着我国对外开放的不断深入、国际交流的日益增多,土木工程领域的国际化趋势日趋明显,已深入行业的各个层面。在专业人才培养方面,各级各类的中外合作办学广泛开展;在专业学术交流方面,每年世界范围内都会召开大量的国际学术会议和工程技术交流会;在工作实践层面,跨国工程公司、国际工程承包、援外工程建设、国际工程咨询与合作等屡见不鲜。

在此背景之下,对于我国土木工程专业的学生而言,他们不仅需要掌握扎实的专业知识和技能以及较好的英语读写技能,亦需要具备一定的运用英语进行专业交流的能力。

实际上,用英语进行专业交流的场合、对象及内容等千差万别,但根据语言交际场合和对象的不同,大体可分为演讲式和问答式两大类。其中,用英语进行演讲式专业交流主要用于各类学术会议、学术报告、专业介绍与演讲等,其语言交际特征为"一对多",即"演讲者对听众";用英语进行问答式专业交流主要用于专业问题的讨论以及国际会议的问答环节等,其语言交际特征通常为"一对一",即"提问者与回答者口语交互"。

本单元着重介绍以上两类口语交流时应注意并掌握的技巧,在其他背景下用英语进行专业交流时亦可灵活借鉴这些技巧。

19.2　用英语进行演讲式专业交流的技巧

19.2.1　撰写演讲稿

在用英语进行专业演讲时,演讲者不应在台上全文朗读自己的论文或研究报告等,这样会使听众觉得演讲枯燥无味,而是应当在演讲之前精心撰写一份演讲稿。

在撰写演讲稿之前,演讲者应首先了解听众的身份、专业知识水平、专业兴趣、听讲需求等,这样可使自己的演讲内容更具针对性。写演讲稿最重要的原则之一就是根据拟定的演讲时间,合理地安排演讲内容,在选定演讲内容时,做到精讲重点内容,弱化甚至是省略非重点内容。

一个出色的演讲应当是语言精练而富于逻辑,因此通常由"演讲内容简介—演讲内容展开—演讲内容总结"三部分依次组成,这一结构安排一般比较容易使听众把握演讲重点。

　　演讲的开场简介部分最好能用幽默性的语言介绍事实或引出问题,以提高听众的兴趣。例如:"Today I am going to describe one of the most important new developments in...".在撰写演讲稿时应注意,演讲的第一分钟不要安排重要的专业内容,因为在演讲之初听众的注意力往往不易集中,此时最好安排具有幽默性的开场白。

　　在撰写演讲稿时,对于关键性的专业内容不要害怕重复,通常重要内容在演讲中只有重复至少两次以上才能被听众接受。当然,需要注意重复的技巧,利用幻灯片或图表来实现重复,既可以取得很好的强化效果,又能避免机械重复。

　　在演讲中每一部分的结尾,最好采用过渡性语句来衔接下一部分内容。例如:在简介某系统的一般特征后,用"Now let us consider some of the more important aspects of this system."引出下一部分将要重点介绍的内容。

　　最后,在撰写演讲稿时应注意,演讲最后一部分结论的句数不应过多,且应尽可能使用短句,以使演讲收场简明有力。

19.2.2　使用过渡词

　　合理使用过渡词和过渡短语可以使得演讲流畅,且有助于听众把握演讲结构。但根据句式和上下文背景的不同,演讲中所用的过渡词也有所不同。现将专业演讲中常用的过渡词和过渡短语归纳如下:

　　1. 用于引起、承接、补充或简介观点

to begin with; let me start by mentioning; first; second; third; finally; last; last but not least; firstly; lastly; then; next; with the above; and; also; another way; a second method; as well as; besides; equally important; further; furthermore; additionally; in addition; moreover; one further remark; remember; subsequently...

　　2. 用于详述(包括强调、举例、具体化)观点

as an application; as an example; basically; for example; for instance; generally; in general; in particular; namely; often; specifically; that is to say...

　　3. 用于比较观点

a similar analysis shows; in like manner; in a similar manner; in this connection; similarly; likewise; in practice; but; however; in comparison; in contrast; after all; alternately; instead; in the later case; nevertheless; nonetheless; notwithstanding; on the contrary; on the other hand; otherwise; still; whereas; while this may be true; yet...

　　4. 用于重复观点

again; as before; a second time; as has been stated; as I have said; in other words; in review; once more; to reiterate...

　　5. 用于表达方位

adjacent to; beyond; here; near; on the other side; opposite to; the former; next to; in front of; in the distance; at the top...

　　6. 用于表达目的

for this purpose; for this reason; in order to; to this end; toward this objective; with this goal...

7. 用于表达肯定或疑惑

admittedly；certainly；conclusively；fortunately；hopefully；indeed；in any event；in fact；in reality；obviously；of course；ordinarily；perhaps；to be sure；truly；undeniably；without any question...

8. 用于表达时间

after an hour；a day；a week；afterward；a little later；as will be seen；at length；at present；at this point；earlier；finally；immediately；in the meantime；in those days；in what follows；meanwhile；presently；recently；so far；then；today；ultimately；will now be developed；will now be illustrated...

9. 用于表达结果

accordingly；a review of the literature reveals that；as a result；consequently；hence；in consequence；in spite of；in view of these considerations；so far；the foregoing discussion illustrates；therefore；thus；to meet this additional difficulty；wherefore...

10. 用于总结观点

in effect；in essence；in other words；in short；in summary；let us briefly review the steps in the analysis；several remarks need to be made at this point；summarizing；the foregoing discussion illustrates；to sum up ...

19.2.3 宣读演讲稿的技巧

在演讲前准备好演讲稿是必要的,但在演讲现场,既可采用脱稿演讲,也可采用宣读演讲稿的方式。若决定宣读演讲稿,则应注意以下读稿技巧:

宣读演讲稿时应做到声音洪亮清晰,面向听众。不可低头喃喃自语,也不要忽视坐在最后一排的听众,可以询问他们"Can you hear me in the back of the room?"。但是,也不要对着麦克风声音过大,这样会使听众觉得是噪音。

在宣读演讲稿时不要忽视听众,尽可能多地用眼神与他们交流。但必须避免因眼观听众而找不到演讲稿中位置的现象,为此,在准备演讲稿时应使用大字体和多倍(两倍或三倍)行距来避免这一问题。

在宣读演讲稿时应避免不停顿的快速朗读,最好读得慢一点,且在撰写演讲稿时就应避免使用在书面语中很优美但在口语表达中很拗口的词汇和短语。

此外,无论演讲现场的气氛多么紧张严肃,演讲者都应不时地用微笑去感染听众。

19.2.4 演讲的常用例句

语法的完美和词汇的优美对于英语书面表达是十分重要的,但对于口语演讲却并不是很重要。专业性演讲中的每一句话都应简明扼要、直截了当,不应模糊难懂或是到演讲结束时方才让听众理解其中含义。为此,应做到每句话只表达一个中心意思;尽可能使用短句,避免使用长句和复杂句。此外,与专业论文、研究报告等书面语不同的是,口头演讲中进行确定性陈述时通常用主动语态而不是被动语态;在表达演讲者自己的观点和经验时,可直接采用第一人称代词。

除上述演讲的句式特点外,专业性演讲中通常还会使用一些固定句式,现将这些常用句式归纳举例如下：

1. 演讲开场时的例句

1) Thank you, Mr. Chairman. Good morning, distinguished members of the panel. It is a pleasure for me to be here. Today I would like to briefly describe to you, and hopefully it will actually be brief, our work over the past few years with X whose name you have heard before. It has lately become of general interest in our field and has attracted special interest of our group.

2) Thank you, Professor Brock, for your very kind introduction. I am particularly honored to have been invited to speak at this conference. Now I am going to present the work that I have been interested in for a long time. I'll lay my stress on the following three aspects. The first...

3) Mr. Chairman, Mr. Co-chairman, fellow colleagues. First of all, let me thank Prof. Wang for his warm introduction in which he mentioned the research work X that I have been carrying on in recent years. Now I would like to make a brief introduction to this study. I hope I can hear comments from every one of you.

4) Good morning, ladies and gentlemen. My name is... I'm a professor with X University. I'm going to discuss with you today... I shall express my own opinion on this, hoping that it will have something to offer you.

5) Good morning. It's an honor to have the opportunity to address such a distinguished audience. Let me start by saying just a few words about my own background. I know I've met some of you, but just for the benefit of those I haven't, my name is... I started out in...

6) Mr. Chairman, ladies and gentlemen. I'm very glad to present my paper today. The title of my presentation is... My presentation will focus on studies from my own laboratory at A university in collaboration with B, as well as studies done in collaboration with Prof. C of D University.

7) Good morning, everyone. I am very delighted to have the opportunity to report my research and to discuss the present status of X on such an occasion. First, I'd like to tell you briefly the background of my study and the findings of my research. Then make a brief comment on it.

8) Good afternoon. Today I'd like to talk about... (to present the recent...; to explain our position on...; to brief you on...; inform you about...; describe briefly...) I shall say what I think, in the hope that it might contain something of interest to you.

9) Friends, it is an honor to have this opportunity to be with you and a pleasure to renew an old friendship with some of you whom I have not met for three years since our last meeting in Paris.

10) Thank you, Professor Smith and Dr. Dowton, for giving me this opportunity to address you today. My presentation will be made from the perspective of a basic scientist

who is intent upon translating basic laboratory findings to meaningful avenues for application. Now, please look at the first slide. Could someone turn off the lights, please? (Lights off, please.) This is a slide showing you what made for... Now look at the second slide.... Could we have the lights on here, please? (Lights on, please.)

2. 引出演讲主题、简介演讲内容时的例句

1) Today I would like to discuss ... and relate this subject to the activities of ... I'd be glad to answer any questions at the end of my talk.

2) I would like to concentrate my discussion on the research and development of ... Please interrupt me if there's something which needs clarifying. Otherwise, there'll be time for discussion at the end.

3) The purpose of this talk is to update you on ... I shall only take 10 minutes of your time. I've divided my presentation into four parts/sections. They are...

4) This talk is designed to put you in the picture about ... I promise to be brief. If you have any questions, please feel free to interrupt. The subject can be looked at under the following headings...

5) My today's presentation will give you the background to... Hopefully, it can act as a springboard for our discussion. My talk should only last five minutes. I'll be grateful if you hold your questions to the end of my talk.

6) We can break this area down into the following fields: Firstly/First of all ... Secondly/Then/Next ... Thirdly/And then we come to ... Finally/Lastly/Last of all ...

7) In discussing these issues, I shall begin with a brief outline of... Next, I shall describe how X functions as... I then shall search for a solution in Y in its broadest sense. Finally, I shall examine at some length the experience of... and arrive at some conclusions which may be of use to those who...

8) The experts are asking numerous questions about this problem: How...; What...; To what extent...; Is there a common factor shared by...; Is this common factor unique? How...? These and other questions are widely discussed in the academic, official and popular literature and the opinions expressed are equally varied. My own assessment of reasons for the... is that...

9) Although the hour is indeed late, and I am sure you are no doubt nearing a saturation point for new information. I hope to discuss with you today what I believe is an exciting new area with considerable clinical impact.

10) I would like to divide my talk into four parts. The first point discusses... The second covers... The third concerns... Last but not least, I will talk about something relating to...

11) It's a great honor to be here and especially to listen to some of the seminars. The one this morning with Robert Johnson and Kansa Rode was truly brilliant, and I hope I can live up to that. What I want to talk about today is... I'm going to divide this

up into two parts. I'm going to talk a bit about democracy, a bit about technology and put them together and give you a bunch of case studies around the world.

12) I want in this presentation to say something about the process of modernization in its contemporary form. Modernization is a process, and it has, in European states, been going on since at least the sixteenth century, but now, at the end of the twentieth century, we can observe quite new dimensions of the process. For one thing, it has now "gone global" and few can escape it. It will evidently have different effects on those like ourselves in the West who have been living with it for centuries and those like the peoples in other civilizations who are experiencing its first highly concentrated impact without any earlier familiarity. Hence I shall make some remarks on the following topics: Firstly, the distinction between modernization and westernization. Secondly, the distinction between identity and instrumentality, which I shall use to give an account of what I take to underlie the process of modernization. Thirdly, to consider the way in which life in consumer cultures is becoming increasingly abstract. Finally, I shall make some remarks on the different ways in which modernization affects traditional societies...

3. 转到新主题时的例句

1) Let us consider the implications of two of these new inventions.

2) I now turn to another basic fact about the problem.

3) I now turn to the subject of...

4) I'd now like to discuss some issues of relevance to this subject.

5) I would also like to take this opportunity to explain some aspects of...

6) Furthermore, in recent years there have been structural changes in...

7) With reference to the..., it should be mentioned that...

8) Another concern of X is to... Additionally...

9) There are two further comments I want to make on this subject. The first is that...the second point I want to make is that...

10) As regards..., we must... This is because...

11) Before I discuss..., I want to go back to the...

12) Before I move on, I'd like to...

13) Now I would like to concentrate on the problem of... As to the point I mentioned just now, I will return to it and talk in detail in a few minutes.

14) My main concern, however, is not with the..., but with its objective character. How has it come about? How has... emerged from the...? The answer is that...

4. 演讲总结、结束演讲时的例句

1) Let me now draw together the threads of argument in this long discourse.... That's all. Thank you.

2) I would like to end on a note of caution.... That concludes my report. Thank

you very much for your attention.

3）It would be reasonable to conclude that... I think that is all what I want to say about the subject. Thank you.

4）In this concluding section, I want to discuss the impact of... My time is running out. I have to stop here. Any questions or comments?

5）As a final word, I would like to suggest for your consideration some incentives for improving... That's all that I want to talk this afternoon. Please don't hesitate to ask me if you have questions.

6）Before closing my presentation, I would like to make a few general remarks about the problem we are discussing... That's all for my presentation. I hope you'll give me your comments and suggestions. Thank you.

7）I am sorry I have talked very long. If there are any points which I didn't make clear, please feel free to point them out, and I would like to make further explanations. Thank you.

8）This concludes my presentation, and I would be pleased to answer any questions you may have.

9）So，let's put all this together as I sum up. You've got knowledge...

10）I have given you my ideas that I have long been thinking about this matter, and now I'd love to hear your comments.

19.2.5　演讲的注意事项

在演讲之前,演讲者应提前去查看演讲现场,了解演讲设备,如麦克风是固定式还是无线式等,并熟悉这些设备的操作,最好还能试一下音响效果。如果需要主持人在演讲之前向观众介绍演讲者,应提前将个人简介文稿交给主持人。

演讲者应自信、稳步地走上演讲台;记住听众是最重要的,应得体的称呼他们"Mr. / Madame Chairperson, Fellow colleagues"等。

在开始演讲之前,演讲者用眼神与听众交流,并在整个演讲过程中保持这种交流。从听众的眼神反馈中,可以获知他们对演讲内容的理解程度——若感觉听众有不耐烦之感,可适当加快演讲速度;如果感觉听众眼神迷惑,则应放慢语速并重复前面的重要内容。

演讲时演讲者应选择一个正常的、易被听众接受的语速,并做到声音洪亮、清晰、充满自信;此外,需注意语言的顿挫,通常在引入一个新的关键内容后适当停顿,或是在目视听众后停顿以留给听众反应时间,合理的停顿是取得完美演讲效果的关键之一。

在演讲中应尽量使用简单明了的句式,避免使用专业性很强的术语,除非听众很熟悉这些术语。演讲者应讲清所用统计、数据、图表等的含义,但要避免背诵长串的数字;若确需这些数据,可将其设计在幻灯片中向听众展示,并选择其中的重要数据进行介绍。此外,演讲中若必须介绍数学公式,则应放慢语速,不要以为听众能在短时间内自行轻易理解公式,应当一步一步地向听众介绍清楚。

在演讲中应重复强调其中的重要内容,但为了避免机械重复,每次尽量使用不同的词汇、短语或不同的方法来表达相同的重复内容。此外,演讲者应充分利用视觉工具来辅助演讲,在用语言告诉听众专业知识的同时,还应充分利用视觉辅助工具让听众看到相关的专业视频、照片、图、表等,以此来增强演讲效果,使听众更好地理解演讲内容。值得注意的是,虽然演讲者的肢体语言对于吸引听众、强调观点非常重要,但也不应过度使用,否则反而会分散听众的注意力。

演讲者不应超过给定的演讲时间,因为演讲超时对演讲者、听众以及主持人而言都是不礼貌的,且会使听众感觉演讲者经验欠缺。但若演讲者发现自己演讲超时了,切记不要用"I think I'll stop here"或"I'm afraid I have to leave it here"之类的话来结束演讲,因为这样会使听众觉得演讲者准备不充分,此时可用一些简短的语句来总结结束演讲。

在演讲过程中演讲者不要离开听众的视线,若演讲稿不慎滑落,演讲者应先向听众表达"Sorry"之类的话,然后再弯腰拾起演讲稿,否则听众会为演讲者突然离开演讲台而感到疑惑。此外,演讲时演讲者不应面向投影幕,而要面向听众。当需要向听众指示演示内容时,不要站在投影幕前用手去指示其中的内容,而应用激光棒指示投影幕,或是直接在电脑幻灯片中用光标指示相关内容。

演讲应设计一个富有震撼力的结尾,结尾可以总结自己演讲中的最重要内容,也可以介绍现有研究的缺陷以及将来的研究方向。注意演讲一定要有明确的总结部分,并欢迎听众提问,否则会让听众不知演讲者已结束演讲,从而不报以掌声或是掌声极少。最后,如同登台时一样,当演讲结束后,演讲者应充满自信地返回自己的座位。

19.3   用英语进行问答式专业交流的技巧

在进行问答式的专业交流时,随着交流场合的不同,如学术大会的问答环节、日常会议的专业讨论、闲暇时带有专业背景的聊天等,语言交流的特征实际上是复杂多样的。为避免空泛介绍,本节主要以学术会议的问答环节为背景,介绍在进行问答式专业交流时应注意事项和常用例句,这些技巧亦可供其他场合下进行问答式专业交流时参考。

19.3.1   专业问答的注意事项

在提问者提问时,回答者应记住提问者的姓名,并在回答时提起,以显示对提问者的热情和尊重;若回答者不知提问者的姓名,可用"The lady in the first row""The gentleman at the far table"等代替。

当提问者所提的问题与所讨论主题明显无关时,回答者可以用"I believe that is another subject; next question, please." "I'm sorry that I can't answer that question, as it is outside my field of research."或"It sounds like an interesting subject. (It's a very interesting question.)But I'm afraid it's beyond the theme of this forum. If you're interested, we may discuss it later."等回答来回避。

当回答者遇到自己无法回答的问题时,此时通常有以下几种回答方法:

1) 直接应对"I don't know",此时回答者通常仍会因为坦诚而受到提问者的尊重。

2) 在回答"I don't know"之后,加上自己对所问问题的见解等。

3) 给予提问者帮助,如回答"I don't know, but I will be happy to find out when I go back and write you the answer."、"That data is not yet available. I'll find out and get

back to you later."或是"Sorry, that's outside my area of expertise. That's Professor Smith's field. You may turn to him for a satisfactory answer after the session."

如果回答者被问到已经回答过的问题，可以回答："I believe we've already covered that. If you have special interest in this, we may talk more about it after the panel.";当然，如果以前回答问题时没有回答完整，也可借助这个重复的提问把相关问题进一步说清。

若想打断对方长篇大论似的讨论，可以礼貌地说："Sorry to interrupt you. But for the time being, we should have time for other questions. Now we want to be clear on what you're asking. So, would you please give us your question now?"，这样对方会感到你是因为时间关系不得已打断他的而不是出于其他原因。

对于学术会议问答环节等正式场合，当回答完每一个提问后，回答者应确认自己的回答是否让提问者满意，例如说："Thank you for your question, Professor Smith. Does that answer your question?"。当需要结束问答时，回答者可以看钟或是询问主持人还剩余多少时间，然后询问听众："We have time for one more question. Whose will it be?"。

19.3.2 专业问答的常用例句

1. 确认对方问题时的反问

1) I am very interested in the subject of your presentation. If I am not mistaken, I remember that you used the abbreviation BLS. To the best of my knowledge, I understand it has slightly different implication if it is compared with the very approach described in your written paper published in a recent issue of optics letters. Could you explain it a little more explicitly?

2) I don't quite understand what you really mean by saying that... Can you explain it again?

3) I would be very glad if you could give definitions to the essential concept of used in your presentation, for I am curious why you use them in this way in your report.

2. 表达自己的兴趣

1) I'm very keen on your mentioning about the... How is it being carried out in your laboratory? What do you mean by..., and how do you think it can be tackled?

2) I'm very much interested in hearing your today's presentation, since we have studied this problem for many years. We would like to ask you to say a few more words about your work. And we would also like to exchange some materials with you.

3) As you mentioned in your talk, you're making the experiments on... I'm not specialized in the subject, but I'm sure it will involve a rather complicated technological process. Would you mind telling us more about that?

3. 提出不同的观点

1) I agree with you about the approaches to solving this problem. But so far as the application aspect is concerned, I'm afraid I can't say that I go along with you on that. According to the result of the experiment we made last year, I think... I would like to hear your opinion on this matter.

2) In your presentation you mentioned XYZ which is also the subject we've been groping for. I am very glad that I got some sidelights from your views. But as to your talking about..., I'm afraid at least the following case seems to have been overlooked. First ... Second ... What do you think of it?

3) I am not an expert in this area. But I'm afraid there are other problems connected with the subject, which have not been mentioned in your presentation. Here are some relevant examples. Can I have your comments on that?

4) Before I give my comments on what the speakers have said with respect to a research agenda for the coming years, I would like to say that I disagree with Professor Smith in that... I think that...

5) Now I'd like to raise a different question...I would like to know a justification in details.

6) I may be wrong, but I think that...

7) Maybe I missed the point, but I don't know whether Mr. Boston is correct about this....

8) Although Mr. A has touched upon the subject, I feel he has not dealt thoroughly with the question.

9) I'm in favor of your saying... However, I disagree with you on the second point. How would you explain if X is taken into consideration?

4. 转移话题时的提问

I am very interested in the new theory you are talking about, but may I ask a question about the application of this theory?

5. 确认对方问题时的回答

1) I'm sorry. I didn't hear you clearly. Would you please repeat your question?

2) I beg your pardon?

3) What was that again?

4) Are you asking me a question about X?

5) I am not quite sure what your question is?

6) I think that's a very good question. But, could you please be more specific?

7) I'm sorry. I didn't quite get your question. Could you clarify your first point, please?

8) Sorry, I didn't follow you. What's your last point, please?

9) I would like to have the second question repeated, for I didn't really understand it.

6. 回答常规的问题

1) It's a good question. To answer your question, I would ...

2) My answer to this question is that...

3) Let me reply to your first question briefly...

4) Thank you for your question. If I understand your question correctly, I can say

that...

5) I'm very glad you asked this question. This is just the point that I want to explain again.

6) Well，what I mean is that...

7) You are right，Prof. Johnson. My saying about X is indeed out of the considerations of...

8) Yes，I do feel that...

9) Indeed，yes. We...

7. 仅回答问题中的一部分

1) Since time is limited. I would like to answer the last point of your question.

2) One of the questions you asked is about X, which I think is very interesting....

3) As far as I know，no one has paid special attention to this subject. At the present time，I can only give you one possible explanation that...

4) I can only provide some examples to partially answer your question.

5) As to your first question，I should say that...

8. 回答困难的问题

1) It's a difficult question. I'm afraid my answer cannot give you a satisfactory explanation.

2) It's hard to answer this question. What I'm going to say is not quite an answer to your question though it is somewhat relevant to it.... The justification for this matter needs more details.

3) I'm sorry，I have had very little experience with this problem，since our emphasis is not laid on this point.

4) I'm afraid I know little about the question you asked. To justify this we need more information for further explanation. I will have to consult about it with other members of my group. Sorry.

5) I am not familiar with the subject，but I suppose you may be right in saying so.

6) Well，of course in a way you might say so，I suppose.

7) Yes，it can also be looked at that way.

8) It is difficult to find the exact expression，but...

9) There is a 50-50 possibility either way，but...

10) There is not enough time to go into details，but...（Time does not allow us to go into details，but...）

9. 同意对方观点时的回答

1) I would agree with you and Mr. Odetts on this viewpoint.

2) I agree with your statement that X is true.

3) I think I am quite in agreement with you and there is nothing to add in this connection.

10. 不同意对方时的回答

1) I'm sorry. I can't agree with you in saying... What I said just now indicates that...

2) You're right in saying that... But this is quite another problem.

3) What I argued in my presentation is that...

4) I'm afraid our different views may result from...

5) That's not what I'm saying.

6) As I told you the other day, ...

7) If I may express it in my own way, ...

8) To avoid being misunderstood, I would like to explain myself for... It goes without saying that... What we should not forget is that...

9) Perhaps it is only to say the same thing in different words, but...

10) I have already touched briefly on the topic, but to go on a step further I would like to...

11) I'm afraid what you asked is beyond the point. It seems that it is not pertinent to our present topic.

12) That's interesting, but just how does your point fit in with the issue being considered?

13) I think that's a good point. But in practice, things just do not work that way. It is very difficult both from the theoretical and from the practical point of view.

14) I'm sorry, I cannot agree to the proposal laid on before us until we have further examined the point that...

15) I'm sorry to say that you seem to have a very much mistaken understanding of our work. To answer your question, I think something must be stated here clearly. First, ...

16) Judging from your question, I could see that your understanding about X is totally different from that of mine. My original intention is that...

17) It seems to me that what you said doesn't hold water, for you couldn't provide sufficient and convincing facts.

18) I agree with most of what you said. I have only one main reservation. You said... But I...

11. 借助例子或建议来回答对方

1) I do not intend to dwell on small details, but I wish to call your attention to the fact that...

2) I call your attention particularly to the following features. The first point which should be taken into consideration is that...

3) Here, I am just showing you a few examples of... The best one is...

4) Examining the matter in the light of past experience, we would like to suggest that...

5）I found the means of ... is extremely useful. You might also want to try it.

6）... of course a communicative-style test should be used and I'd suggest the teacher establish contacts with other teachers in an attempt to bring about a change in this area.

7）Your question on testing... is a complex one. I would suggest... It would be a rich source of material for you and your colleagues.

12. 回答完问题后确认对方是否满意

1）Are you satisfied with my answer?

2）Does that answer your question?

3）I don't know if this is a satisfactory answer?

4）Is this enough for your question?

5）I think that might be the answer to your question.

6）This is my answer to your question. What do you think of it?

7）This is my own opinion. Can I have your comments on this subject?

19.4　技能训练

1）请选择一篇关于土木工程专业的英文论文，将其改写为一篇口头演讲稿。

2）请选择一篇关于土木工程专业的英文研究报告，将其改写为一篇口头演讲稿。

3）请用英语为班级同学做一次有关土木工程的专业性演讲，要求在演讲过程中不照读演讲稿。演讲之后请回答同学们提出的相关问题。

4）请用英语为班级同学宣读一篇有关土木工程专业的演讲稿，注意体会并运用本单元所学习的读稿技巧。宣读演讲稿后请回答同学们提出的相关问题。

5）两位同学一组，围绕土木工程中的某一专业问题进行问答并展开讨论（提问者注意体会如何发问、如何提出不同意见、如何对某一问题表达兴趣等；回答者注意体会未听清问题时如何处理、遇到无法回答的问题时如何处理等）。

Unit 20　Skills of Special Translation in English

土木工程专业英语的翻译除了要具备一定量的专业词汇和语法基础外,还必须掌握必要的翻译知识和技巧。本单元主要介绍在土木工程专业英语翻译过程中常用的一些技巧。

20.1　词类转译法

在翻译过程中,原文中有些词需要转换词类,才能使译文通顺自然。例如:有时根据上下文表达意义的需要,把英语名词译成汉语动词、英语动词译成汉语名词等。词类转译的情况主要有转译成动词、转译成名词、转译成形容词和转译成副词四种。

20.1.1　转译成动词

与汉语相比,英语句子中大多只有一个谓语动词;而汉语动词则用得比较多,往往可以几个动词连用。汉语中使用动词的场合,在英语中有时常用介词、分词、不定式、动名词和抽象名词等来表达。

1. 介词转译成动词

许多含有动作意味的介词,如 across、past、toward 等,汉译时通常转译成动词;一些仅表示时间、地点、方式的介词,如 in、at、on 等,虽然没有动作意味,但汉译时根据汉语的行文习惯有时也需转译成动词。

例 20-1　Rigid pavements are made from rigid Portland cement concrete.

刚性路面是用具有刚性的水泥混凝土修筑的。(句中的介词 from 汉译为"用")

例 20-2　The Eads bridge was completed in 1874 over the Mississippi at St Louis, Missouri.

伊兹桥于 1874 年建成,它位于密苏里州圣路易斯,横跨密西西比河。(句中的介词 over 汉译为"横跨";介词 at 汉译为"位于")

2. 名词转译成动词

英语中有大量从动词派生的名词和具有动作意味的名词,这类名词在英译汉时常能转译成汉语动词。

例 20-3　A heavy steel rolling mill is under construction.

一座大型轧钢厂正在兴建中。(句中的名词 construction 汉译为"兴建")

例 20-4　Since the advent of freeway, travel has been speeded up.

自从出现了高速公路,旅行的速度就加快了。(句中的名词 advent 汉译为"出现")

3. 形容词转译成动词

英语中表示知觉、感觉、情感、欲望等心理状态的形容词在系动词后作表语时,常常可转译成汉语动词。

例 20-5　Steel is widely used in engineering, for its properties are most suitable for construction purposes.

因为钢材的性能非常适合于工程建设,所以钢材广泛应用于工程中。(句中的形容词 suitable 汉译为"适合")

例 20-6　They are confident that they will be able to build the modern bridge in a short time.

他们确信在短时间内能建成这座现代化桥梁。(句中的形容词 confident 汉译为"确信";形容词 able 汉译为"能够")

4. 副词转译成动词

英语中有些副词本身含有动作的意味,如 on、back、off、in、behind、over、out 等,这些副词在英译汉时往往需转译成动词。

例 20-7　The construction of the highway was two months behind.

这条公路施工工期延误了两个月。(句中的副词 behind 汉译为"延误")

例 20-8　The broken bridge is now back to traffic.

这座被破坏的大桥现在已恢复通车。(句中的副词 back 汉译为"恢复")

20.1.2　转译成名词

1. 动词转译成名词

英语中有很多名词派生的动词和由名词转用的动词,在英译汉时不易找到适当的汉语对应词,因而常将其转译成汉语的名词。另外,有些英语被动句子中的动词,宜译成"受到(……)+名词"或"加以(……)+名词"这类结构。

例 20-9　This tool weighs about sixteen pounds.

这种工具的重量约为 16 磅。(句中的动词 weighs 转译为名词"重量")

例 20-10　Mechanical Stabilization is chiefly characterized by its maximum usage of locally available materials in highway embankment.

力学稳定法的主要特点是它能最大限度地利用当地现成材料修筑路堤。(句中的动词 characterized 转译为名词"特点")

2. 形容词转译成名词

英语形容词转译成汉语名词大致有以下三种情况:

有些形容词加上定冠词表示某一类人或事,汉译时常译成名词,如 the rich(富人)、the poor(穷人)等。

英语的关系形容词在汉语中没有相应的形容词,其相当于汉语的名词,汉译常作名词处理,如 machinable steel(机械钢)、rural highway(农村公路)、ideal structure(理想结构)等。

科技英语中往往习惯用形容词来表示物质的特性,但汉语却习惯用名词,通常在这类形容词后加上"度""性"等词而转换成汉语名词。

例 20-11　Of those stresses the former is compressive stress and the latter is tensile stress.

在两种应力中,前者是压应力,后者是拉应力。(句中的形容词 former 转译为名词"前者";形容词 latter 转译为名词"后者")

例 20-12   The cutting tool must be strong, tough, hard and wear resistant.

刀具必须有足够的强度、韧性、硬度,而且要耐磨。(句中的形容词 strong、tough 和 hard 分别转译为名词"强度""韧性"和"硬度")

3. 副词转译成名词

英语中由名词派生的副词汉译时常可作名词处理,少数不是名词派生的副词有时也可译成名词。

例 20-13   Structural drawings must be dimensionally correct.

结构图的尺寸必须准确。(句中的副词 dimensionally 转译为名词"尺寸")

例 20-14   There seems to be no other competitive technique which can measure range as well or as rapidly as can a laser.

就测量的精度和速度而论,似乎还没有其他技术能与激光相比。(句中的副词 well 和 rapidly 分别转译为名词"精度"和"速度")

20.1.3   转译成形容词

1. 名词转译成形容词

形容词派生的名词常可转译成形容词,有些名词加不定冠词作表语时可转译成形容词。

例 20-15   The lower stretches of rivers show considerable variety.

河流下游的情况是多种多样的。(句中的名词 variety 转译为形容词"多种多样的")

例 20-16   The experiment was a success.

试验很成功。(句中的名词 success 转译为形容词"成功")

2. 副词转译成形容词

当英语动词或形容词转译成汉语名词时,原来修饰该动词或形容词的副词通常转译成汉语的形容词。

例 20-17   The engineer had prepared meticulously for his design.

工程师为这次设计作了十分周密的准备。(句中的副词 meticulously 转译为形容词"周密的")

例 20-18   It is a fact that no structural material is perfectly elastic.

事实上,没有一种结构材料是完全的弹性体。(句中的副词 perfectly 转译为形容词"完全的")

20.1.4   转译成副词

1. 形容词转译成副词

英语名词转译成汉语动词时,修饰该名词的形容词自然就转译成副词。此外,由于英汉两种语言的表达习惯不同,英语形容词有时需转译成汉语副词。

例 20-19   We have made a careful study of the properties of these structures.

我们仔细地研究了这些结构物的特性。(句中的形容词 careful 转译为副词"仔细地")

例 20-20   Although the basic materials for bridge building remain either prestressed concrete or steel, there have been some recent changes.

尽管造桥的基本材料仍然是预应力混凝土或钢材,但最近已有某些变化。(句中的形容词 recent 转译为副词"最近")

2. 名词转译成副词

例 20-21 We find difficulty in solving this problem.

我们觉得难以解决这个问题。(句中的名词 difficulty 转译为副词"难以")

例 20-22 It is our great pleasure to note that our highway net is being perfected.

我们很高兴地看到,我国道路网正日趋完善。(句中的名词 pleasure 转译为副词"高兴地")

20.2 增词法

增词法就是在翻译时根据句法、意义或修辞上的需要增加一些词,以便能更加忠实、通顺地表达原文的思想内容。当然,增词不是随意的,而是基于英汉两种语言表达方式的差异,增加原文中虽无其词但内含其意的一些词,以使译文忠实流畅,这种情况在科技英语文献翻译时比较常见。

20.2.1 根据句法上的需要增词

由于英汉两种语言表达方式存在差别,有些在英语中需要省略的成分在汉语中只有补出才能符合汉语表达的习惯。

例 20-23 Hence (...) the reason why regulations to control parking in towns are so often viewed with suspicion by Chambers of Commerce.

因此,这正是商会总是以怀疑的眼光对待城市管理停车条例的原因。(本例为省略句,句中在 Hence 后省略了 that is,翻译时应补出)

例 20-24 He first repaired and built dry stone walls, then (...) small building and, at the age of about twenty, (...) a large forge in Cardiff, twelve miles from his home.

最初他修建干砌石墙,后来便开始修建小型建筑物,20 岁左右时他在离家 12 英里外的加得夫修建了一座大型锻工厂房。(句中在"small building"和"a large forge"前均省略了 repaired and built,翻译时应补出)

20.2.2 根据意义上的需要增词

1. 英语复数名词的增译

复数名词前后增译"许多""一些""一批""各种""类"等词,可使复数意义更明确。

例 20-25 Important data have been obtained after a series of experiments.

在一系列的实验之后,得到了许多重要的数据。[句中的 Important data(重要的数据)在汉译时前面增译了"许多"]

例 20-26 As is known to all, air is a mixture of gases.

大家知道,空气是多种气体的混合物。[句中的 gases(气体)在汉译时前面增译了"多种"]

2. 英语中表示动作意义名词的增译

翻译时可根据上下文的逻辑意义,补充一些表示动作意义的名词,如"作用""现象""效应""方案""过程""情况""设计""变化"等。

例 20-27　Oxidation will make iron and steel rusty.

氧化作用会使钢铁生锈。［句中的 Oxidation（氧化）在汉译时后面增译了"作用"］

例 20-28　The type of traffic volume data collected at any given time and location depends upon the use to which the data will be put.

在一定时间和地点所收集的交通量数据类型，要根据该数据的使用情况而定。［句中的 use（使用）在汉译时后面增译了"情况"］

3. 在英语名词或动名词前后增译汉语动词

根据意义上的需要，可以在名词或动名词前后增加动词。通常增译的汉语动词有"进行""出现""生产""引起""发生""遭受""使"等。

例 20-29　Testing is a complicated problem and long experience is required for its mastery.

进行试验是一个复杂的问题，需要有长期的经验才能掌握。［句中的动名词 Testing（试验）在汉译时前面增译了"进行"］

例 20-30　Ideal conditions for passenger cars are considered to include traffic lanes 3.7 m wide, hard shoulders along both edges, at least 1.8 m wide, a minimum sight distance of 45 m, and no commercial or other extra wide vehicles.

对于轿车来说，一般认为理想的条件应包括 3.7 米宽的通道，路两侧设有至少宽 1.8 米的硬质路肩，最小视线距离为 45 米，并且不得有运货汽车或其他非常宽的车辆通行。［句中的名词 vehicles（车辆）在汉译时后面增译了"通行"］

4. 增译解说性的词

英语中常因惯用法或上下文关系，省去了不影响理解全句意义的词语。为了使译文清晰易懂，翻译时必须增译一些词语。

例 20-31　Air pressure decreases with altitude.

气压随海拔高度的增加而下降。（汉译时增译了"海拔……的增加"）

例 20-32　In contrast with a concrete train, a slip-form paver is small and cheap.

与混凝土铺路机组相比，滑模摊铺机体积小、价格便宜。（汉译时增译了"体积"和"价格"）

5. 增译概括性的词

在句子列述几种情况后，随即增加"三个方面""四个项目""如此种种"等概括词。

例 20-33　Both water waves and electromagnetic waves have the characteristics of velocity, frequency and amplitude.

水波和电磁波都有速度、频率和波幅这三项特性。（汉译时增译了"三项"）

例 20-34　A designer must have a good foundation in statics, kinematics, dynamics and strength of materials.

一个设计人员必须在静力学、运动学、动力学和材料力学这四个方面有良好的基础。（汉译时增译了"四个方面"）

20.2.3　根据修辞上的需要增词

英译汉时，有时需要在译文中增加一些起连贯作用的词，主要是连词、副词和代词，以达到使句子连贯、行文流畅的修辞目的。

例 20-35　The question is really a route selection rather than a drainage problem.

这个问题实在是属于选线方面的、而不是排水的问题。（汉译时增译了"属于……方面"）

例 20-36　The Japanese have developed a new type of machine called moles, which can bore through soft and hard rock by mechanical means.

日本人已研制出一种名叫鼹鼠掘进机的新型机械,这种掘进机使用机械方法既可挖掘软岩又可挖掘硬岩。（汉译时增译了"既……又"）

20.3　重复法

重复法实际上也是一种增词法,只不过所增添的词是上文出现过的词。重复法是指在译文中重复原文中重要的或关键的词,以期达到两个目的:一是清楚、二是强调,从而使译文生动有力,清晰流畅。

20.3.1　重复代词

1. 英语中代词代替名词的地方,翻译时往往按汉语习惯译出代词所指代的对象。

例 20-37　An alternative way to use reinforcement is to stretch it by hydraulic jacks before the concrete is poured around it.

使用钢筋的一种方法是先用液压千斤顶把钢筋拉长,然后在钢筋周围浇灌混凝土。（句中的代词 it 汉译为"钢筋"）

例 20-38　The conductor has its properties, and the insulator has its properties. Their properties are different from each other.

导体有导体的特性,绝缘体有绝缘体的特性。它们的特性是互不相同的。（句中的第一个 its,汉译为"导体的";第二个 its,汉译为"绝缘体的"）

2. 英语中强调关系代词或关系副词,翻译时往往需要使用重复的方式处理,这些词有 whoever、whichever、whenever、wherever 等。

例 20-39　You may solve the problem in whichever way you know well.

你熟悉哪种方法,就用哪种方法去解答这个问题。（句中的代词 whichever 汉译为"哪种方法"）

例 20-40　You may do the experiment whenever you have time.

什么时候你有时间,你就在什么时候做这个实验。（句中的代词 whenever 汉译为"什么时候"）

20.3.2　重复名词

1. 重复英语中的宾语

例 20-41　Operators should inspect and oil their machines before work.

操作人员在操作前应当检查机器,并给机器加油润滑。（句中的名词 machines 汉译为"机器"时重复了两次）

例 20-42　I had experienced oxygen and/or engine trouble.

我曾碰到过,不是氧气设备出故障,就是引擎出故障,或两者都出故障。(句中的名词 trouble 汉译为"故障"时重复了三次)

2. 重复英语中的先行词

例 20-43　A synthetic material equal to that alloy in strength has been created, which is very useful in civil engineering.

一种在强度上和那种合金相等的合成材料已经制造出来了,这种合成材料在木土工程中很有用。(汉译时重复了"合成材料")

例 20-44　Water can be decomposed by energy, a current of electricity.

水可由能量来分解,所谓能量也就是电流。(汉译时重复了"能量")

20.3.3　重复动词

1. 重复英语中被代替或共有的动词

例 20-45　Atmospheric pressure decreases with increase in altitude and so does the density of the atmosphere.

大气压力随高度增加而降低,大气密度也随高度增加而降低。(句中的 does 代替 decreases,汉译为"降低")

2. 重复英语中省略的动词

英语句中常在助动词后或在不定式符号后,省略动词。译成汉语时,有时需重复省略的动词,以达到通顺的目的。

例 20-46　Some of the gases in the air are fairly constant in amount, while others are not.

空气中有些气体的含量相当稳定,有些就不稳定。(汉译时重复句中省略的"稳定")

20.3.4　汉译时采用一种在内容上、而不是在形式上的重复

例 20-47　There has been no radio, TV or press mention of this information.

这则消息既没有电台广播,也没有上电视节目,在报刊上也没有报道。

20.4　省略法

省略就是将原文中的某些词语略去不译。有些词在英语中经常出现,如介词、冠词等,但在译成汉语时如果逐词死译,译文会显得不伦不类,甚至不通顺,这时,就有必要采用省略手法。但省略绝不能损害原文的思想内容,删减的应该是一些可有可无的词。或者译文中虽无其词却有其意,或者在译文中是不言而喻的,硬译出来反嫌累赘,有悖于汉语的行文习惯。

20.4.1　省略冠词

英语中冠词用得比较多,而汉语中没有这一词类。虽然在词典中可以找到 the 或 a 的汉语对应词语,但在具体的翻译实践中,有许多情况是可以而且有必要省去不译的。

1. 不定冠词 a 或 an 的省略

不定冠词 a 或 an 表示某一类事物中的"任何一个"或泛指时,译文中往往需要省略;此外,短语或固定搭配中的不定冠词需省译。

例 20-48 A beam bridge is the simplest type of span.

梁桥是最简单的桥孔结构。(句中的不定冠词 A 在汉译时省略)

例 20-49 A central barrier alters the types of accidents which occur, but does not necessarily reduce the overall number.

中央防撞栏可以改变事故发生的类型,但不一定能减少事故发生的总数。(句中的不定冠词 A 在汉译时省略)

2. 定冠词 the 的省略

定冠词 the 的省译主要用于以下几种情况:用在单数可数名词前泛指一类事物;用于某些形容词前表示某一类人或物;其后面的名词为世界上独一无二的事物以及报纸杂志和机关团体等专用名词。

例 20-50 At constant temperature, the pressure of a gas is inversely proportional to its volume.

温度不变,则气体压力与其体积成反比例。(句中的定冠词 the 在汉译时省略)

例 20-51 The moon was slowly rising above the sea.

月亮慢慢地从海上升起。(句中的定冠词 The 在汉译时省略)

值得注意的是,不是在任何情况下冠词都可省译。在一些英语词组中,冠词可使词组的意义发生很大的变化,翻译时,虽无需译出冠词本身的意义,但要重视冠词在整个词组中所起的改变词组意义的作用。

例 20-52 In the past, construction of such a tall building was out of the question.

在过去,施工如此高的一座大楼是不可能的。

例 20-53 In these days, construction of such a tall building was out of question.

现在,施工如此高的一座大楼是不成问题的。

20.4.2 省略代词

英语里代词使用得较多,而汉语中却较少使用代词。在英译汉时,许多情况下代词可省略不译。

例 20-54 If you know the frequency, you can find the wave length.

如果知道频率,就能求出波长。(泛指的人称代词 you 可省略不译)

例 20-55 The effect of commercial vehicles is of course much greater when the road has a few lanes than when it has many.

当道路的车道较少时,运货汽车的影响当然要远远大于车道多时对道路的影响。(省译人称代词 it)

例 20-56 A gas distributes itself uniformly throughout a container.

气体均匀地分布在整个容器中。(反身代词 itself 省略不译)

例 20-57 A fine bridge engineer, Shirley Smith, who was the contractor's agent for the Forth Road Bridge, has written very well about his love of bridge work in his book *Great Bridges of the World*.

曾担任福斯公路桥承包代理人的优秀桥梁工程师雪利·斯密斯在他的著作《世界大桥》一书中出色地表达了他对桥梁工程的热爱。(省译关系代词 who)

20.4.3　省略动词

英语中的系动词以及一些抽象的行为动词(have、make、take)等,翻译时往往可以省略。

例 20-58　The decking is usually between and monolithic with the tie members.

桥面通常在系杆之间,并与之成为整体。(省略动词 is)

例 20-59　All kinds of excavators perform basically similar functions but appear in a variety of forms.

各种挖土机的作用基本相同,但形式不同。(省略动词 perform 和 appear)

20.4.4　省略介词

介词在英语中使用的频率比汉语要高得多,英译汉时,许多情况下要根据汉语的习惯省略介词。

例 20-60　The new type of rope-operated scraper has worked well for several years.

这台新型的缆索式铲运机已正常工作了好几年。(省略表示时间的介词 for)

例 20-61　Most substances expand on heating and contract on cooling.

大多数物质热胀冷缩。(省略动词后面搭配的介词 on)

例 20-62　With most bulldozers maintenance is necessary.

大部分推土机需要维修。(省略介词短语中的介词 with)

20.4.5　省略连词

英语中两个句子成分或两个句子之间必须有连词或关联词,而汉语较少使用连词。因此,英译汉常可将连词省略。

例 20-63　The instrument is very small in size and very light in weight.

这种仪器尺寸极小,重量很轻。(省略并列连词 and)

例 20-64　A gas becomes hotter if it is compressed.

气体受压缩,温度就升高。(省略从属连词 if)

例 20-65　In downstream areas where the river passes through a broad gentle flood plain, civil engineers may be asked to build flood protection works.

在下游地区,河流经过的广阔平缓的洪泛平原,需要土木工程师修建防洪工程。(省略起连接作用的关系副词 where)

20.4.6　省略意义上重复的词

英语中常用 or 引出同位语,这些同位语有的可分别译出,有的具有相同的译名,只能译出一个省略一个。有时句子里个别词与其他词意义重复,翻译时也应予以省略。

例 20-66　The mechanical energy can be changed back into electrical energy by means of a generator or dynamo.

机械能可利用发电机再转变成电能。(省略同位语 dynamo)

例 20-67 Insulators in reality conduct electricity but, nevertheless, their resistance is very high.

绝缘体实际上也导电,但其电阻很高。(省略 nevertheless)

20.5 被动句型的译法

与汉语相比,英语中被动语态使用的范围要广泛得多。英语被动语态的句子,译成汉语时,很多情况下都可译成主动句,但也有一些可以保持被动语态。

20.5.1 译成汉语主动句

1. 原文中的主语在译文中仍作主语,将被动语态的谓语译成"加以……""是……的"等

例 20-68 Other problems will be discussed briefly.

其他问题将简单地加以讨论。

例 20-69 Distances between elevations are measured in a horizontal plane.

高地之间的距离是用水平测量法来测量的。

2. 原文中的主语在译文中作宾语,将英语被动句译成汉语的无主语句或加译"人们""我们""大家""有人"等词作主语。

例 20-70 The wearing surface is then laid.

之后才铺磨耗层。(无主语句)

例 20-71 Attempts are also being made to produce concrete with more strength and durability, and with a lighter weight.

目前仍在尝试生产强度更高、耐久性更好,而且重量更轻的混凝土。(无主语句)

例 20-72 The principle on which the engine works has been known for hundreds of years.

人们知道发动机的工作原理已有几百年了。(加译"人们")

例 20-73 Mr. Wang was seen making a test in the laboratory.

有人看见王先生在实验室做实验。(加译"有人")

3. 用英语句中的动作者(通常放在介词 by 后)作汉语句中的主语

例 20-74 Matter can be moved and changed by energy.

能量可以使物质运动并发生变化。

例 20-75 The top layers were bound together more firmly by mixing the crushed rock with asphalt.

用沥青掺拌碎石能使表层更坚固地黏结在一起。

4. 将英语句中的一个适当成分译成汉语句中的主语

例 20-76 Much progress has been made in civil engineering in less than one century.

不到一个世纪,土木工程学取得了许多进展。

例 20-77　Considerable use will be made of these experimental data.

这些实验数据将得到充分的利用。

20.5.2　译成汉语被动句

原句中的主语仍译成主语,而原句中的被动意义用"被……""受到……""使……"等词来表达。

例 20-78　The amount of earth to be moved has been previously calculated.

运土量被预先计算出来。

20.5.3　把原句中的被动语态谓语动词分离出来,译成一个独立结构

此类情况主要包括:"It is asserted that..."(有人主张……)、"It is believed that..."(有人认为……)、"It is suggested that..."(有人建议……)、"It is stressed that..."(有人强调说……)、"It is generally considered that..."(大家认为……)、"It is told that..."(有人曾经说……)、"It is well known that..."(众所周知……)等。

例 20-79　It is well known that silver is the best conductor.

众所周知,银是最好的导体。

20.5.4　某些固定的被动句型翻译时不加主语

此类情况主要包括:"It is estimated that..."(据估计……)、"It is reported that..."(据报道……)、"It is hoped that..."(希望……)、"It is supposed that..."(据推测……)、"It is said that..."(据说……)、"It must be admitted that..."(必须承认……)、"It must be pointed out that..."(必须指出……)、"It will be seen from this that..."(由此可见……)等。

例 20-80　It is estimated that about one third of all accidents happen when it is dark, although obviously there is more traffic during daytime.

据估计大约有 1/3 的事故发生在晚上,尽管白天交通运输显然要繁忙得多。

20.6　否定句型的译法

20.6.1　部分否定

英语中 all、both、every、each 等词与 not 搭配使用时,表示部分否定。

例 20-81　All these building materials are not good products.

这些建筑材料并不都是优质产品。(不能译成"所有这些建筑材料都不是优质产品")

例 20-82　Both of the dams are not gravity dams.

这两座水坝并不都是重力坝。(不能译成"这两个水坝都不是重力坝")

例 20-83　Every machine here is not imported from abroad.

这里的机器并非每台都是从国外进口的。(不能译成"这里的每台机器都不是从国外进口的")

表示部分否定的常用结构还有："not ... many"(不多)、"not ... much"(一些)、"not ... often"(不经常)、"not ... always"(不总是)等。

20.6.2 意义否定

英语里有些句子包含着带否定意义的词或词组,虽然在形式上是肯定句,实际上却是否定句。翻译时要恰当地表达出其否定意义来。

例 20-84 The book is too difficult for us to read.

这本书太难,我们读不懂。

例 20-85 Such glass would bend like metal when dropped rather than shatter into bits.

这种玻璃掉在地上时会像金属一样弯曲而不碎裂。

例 20-86 He would be the last man to say such things.

他绝不会说这种话。

表示意义否定的常用词组还有："but for"(如果没有)、"in the dark"(一点也不知道)、"free from"(没有,免于)、"safe from"(免于)、"short of"(缺少)、"far from"(远非,一点也不)、"in vain"(无效,徒劳)、"but that"(要不是,若非)、"make light of"(不把……当一回事)、"fail to"(没有)等。

20.6.3 双重否定

结构上是双重否定形式的情况很容易理解,不属于特殊句型。我们在这里要讨论的是形式上是否定、但意思却是双重否定(或肯定)的情况。

例 20-87 We cannot be too careful in making a microanalysis.

我们做微量分析时无论怎样小心也不过分。(不能译成"我们在做微量分析时不能太小心")

例 20-88 He finds this no other than a crane.

他发现这不是别的,正是一台起重机。

例 20-89 There is no material but will deform more or less under the action of force.

在力的作用下,没有一种材料不或多或少地发生变形。(此处 but 作关系代词相当于 that ... not)

表示双重否定的常用搭配还有："not ... until"(直到……后,才……)、"never ... without"(每逢……总是)、"not (none) ... the less"(并不……就不)、"not a little"(大大地)等。

20.7 强调句型的译法

1. 英语强调句型"it is (was) ... that (which/who) ..."几乎可用于强调任何一个陈述句的主语、宾语或状语。

例 20-90 It was on wild nights like these that travelers from Edinburgh to Dundee were thankful for the wonderful new bridge.

正是在这种狂风暴雨的夜晚,从爱丁堡到丹迪去的旅客们会非常感激这座令人叹服的新桥。

2. "It is (was) not until ... that ..."是强调时间状语常见的一种句型,可译成"直到……才……"

例 20-91　It was not until 1936 that a great new bridge was built across the Forth at Kincardine.

直到 1936 年才在肯卡丁建成一座横跨海口的新大桥。

3. 在强调句中,被强调的部分不仅可以是一个词或词组,而且还可能是一个状语从句,翻译时应注意这点。

例 20-92　It is only when piers for long span bridges is built across wide rivers that cellular cofferdams are often used.

只有当需要在宽阔的河面上构筑大跨度桥的桥墩时,才经常使用格型围堰这种方法。

例 20-93　It is not until the stiff concrete can be placed and vibrated properly to obtain the designed strength in the field that the high permissible compressive stress in concrete can be utilized.

只有做到工地上正确地灌筑与振捣干硬性混凝土,使之达到设计的强度时,才能充分利用混凝土较高的容许压缩应力。

20.8　长句的译法

按照英语的语法结构和修辞手段,利用介词短语、分词短语、不定式短语和各种并列结构,可以构成很长的简单句;利用适当的连词又可将许多句子组合成更长的复杂句,如并列复合句和主从复合句等,这样,往往是从句之中有分句,分句之中又有从句。相比之下,汉语的特点是使用较多的动词和较少的连词,靠词序和逻辑关系来组织句子,句子中的各部分(分句或独立短语)一般不太长。因此,翻译时首先要弄清楚原文的句法结构,找出整个句子的中心内容及其各个部分意思,然后分析相互间逻辑关系,再根据汉语的行文习惯,重新加以组织,力求在"明确"的基础上,做到"通顺"和"简练"。

英语长句汉译通常采用顺译、倒译、拆译重组等三种方法,而且往往需要同时并用几种方法。

20.8.1　顺译法

对科技英语而言,只要不太违反汉语的行文习惯和表达方式,一般应尽量采用顺译法。顺译有两个长处:一是可以基本保留英语语序,避免漏译,力求在内容和形式两方面贴近原文;二是可以顺应长短句相替、单复句相间的汉语句法修辞原则。

1. 在主谓连接处切断

例 20-94　The rapid growth from 1945 onwards in the prestressing of concrete shows that there was a real need for this high-quality structural material.

1945 年以来预应力混凝土的迅速发展,反映了对于这种高质量结构材料的实际需要。

例 20-95 The main problem in the design of the foundations of a multi-storey building under which the soil settles is ｜ to keep the total settlement of the building within reasonable limits, but specially to see that the relative settlement from one column to the next is not great.

在土壤沉降处设计多层建筑基础的主要问题,就是要使建筑物的总沉降保持在合理的限度内,而且特别要注意使一个柱子对另一个柱子的相对沉降量不能过大。

2. 在并列或转折连接处切断

例 20-96 Applied science, on the other hand, is directly concerned with the application of the working laws of pure science to the practical affairs of life, ｜ and to increase man's control over his environment, thus leads to the development of new techniques, processes and machines.

另一方面,应用科学则直接研究如何将理论科学中的定律用于生活实践,用于加强人类对周围世界的控制,从而导致新技术、新工艺和新机器的产生。

例 20-97 Jointly or separately Council Committees and subsidiary commissions study the problems ｜ and submit recommendations to the Council in due time.

理事会各委员会及辅助委员会可单独亦可共同研究此类问题,并适时向理事会提出建议。

3. 在从句前切断

例 20-98 In the course of designing a structure, you have to take into consideration ｜ what kind of load structure will be subjected to, ｜ where on the structure the said load will do what is expected and ｜ whether the load on the structure is applied suddenly or gradually.

你在设计一个结构物时,必须考虑到:结构物将承受什么样的荷载,荷载将在结构物的什么地方起作用,以及这荷载是突然施加的,还是逐渐施加的。

例 20-99 It will be found that similar soil conditions are encountered where similar soil profile characteristics exist; ｜ that changes in soil conditions accompany similar changes in relief and parent material; ｜ and that similar subsurface conditions have similar air photo soil patterns for regions where no appreciable changes occur in climate or vegetation.

人们会发现相同的土壤条件往往是出现在土壤剖面特征相同的地方;土壤条件的变化会伴有相同的地形起伏变化和原材料变化;对于气候或植被无明显变化的地区来说,地下条件相同,所拍摄的土样航测照片也必然相同。

20.8.2 倒译法

在英译汉时,常常需根据汉语的习惯表达方式将英语长句进行全部倒置或局部倒置。当然,翻译时只要能做到顺译,就不一定非倒置不可,在大多数情况下,倒置也只是一种变通手段,并不是唯一可行的办法。

1. 将英语原句全部倒置

例 20-100 Drivers have sometimes been lulled to sleep by the regular thudding noise that tires make when rolling over the expansion joints on concrete roads.

当轮胎驶过水泥混凝土路面伸缩缝时,发出的有规律的声响,有时使驾驶员昏昏欲睡。

例 20 - 101　About one third of all accidents happen when it is dark, although obviously there is more traffic during daytime.

虽然在白天交通运输显然繁忙得多,但大约 1/3 的事故发生在晚上。

2. 将英语原句部分倒置

所谓部分倒置实际上是将句首或首句置于全句之尾。

例 20 - 102　The importance of this point cannot be over-emphasized, in which parking policy must always be considered in the light of its effects on land use and transport policy, and in fact parking control is the key to proper traffic control in many towns.

汽车停放规则始终要根据其对土地利用和交通规则的作用来考虑。事实上,在许多城市,汽车停放管理是进行有效交通管理的关键,这一点的重要性无论如何强调也不过分。

例 20 - 103　It is most important that the specifications should describe every construction item which enters into the contract, the materials to be used and the tests they must meet, methods of constructions in particular situations, the method of measurement of each item and the basis on which payment should be calculated.

对于合同所列的各项施工项目、需要的材料及其检验要求、具体条件下的施工方法、每个施工项目的检定方法以及付款计算的依据等,说明书中都应加以详细说明,这一点是很重要的。

20.8.3　拆译重组法

为汉语行文方便,有时可将英语原文的某一短语或从句先行单独译出,并利用适当的概括性词语或通过一定的语法手段把它同主语联系在一起;或者用将后面重组法。

例 20-104　The loads a structure is subjected to are divided into dead loads, which include the weights of all the parts of the structure, and live loads, which are due to the weights of people, movable equipment, etc.

结构物的荷载可分为静载和活载两种。静载包括该结构物各部分的重量,活载是由人的重量、可移动设备的重量等等所引起的。

例 20-105　The integrated products quality control system used by thousands of enterprises in Russia is a combination of controlling bodies and objects under control interacting with the help of material, technical and information facilities when exercising QC at the level of an enterprise.

俄国成千上万家企业采用的产品质量综合管理体系,是通过在整个企业范围内实行质量管理、把企业内各个管理机构和各种管理对象联结一起的综合体,这种联结是借助于材料部门、技术部门和信息部实现的。

20.9 定语从句的处理方法

定语从句是句中的一种扩展的定语形式,是构成英语长句的一个重要因素。定语从句既具有本身的独立性,又具有对限定修饰词的引申叙述、补充说明、分层分析的作用;此外,定语从句在逻辑意义上往往与所限定的词有着表示"目的""结果""原因""让步"等含义。因此,在英译汉时,需要先弄清楚定语从句与先行词的逻辑关系。下面介绍定语从句的几种翻译方法。

20.9.1 译成前置定语

限定性定语从句往往译成前置定语结构,但有些非限定性定语从句有时也可作前置处理,尤其是当从句本身较短、或与被修饰词关系较为密切、或因拆译造成译文结构松散时。

例 20‒106 In this cause, the third world, where the largest part of world population live and where abundant natural resources still lie hidden, must play its due role.

在这一行动中,集中了世界最大多数人口并蕴藏着丰富资源的第三世界应当发挥自己的作用。

例 20-107 This is generally the only of the examination from which candidates are never excused, whatever their civil engineering degree.

不论申请入会者的土木工程学位是什么,该项口试通常是入会者唯一不能免试的考试项目。

20.9.2 译成并列句

非限定性定语从句往往需要拆译成并列句。有时,限定性定语从句因从句本身太长,前置会使句子显得臃肿,故也可采用拆译分列。

例 20‒108 Their main interest is in large dams, where they may reduce the heat given out by the cement during hardening.

它们主要用于大型水坝,在大坝中它们能减少水泥硬化时放出的热量。

例 20‒109 The main US lightweight aggregate, also made in Britain, is called expanded-shale aggregate because it is made from clay or shale which swells and thus reduces its density during firing.

美国的主要轻骨料在英国也有生产,这种骨料称为膨胀页岩骨料,由黏土或页岩制成。它们在煅烧过程中膨胀,因此密度减少。

20.9.3 译成状语从句

定语从句有时与主语之间的关系,实际上是原因、条件、目的、让步、结果、转折等隐含逻辑关系。因此,英译汉时应以逻辑关系为基础,以忠实表达原文的意思为前提,将定语从句转译成汉语中的状语从句。

例 20‒110 This is particularly important in fine-grained soils where the water can be sucked up near the surface by capillary attraction.

　　在细颗粒土壤中这一点尤其重要,因为在这种土壤中,由于毛细管吸力作用,水能被吸引到靠近道路表面的地方。(译成原因状语从句)

　　例 20-111　There is a minimum size for the reactor at which the chain reaction will just work.

　　要使链式反应有效发生,反应堆要有一个最起码的尺寸。(译成目的状语从句)

20.9.4　译成单句中的一部分

　　限定性定语从句有时在翻译时可压缩成宾语、谓语、表语或同位语。

　　例 20-112　Obtaining a photograph of uniform density from center to edge requires a copy-board which is uniformly illuminated.

　　要获得中心到边缘的密度均匀照片,就要求拷贝板的照明均匀。(译成宾语)

　　例 20-113　Fig. 1 incorporates many of the factors which must be considered in developing a satisfactory system.

　　图1所示的许多因素,在研制一个性能良好的系统时必须予以考虑。(译成谓语)

　　例 20-114　Soft-rock tunneling has as its main characteristic the tunneling process which needs no explosives.

　　软岩隧道施工的主要特点是在施工过程中不需要使用炸药。(译成表语)

　　例 20-115　Pure aluminum is used extensively as foil, that is, metal that has been rolled into sheets or strips.

　　纯铝广泛用于制造铝箔,即压制成薄片或薄带。(译成同位语)

20.10　技能训练

1. 应用词类转换法翻译技巧将下列英文句子翻译成中文

1) This steam engine is only about 15 percent efficient.(提示:adj. 转译为 n.)

2) Bituminous seals are placed on the joints between concrete slabs to prevent the ingress of water.(提示:n. 转译为 v.)

3) Cement stabilization is still considered of great value in our highway construction.(提示:prep. 转译为 v.)

4) Work on that freeway project is already fifteen days behind schedule.(提示:adv. 转译为 v.)

5) This communication system is chiefly characterized by its simplicity of operation and the ease with which it can be maintained.(提示:adv. 转译为 adj. , v. 转译为 n.)

6) In their work they pay much attention to the combination of theory with practice.(提示:n. 转译为 v.)

7) If we were ignorant of the safety on highways, it would be impossible for vehicles to travel in high-speed.(提示:adj. 转译为 v.)

8) This paper aims at discussing the properties of the newly discovered material.(提

示:v. 转译为 n.)

9) These modern equipment and installations <u>affect</u> <u>tremendously</u> the entire project. (提示:v. 转译为 n. ，adv. 转译为 adj.)

10) The <u>combination</u> of mechanical properties of this alloy can be well achieved by heat treatment.（提示:n. 转译为 adj.)

2. 应用增词法或重复法翻译技巧将下列英文句子翻译成中文

1) It is known to all that light travels much more quickly than sound does.

2) Someone wants to stop this experiment，but she prefers not to.

3) He is familiar with engineering technology and management.

4) That modification of route selection has been carefully checked.

5) Roads built here are narrower than those built in Japan.

6) The frequency，wave length，and speed of sound are closely related.

7) The company was hopeful that the start-up period would go without incident.

8) Although machines do work better and faster than a person does，they will never be able to replace him because they must be controlled by man.

9) Machining is suited to either brittle or ductile materials，preferably the former.

10) You can't build a ship, a house, or a machine tool if you don't know how to make a design or how to read it.

3. 应用省略法翻译技巧将下列英文句子翻译成中文

1) Liquids have no definite shape，yet they have a definite volume.

2) Different materials differ in their mechanical behavior.

3) The new type electrical machine has worked well for several years.

4) All substances expand when heated and contract when cooled.

5) Let us consider a plain concrete beam，that is，one having no steel embedded in it.

4. 将下列英文句子翻译成中文,注意其中特殊句型的译法

1) All the structural elements here are not beams.

2) We cannot estimate the value of computers used in structural design too much.

3) But that the careful check was made，many errors in the blue print could not be discovered.

4) The working plan will be examined by a special committee first.

5) Before any civil engineering project can be designed，a survey of the site must be made.

5. 将下列英文长句翻译成中文

1) The Congressman tends to be very interested in public works — such as new government buildings, water projects, highways and bridges, etc — that will bring money to the area or improve living conditions.

2) Most motorways have three lanes each way in addition to the central strip which of course carries no traffic and the hard shoulders along each edge which are some 2 m

wide and also normally carry no traffic.

　　3）For any unusual structure the tasks of design and analysis will have to be repeated many times until, after many calculations, a design has been found that is strong, stable and lasting.

　　4）Manufacturing processes may be classified as unit production with small quantities of parts being made and mass production with large numbers of identical parts being produced.

　　5）As the correct solution of any problem depends primarily on a true understanding of what the problem really is and where lies its difficulty, we may profitably pause upon the threshold of our subject to consider first, in a more general way, its real nature; the causes which impede sound practice; the conditions on which success or failure depends; and the direction in which error is most to be feared.

6. 将下列中文句子翻译成英文,注意其中中、英文表达方法的差别

　　1）金属是热的最佳导体,而气体则是热的最差导体。

　　2）物质可以转化为能,能也可以转化为物质。

　　3）已对几个桥址作了初步踏勘,并已完成了对每一个桥址所作的室内研究和估算。

　　4）我们已经解决了这个问题,并且得出了和他们相同的结论。

　　5）一切准备工作已经做好了。

　　6）电阻是以欧姆为单位来度量的。

　　7）已经采取措施来阻止混凝土开裂。

　　8）众所周知,混凝土是一种重要的建筑材料。

　　9）太阳到地球的距离大得难以想象。

　　10）我们必须想各种办法克服的正是由于摩擦所引起的损失。

Unit 21   Skills of Special Writing in English

对于土木工程专业的本科学生而言,在《中学英语》和《大学英语》课程中已经系统学习过英语写作方面的知识,并进行了一定数量的英语写作训练,这些积累将是写好土木工程专业英语文章的重要基础。但是,并不是具备了较好的英语写作基础就一定能够写好专业英语文章,许多通过大学英语六级考试的本科生所撰写的毕业论文的英文摘要仍会让人感觉其语言表达很不规范。造成这一问题的重要原因在于专业英语文章具有其自身的特点,若没有注意并掌握这些语言特点,很难写出像样的专业英语文章。

本章不拟重复介绍公共英语的写作知识,仅从科技英语的特点出发,介绍用英语进行专业写作时应注意的技巧以及如何进行专业论文写作。

21.1 用英语进行专业写作的技巧

21.1.1 多用名词化结构

专业英语文章通常要求行文简洁、表达客观、内容确切、信息量大,强调存在的事实、而非某一行为。因此,在专业英语写作中,应尽可能用名词化结构来精简句式。

例 21-1 Archimeds first discovered the principle that water would be displaced by solid bodies.

若将其中的同位语从句简化为名词化结构,将更符合专业英语的文体特点。原句修改为:

Archimeds first discovered the principle of displacement of water by solid bodies.

例 21-2 When we had completed the experiment, we immediately recorded the result.

若将其中的时间状语从句简化为名词性短语,将使专业英语句式更加简洁。原句修改为:

On completion of the experiment, we immediately recorded the result.

21.1.2 多用被动语态

根据英国利兹大学 John Swales 教授的统计,英语专业文章中的谓语至少有三分之一采用被动语态。造成这一现象的原因主要有两点:其一,专业文章的内容侧重叙事推理,强调客观准确,若第一、第二人称使用过多,则会造成主观臆断的不良印象;其二,英语专业文章的语句倾向于将主要信息前置,放在主语部分,这便在客观上造成了全句使用被动语态。

因此,我们在用英语进行专业文章写作时,应尽量使用第三人称叙述,采用被动语态,以使文章简洁客观;特别是在遇到不必说出动作的行为者、无从说出动作的行为者或是不便说出动作的行为者时,通常均用被动语态来表达。

1. 当写作中想要表达意思的主语为"人们""我们""大家""有人"等时,往往可考虑使用被动语态

例 21-3　写作中若要表达"我们应注意混凝土试件的养护温度和养护湿度",可用主动语态 "We should pay attention to curing temperatures and humidity of concrete specimens.";但若用被动语态来表达"Attention should be paid to curing temperatures and humidity of concrete specimens.",则使得表达效果更加简洁客观。

例 21-4　写作中若要表达"人们认为这种合金钢是这里能提供的最好的合金钢",可用被动语态表达为"This steel alloy is believed to be the best available here."。

2. 当写作中若要表达"被……""受到……""使……"等被动意义时,大都使用被动语态

例 21-5　若写作中要表达"这位老科学家广受尊敬",用被动语态表达为"The old scientist was widely respected."。

例 21-6　若写作中要表达"通过数学计算使模型方程符合实际情况",用被动语态表达为"The model equation is reconciled by mathematical calculations with the actual situation."。

3. 写作中若要表达"有人主张……""有人认为……"等意思时,多用被动语态固定句型 "It is asserted that..." "It is believed that..."等

例 21-7　若写作中要表达"有人主张,在城市 A 与城市 B 之间修建一条高速公路",用被动语态表达为 "It is asserted that a freeway should be built between city A and city B."。

例 21-8　若写作中要表达"有人认为汽车造成一系列问题,如城市膨胀、土地浪费、市中心的拥挤、油污遍地、空气污染和噪声污染等",用被动语态表达为"It is believed that the automobile is blamed for such problems as urban area expansion and wasteful land use, congestion and slum conditions in the central areas, and air and noise pollution."。

此类情况类似的还包括:

有人建议……　　　It is suggested that...

有人强调说……　　It is stressed that...

大家认为……　　　It is generally considered that...

有人曾经说……　　It is told that...

众所周知……　　　It is well known that...

4. 写作中若要表达"据估计……""据报道……"等意思时,多用被动语态固定句型"It is estimated that..." "It is reported that..."等

例 21-9　若写作中要表达"据估计大约有 1/3 的事故发生在晚上",用被动语态表达为 "It is estimated that about one third of all accidents happen when it is dark."。

例 21-10　若写作中要表达"据报道,滑模摊铺机已经研制出来,使用这些机械铺筑混凝土板,可以省去准确安装筑路模板这一费时的工序",用被动语态表达为"It is reported that slip form machines have been developed and by using these machines to form the concrete slab it is possible to eliminate the length process of accurately laying out road forms."。

此类情况类似的还包括:

希望……　　It is hoped that...

据推测……　　It is supposed that...

据说……　　It is said that...

必须承认……It must be admitted that...

必须指出……It must be pointed out that...

由此可见……It will be seen from this that...

21.1.3　多用非限定动词

科技文章要求行文简练、结构紧凑。因此,在专业英语写作时,通常用分词短语代替定语从句或状语从句;用分词独立结构代替状语从句或并列分句;用不定式短语代替各种从句;用"介词＋动名词"短语代替定语从句或状语从句等。这样既可缩短句子,又比较醒目。

例 21-11　A direct current is a current which flows always in the same direction.

用现在分词短语代替原句中的定语从句,原句修改为:A direct current is a current flowing always in the same direction.

例 21-12　Materials which would be used for structural purposes are chosen so as to behave elastically in the environmental conditions.

用不定式短语代替原句中的定语从句,原句修改为:Materials to be used for structural purposes are chosen so as to behave elastically in the environmental conditions.

21.1.4　多用后置定语

在科技文章中,后置定语被大量使用。因此,我们在撰写专业英语文章时,应注意到下列五种后置定语的结构:

1. 介词短语作后置定语

例 21-13　The forces due to friction are called frictional forces.(由于摩擦而产生的力称之为摩擦力。)

例 21-14　A call for paper is now being issued.(征集论文的通知现正陆续发出。)

2. 形容词或形容词短语作后置定语

例 21-15　In this factory the only fuel available is coal.(该厂唯一可用的燃料是煤。)

例 21-16　In radiation, thermal energy is transformed into radiant energy, similar in nature to light.(热能在辐射时,转换成性质与光相似的辐射能。)

3. 副词作后置定语

例 21-17　The air outside pressed the side in.(外面的空气将桶壁压得凹进去了。)

例 21-18　The force upward equals the force downward so that the balloon stays at the level.(向上的力与向下的力相等,所以气球就保持在这一高度。)

4. 保持较强动词意义的单个分词作后置定语

例 21-19　The results obtained must be cheeked.(获得的结果必须加以校核。)

例 21-20　The heat produced is equal to the electrical energy wasted.(产生的热量等于浪费了的电能。)

5. 使用定语从句

例 21-21　During construction, problems often arise which require design changes. (在施工过程中,常会出现需要改变设计的问题。)

例 21-22　The molecules exert forces upon each other, which depend upon the distance between them. (分子相互间都存在着力的作用,该力的大小取决于它们之间的距离。)

21.1.5　注意使用特定句型

英语专业文章中经常使用一些特定的句型,从而形成了科技文体区别于其他文体的标志,除前面已经详细介绍的被动语态句型外,这些特定句型主要还包括强调句型、比较结构的句型、分词短语结构的句型等。我们在用英语进行专业文章写作时,应当注意并应用这些特定句式。

1. 强调句型

英语强调句型"it is (was) ... that (which/who) ..."几乎可用于强调任何一个陈述句的主语、宾语或状语。

例 21-23　写作中若要表达"正因为有这些缺点需要消除,才导致了对新方法的研究探求",并强调主语"缺点",可用强调句型表达为"It is these drawbacks which need to be eliminated and which have led to the search for new processes."。

例 21-24　写作中若要表达"正是这里决定着整个作业线的相对生产速度",并强调状语"这里",可用强调句型表达为"It is here that the relative speed of the whole line is determined."。

例 21-25　写作中若要表达"建筑工地最急需的正是这种钢",并强调宾语"这种钢",可用强调句型表达为"It is this kind of steel that the construction worksite needs most urgently."。

例 21-26　写作中若要表达"正是在这种狂风暴雨的夜晚,从爱丁堡到丹迪去的旅客们会非常感激这座令人叹服的新桥",并强调状语"在这种狂风暴雨的夜晚",可用强调句型表达为"It was on wild nights like these that travelers from Edinburgh to Dundee were thankful for the wonderful new bridge."。

2. 比较结构的句型

例 21-27　Electromagnetic waves travel at the same speed as light. (电磁波传送的速度和光速相同。)

例 21-28　In water sound travels nearly five times as fast as in air. (声音在水中的传播速度几乎是在空气中传播速度的五倍。)

3. 分词短语结构的句型

例 21-29　The resistance being very high, the current in the circuit was low. (由于电阻很大,电路中通过的电流就小。)

例 21-30　Ice keeps the same temperature while melting. (冰在融化时,其温度保持不变。)

21.2 用英语进行专业论文的写作

21.2.1 论文体例(Structure of Thesis)

目前,世界上绝大多数学术组织都对科技论文的写作体例做出了规定,虽然存在一定的差别,但基本内容大同小异。

1. 期刊类论文的体例

期刊类论文通常由以下部分组成:Title(标题),Abstract(摘要),Keywords(关键词)或Subjects(主题词),Main text(正文),Acknowledgments(致谢),References(参考文献)或Appendix(附录)。

其中,论文的正文 Main text 一般包含:Introduction(引言),Analysis of the theory(理论分析)或 Test procedure(试验过程),Results(结果),Discussions(Summary,Conclusions,Suggestion and Development)[讨论(总结,结论,建议和发展)]。论文的Acknowledgments部分可能没有。

2. 长篇科技报告的体例

长篇科技报告主要包括科研成果、学位论文等,其组成如下:

1) Front(前部)

① Front cover(封面)

一般包含:Title(标题),Contract or job number(合同或任务号),Author or authors(作者),Date of issue(完成日期),Report number and serial number(报告编写和系列编号),Name of organization responsible for the report(研究单位名称),Classification notice(密级)。

② Title page(扉页)

③ Letter of transmittal(Forwarding letter)(提交报告书)

④ Distribution list(分发范围)

⑤ Preface or foreword(序或前言)

⑥ Acknowledgments(致谢)(可能没有)

⑦ Abstract(摘要)

⑧ Table of contents(目录)

⑨ List of illustration(图表目录)

2) Main Text(正文)

① Introduction(引言)

② Analysis of the theory or Test procedure(理论分析或试验过程)

③ Discussions(Summary,Conclusions,Suggestion and Development)[讨论(总结,结论,建议和发展)]

3) Back(后部)

① References(参考文献)

② Appendix(附录)

③ Tables(表)

④ Graphics(图)

⑤ List of abbreviations,signs and symbols(缩写、记号和符号表)

⑥ Index(索引)

⑦ Back cover(封底)

在实际写作中,有时也不一定完全按上述内容编写,可视实际情况适当调整。

21.2.2　论文的标题和署名(Title and Signature)

论文标题属于特殊文体,一般不采用句子,而是采用名词、名词词组或名词短语的形式,通常省略冠词。从内容上,要求论文标题能突出地、明确地反映出论文主题。具体而言,在拟定论文标题时应注意以下几点:

1) 恰如其分而又不过于笼统地表现论文的主题和内涵;

2) 单词的选择要规范化,要便于二次文献编制题录、索引、关键词等;

3) 尽量使用名词性短语,字数控制在两行之内。

例 21-31　Bayesian Technique for Evaluation of Material Strengths in Existing Structures(采用贝叶斯技术评估既有结构的材料强度)

1. 标题书写的常用格式

1) 标题中实词的首字母大写、其余字母均小写

例 21-32　Bridge Live-Load Models

例 21-33　Nonlinear Analysis of Space Trusses

2) 标题中所有字母全部大写

例 21-34　RELIABILITY ASSESSMENT OF PRESTRESS CONCRETE BEAMS

3) 标题中首单词的首字母大写,其余字母均小写

例 21-35　Sustainable development slowed down by bad construction practices and natural and technological disasters

此外,值得一提的是,若论文标题的中文含义为"……的研究",中国的作者通常使用"Study on ...",实际上,正确的表达应是"Research on ..."。

2. 署名和作者信息

一般紧跟在论文标题之后的是论文署名和作者的信息,如作者单位、通信地址(近年来还包括 E-mail 地址)、职称、头衔或会员情况等。按照英语国家的习惯,论文署名时名在前(可缩写),姓在后;但为了便于计算机检索,也有姓在前、名在后的情况(参考文献中的作者姓名排列就是这样)。有关作者的信息有时放在署名之后,有时放在论文第一页的页脚,有时放在论文的末尾,有时还分开编排,这要视论文载体的具体要求而定。

例 21-36　作者信息紧接在署名之后

Developing Expert Systems for Structural Diagnostics and Reliability Assessment at J. R. C

A. C. Lucia

Commission of the European Communities, Joint Research Center, ISPRA Establishment,21020 ISPRA(VA),Italy

例 21-37 作者信息放在论文第一页页脚

BRIDGE RELIABILITY EVALUATION USING LOAD TESTS

By Andrzej S. Nowak[1] and T. Tharmabala[2]

在论文第一页的页脚：

[1] Assoc. prof. of Civ. Engrg. , Univ. of Michigan，Ann Arbor，MI 48109

[2] Res. Ofcr. , Ministry of Transp. and Communications，Downvsview，Ontario，Canada M3M IJ8

21.2.3 论文的摘要(Abstract)

摘要是一篇科技论文的核心体现，直接影响读者对论文的第一印象。一篇学术价值较高的论文，若摘要撰写得不理想，会使论文价值大打折扣。因此，掌握英文摘要的特点至关重要。

1. 摘要的基本特点

1）能使作者理解全文的基本要素，能脱离原文而独立存在；

2）摘要是对原文的精华提炼和高度概括，信息量大；

3）具有客观性和准确性。

2. 摘要的形式和内容要求

摘要的基本形式和内容表现在以下几个方面：

1）若无特殊的规定，一般摘要位于论文标题和正文之间，但有时也要求接在正文之后。

2）对于一般篇幅的论文，摘要的篇幅控制在 80～100 个单词左右；对于长篇报告或学位论文，摘要的篇幅控制在 250 个单词左右，一般不超过 500 个单词。

3）一般篇幅的论文摘要不宜分段，长篇报告或学位论文可分段，但段落不宜太多。

4）与标题写作相反，摘要需采用完整的句子，不能使用短语。另外，要注意使用一些转折词连接前后语句，避免行文过于干涩单调。

5）避免使用大多数人暂时还不熟悉或容易引起误解的单词缩写和符号等，不可避免时，应在这些单词缩写和符号在摘要中第一次出现处加以说明。例如：TM（Technical Manual）、CCES（Chinese Civil Engineering Society）等。

6）摘要的句型少用或不用第一人称，多采用第三人称态，以体现客观性。

7）避免隐晦和模糊的语句，采用准确、简洁的语句概括全文所描述的目的、意义、观点、方法和结论等。

8）注意体现摘要的独立性和完整性，使读者在不参看原文的情况下就能基本了解论文的内容；摘要的观点和结论必须与原文一致，忌讳把原文没有的内容写入摘要。

9）通常摘要采用一个主题句开头，以阐明论文的主旨，或引出论文的研究对象，或铺垫论文的工作等，避免主题句与论文标题的完全或基本重复。

10）在摘要之后，通常要附上若干个表示全文内容的关键词、主题词或检索词，应选用规范化的、普遍认可的单词、词组或术语作为关键词，不宜随性编造。

3. 摘要的常用句型

在撰写摘要时，可套用一些固定句型。以下列几个句型供参考：

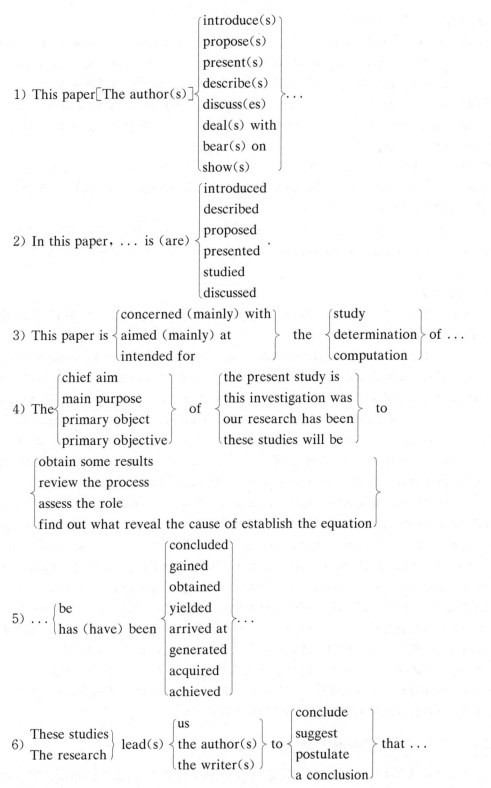

1) This paper[The author(s)]
{ introduce(s)
propose(s)
present(s)
describe(s)
discuss(es)
deal(s) with
bear(s) on
show(s) } ...

2) In this paper, ... is (are)
{ introduced
described
proposed
presented
studied
discussed } .

3) This paper is
{ concerned (mainly) with
aimed (mainly) at
intended for }
the
{ study
determination
computation }
of ...

4) The
{ chief aim
main purpose
primary object
primary objective }
of
{ the present study is
this investigation was
our research has been
these studies will be }
to
{ obtain some results
review the process
assess the role
find out what reveal the cause of establish the equation }

5) ...
{ be
has (have) been }
{ concluded
gained
obtained
yielded
arrived at
generated
acquired
achieved } ...

6)
{ These studies
The research }
lead(s)
{ us
the author(s)
the writer(s) }
to
{ conclude
suggest
postulate
a conclusion }
that ...

例 21-38　This paper describes the objects, contents, significance and impact of Information Super Highway project being constructed. (本文阐述了建设"信息高速公路"

的目标、内容、意义及其影响。)

例 21-39 The main purpose of this paper is to contribute to the development of more rational system reliability-based structural design and evaluation specifications. (本文的主旨是:采用更为合理的系统可靠度理论,促进结构设计和评估规范的发展。)

例 21-40 The method is based on a radial-space division technique in conjunction with automatic generation of unit centerline vectors. (该方法基于极坐标空间分割以及单位中线向量的自动生成技术。)

例 21-41 Results of numerical examples indicate that the proposed method has better accuracy comparing with the Monte-Carlo simulation method. (数值算例的结果表明,与蒙特卡罗模拟方法相比,建议方法的精度较高。)

例 21-42 The proposed approach may be used as a basic for the analysis of distortion-induced stresses in the concrete box girders. (建议的方法可作为分析混凝土箱梁畸变应力的基础。)

4. 摘要的实例

以下是一篇题为"Ultimate loads of continuous composite bridges(连续结合桥的极限荷载)"的论文摘要。

Abstract: The prediction of the ultimate load capacity of composite bridges of slab-on-I-steel girder construction is necessary. This is dictated by design requirements for the ultimate limit states of such bridges. In this paper, the prediction of the most probable yield-line patterns of failure for relatively wide composite bridges is presented. The prediction is based on a parametric study as well as on laboratory test results on composite bridge models. The degree of fixity between the transverse steel diaphragms and the longitudinal steel girders is considered with respect to influence on the ultimate load capacity of the bridge. Good agreement is shown between the theoretical and experimental results. A method of relating AASHTO truck loading to the collapses load is presented. The derived equations can be used either to predict the ultimate load capacity or the required ultimate moments of resistance for design of simple-span and of continuous-span composite bridges. An illustrated design example is presented.

摘要:由于设计需求,对混凝土桥面板与工型钢梁结合的桥跨结构,有必要预测其极限承载力。本文针对相对较宽的结合桥,基于参数研究以及实验室模型试验结果,预测出最可能出现的失效屈服线模式。对钢横隔板与纵向主梁的连接紧固程度以及其对桥梁极限承载力的影响也加以考虑。实验结果与理论分析结果很接近。(文中)提出了一种把 AASHTO 车辆荷载与破坏荷载相关联的方法。所推导的方程可用来预测极限承载力,或用来设计简支或连续结合桥的极限抵抗弯矩。(本文)给出了设计算例。

21.2.4 论文的正文(Main Text)

正文是论文的主体部分。由于学科、论题、方法和手段的差异,正文的组织和写作也不可能千篇一律。总的原则应该是:论文的结构层次分明、逻辑关系清晰、研究重点突出、语言文字简练。通常,正文包括以下几部分:

一、介绍与论题相关的背景情况和研究现状,并提出问题;

二、对理论分析过程、应用材料、计算方法、应用软件、实验设备、研究过程等的详细描述;

三、对计算、分析或实验研究结果进行分析讨论,提出结论、建议、发展方向等。

1. 常用语法

常用语法主要包括一般现在时、现在完成时、无人称被动语态、条件语句、祈使语句等。

1) 一般现在时和现在完成时

在科技类英语的写作中,一般现在时(包括被动语态)用得最多,它常用于描述不受时间限制的客观事实和真理,表达主语的能力、性质、状态和特征等。用得较多的还有现在完成时,但主要是被动语态。它主要用来表述过去发生的(无确切时间),或在过去发生而延续到现在的事件对目前情况的影响。常与现在完成时连用的词有:already、(not) yet、for、since、just、recently、lately 等。

例 21 - 43 By doing so, the results of the nondestructive tests increase the confidence level of the ultimate strength results.

例 21 - 44 Most of the research works to date has been related to undamaged structures, and not to the structure as already by damage to its components.

此外,在论文的引言部分论述某一研究课题的过去情况和目前进展时,时常会用到其他时态,如过去时、现在进行时等。

2) 无人称被动语态

对无需说明或难以说明动作发出者的情况,就用无人称被动语态。

例 21-45 This is shown in Fig. 1.

例 21-46 Material properties, dimensions, and accuracy of the analytical model are treated as random variables.

3) 条件语句

在理论描述中,常常用到一些条件句,说明一种假设情况。最常用的条件句为 if 语句;此外还有其他一些条件表达方式,如:unless(= if ... not)、providing (that)、provided (that)、only if、given + 名词、in case、so(as) long as、suppose (that)、assume (that)、with ... 等。

例 21-47 Hence, if the combination of stresses which cause yield is known, the direction of the strain vector can be determined uniquely from the flow rule.

例 21-48 Given wind speed and environmental conditions, it is possible to predict the actions by wind on buildings.

例 21-49 Provided that the load conditions are known, the forces on structural members can be analyzed easily.

例 21-50 With the equipment, the experiment would be readily conducted.

例 21-51 I will represent alone the length of the beam in the paper unless otherwise stated.

例 21-52 These equations will hold as long as x<0.

4) 祈使语句

在理论解释、公式说明和试验分析中,经常会用到祈使语句。它表示指示、说明、建议或表示条件、假设、设想等。

例 21-53 <u>Note that</u> concrete is a porous material and the carbon dioxide from the air can penetrate into the interior of it.

例 21-54 <u>Let</u> x equal in the Equation (1).

例 21-55 <u>Be</u> sure to fix the mould board in right position.

例 21-56 <u>Suppose that</u> the influence of temperature is negligible and the equation can be rewritten as follows.

2. 常用句型

一般,在科技论文撰写中采用符合语法的何种句型,并无统一的规定。大量采用的仍是"主+谓+宾"和"主+系+表"结构及复合句型等。不过,有些句型简单明了,适应性广,使用频率较高,特举例如下:

1) It $\left\{\begin{array}{l}系动词+形容词\\动词\end{array}\right\}$ + that 从句

例 21-57 <u>It is</u> recommended <u>that</u> waterproof membranes should cover the entire deck surface between parapets.

2) It + 系动词 + 形容词 + to

此类句型主要有:

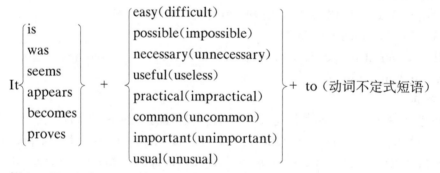

例 21-58 <u>It is</u> convenient <u>to</u> use non-dimensional parameters in the analysis of the test results.

3) 主语 + 系动词 + to(动词不定式短语)

例 21-59 The objective of this paper <u>is to</u> present a practical procedure to calculate the load responses for railway roadbed.

3. 省略形式

专业英语中省略的情况较多,较为常见的几种形式如下:

1) 用分词独立结构代替从句

当主句和从句的主语相同时,从句的主语可以省略,用分词独立结构代替从句。句型为:从属连词(Before、After、When、While、On、By、In 等)+分词,主语······

例 21-60 Before measuring the stresses, the testing equipment should be checked first. (= Before the testing equipment measures ...)

例 21-61 When used with the balanced cantilever construction, the approach works best.

（= When the approach is used with …）

2）用过去分词作后置定语代替定语从句

英语中常用 which、where、what、that 等引导后置定语从句，修饰前面的名词。当后置定语从句中的动词为被动时态时，可省略引导部分，直接用过去分词作后置定语，使句子更简练。下面几个例句括号中为省略部分。

例 21-62　The research（which is）being carried on this subject is extensive.

例 21-63　Currently, several methods（which are）used for simulating the entire process of construction of cable-stayed bridges are available.

3）并列复合句的句子成分的省略

在并列复合句子中，其第二分句（或后续分句）里常省略与第一分句相同的句子成分（主语、谓语、宾语或状语），见下两例的括号部分。

例 21-64　The bending moment is positive if the beam bends downwards, and（the bending moment is）negative if（the beam bends）upwards.

例 21-65　In Fig. 2, R is the resistance, S（is）load effect and K（is）the safety factor.

4）状语从句中句子成分的省略

在表示时间、地点、条件、让步、方式的状语从句中，若其主语与主句的主语一致且谓语含有动词 be，或其主语是 it，就可省略从句中的主语和作助动词或者连系动词的 be。

例 21-66　Steel girders expand when（they are）heated and contract when（they are）cooled.

例 21-67　The flood scored the foundation seriously as（it had been）expected.

4. 写作例句与说明

写作情况多种多样，可采用的句型也不少。下面根据具体对象的不同，介绍一些句型、短语和词汇供参考。

1）进展和评述

在科技论文中，尤其在引言部分，往往首先需要对目前的进展和他人研究进行评述。对这种情况，通常采用现在完成时态。若干句型如下：

① A substantial review of … has been given by …

② An extensive list of references can be found in the review paper by …

③ … have attracted researchers' attention since …

④ There has been theoretical interest in the field of … for the last decade.

⑤ …have been a major concern in the development of …

⑥ Recently this topic has seen tremendous growth in the theory and methods of …

⑦ Much progress has been made in …

⑧ The last decade has seen tremendous growth in the theory and methods of …

⑨ However, attention was just focused on …, not on …

⑩ Since … have been described in detail elsewhere, only a brief outline of the important aspects of … is presented here.

⑪ … is far from simple and it is therefore desirable to …

⑫ There is a growing need for ...

⑬ The problems of ... are issues which have gained increasing importance in 1990s.

⑭ Some attempts has been made to apply ... to ...

⑮ It has been shown by ... that ... have a significant effect on ...

⑯ However, it has been observed that ...

例 21-68 The simulation, including analysis and design on site, of entire process for highway tunnel construction has attracted researches' attention since 1970s.

例 21-69 Recently, there has been an interesting interest in and concern about dynamic loads, such as moving vehicles, earthquake and wind, for bridge design.

例 21-70 However, it has been observed that the previous researches could not show clearly the relationship ... because of lacking of measured data.

2) 定义和描述

在理论分析和公式推导中,常需要对一个事物或概念做出定义,并进行解释和描述。常见语句结构如下:

① Define ... to be ...

② ... is $\begin{cases} \text{defined as} \\ \text{called} \\ \text{said to be} \end{cases}$ ...

③ ... $\begin{cases} \text{is} \\ \text{means} \\ \text{signifies} \\ \text{is considered to be} \\ \text{is taken to be} \\ \text{refers to} \end{cases}$ ...

例 21-71 Here, N is defined as the number of circles required to break a specimen at a particular stress level under a fluctuating load.

例 21-72 B is called safety index, and is taken to be a measurement of safety level for all similar components of structures.

3) 假说和假设

假说(hypothesis)是在事实基础上根据类比推理、归纳推理和演绎推理提出的。假设(assumption)可用来预测事物发展趋势,简化分析和计算过程。常见语句结构如下:

① $\begin{matrix} \text{This hypothesis} \\ \text{The hypothesis of ...} \end{matrix}$ + $\begin{cases} \text{was developed} \\ \text{was put forward} \\ \text{was suggested} \end{cases}$ + $\begin{cases} \text{in 1980.} \\ \text{in the early 1970s.} \\ \text{in the late 1970s.} \end{cases}$

② $\begin{matrix} \text{This hypothesis that ...} \\ \text{This hypothesis of ...} \end{matrix}$ + $\begin{cases} \text{has been checked} \\ \text{has not yet been verified} \\ \text{remains to be supported} \end{cases}$ + by ...

③
$$\left.\begin{array}{l} \text{The theory is based on an assumption} \\ \text{These data lead us to assume} \\ \text{The author proceeds from an assumption} \end{array}\right\} \; + \quad \text{that} \ldots$$

④
$$\left.\begin{array}{l} \text{What we obtained} \\ \text{What the authors report} \end{array}\right\} \; + \left\{\begin{array}{l} \text{(does not) contradict(s)} \\ \text{(does not) agree(s) with} \\ \text{is in agreement with} \\ \text{is not consistent with} \end{array}\right\} \; + \quad \text{the assumption}$$

(that . . .)

例 21-73   These data lead us to <u>assume that</u> it is of little importance to take the vibration caused by wind into account.

例 21-74   <u>What we obtained</u> in the test is in good agreement with the theoretical assumption.

4) 分类与比较

分类与比较是根据事物的特点、属性进行归纳区别,并对两种以上同类事物的异同点或优缺点进行对比,以加深对事物本质的认识。

① 常用分类语句结构如下:

A.   There are N $\left\{\begin{array}{l} \text{types} \\ \text{kinds} \\ \text{varieties} \\ \text{classes} \\ \text{sorts} \end{array}\right\}$ of . . .

B.   . . . $\left\{\begin{array}{l} \text{may be} \\ \text{might be} \\ \text{can be} \end{array}\right\} \; + \left\{\begin{array}{l} \text{classified} \\ \text{divided} \\ \text{categorized} \\ \text{grouped} \end{array}\right\}$ into . . .

C.   There are . . . , the first . . . , the second . . . , . . .

例 21 - 75   A structural engineering project can be divided into three phases: planning, design and construction.

② 常用比较语句结构如下:

A.   as + adj./adv.的原级 + as . . . 和 not so(as) + adj./adv.的原级 + as . . .

B.   倍数 + as . . . as . . .

C.   adj./adv.的比较级 + than . . .

D.   比……好,优于……:

. . . (be) superior to. . .

(be) in advance of. . .

(have) superiority over. . .

(have) advantages over. . .

E.   比……差,不如……:

. . . (be) inferior to. . .

（be) nothing to...

F. 比得上……,与……相比:

...bear comparison to(with)...

stand comparison to...

（be) equal to...

（be) comparable to...

G. 表达"相对于;与……相比;对比起来"的常用短语:

with respect to; as against; in comparison to; as compared to; compared to(with); by the side of; in contract to; in(by) contract 等。

例 21-76 In comparison, the scheme seems superior to the previous one.

5) 方法和方式

在阐述研究过程时,总要论及所采用的方法。在专业英语中,对方法的描述往往是句子的状语成分,内容涉及描述方法的类型、途径、意义、范围、方式等。

① 用短语或结构来表达"使用,采用(某方法)"

by means of; by; with(by) the aid of; by virtue of; in terms of; by the use of; using 等。

② 用副词来表达"使用,采用(某方法)"

mathematically(用数学方法);theoretically(通过理论探讨,理论上);statistically(用统计方法);empirically(用经验方法);experimentally(用实验方法)等。

③ 表达"以……方法"

one way or another(以某种方法/式);in a similar way(以类似方法);in all manner of ways(以各种方法);by some means or other(以某种方法);in much the same way(以基本相同的方法);in a regular manner(以常用方法);in the usual manner(以常用方法)等。

④ 表达"在……意义上"

in a sense(在某种意义上);in all sense 或 in every sense(在各种意义上);in the same sense(在同样意义上);in this sense(在这个意义上);in a narrow sense(狭义上);in a broad sense(广义上)等。

⑤ 各种方式的表达

without break(intermission)(不间断地);intermittently 或 with intermittence(间断地);on and on 或 continuously(持续不断地);in fits and starts 或 on and off(断续地);in combination(conjunction) with(与……结合);in isolation(孤立地);independently(独立地);in a discrete fashion(以离散的方式);in an analogous manner(以类似的方式);after the manner(pattern, fashion) of(仿效,仿照);in chronological order(按时间顺序);in descending(ascending) order[以递降(升)顺序];clockwise 或 in a clockwise sense(顺时针地);counterclockwise(逆时针地);in groups(成群地);in a line(成直线地);in pairs(成对地);in rows(成排地);in a circle(成圆圈地);upside down(上端朝下);downside up(下端朝上);inside out(里面朝外);outside in(外面朝里);the right side up(正面朝上)等。

例 21-77 The change in strain can be recorded by means of strain gauge or an analogous apparatus.

例 21-78 Attempts have been made to maintain and rehabilitate the existing building structures <u>one way or another</u>.

6) 程度和度量

在论文写作中,需要对事物某一方面的水平进行修饰,以具体或不具体的方式进行度量。这部分内容通常作为句子的定语、表语或状语成分。可参考的词汇或词组如下:

① 显著的(用作定语和表语)

pronounced; appreciable; noticeable; conspicuous; considerable; remarkable; marked; significant; substantial 等。

② 显著地,远……得多(作状语)

significantly; substantially; considerably; a great deal; much; a lot; far 等。

③ 稍微(作定语或状语)

a little; a bit; somewhat; slightly; more or less 等。

④ 大小

predominantly(主要地); to a less(slight) degree(extent)(在较小程度上); in a greater degree(在较大程度上); in a considerable degree(在很大程度上); to a high degree(在很高程度上); to a certain extent(在某种程度上); in some degree(在某种程度上)等。

⑤ 粗细

at large(笼统地); in detail、in full length(详细地); in a more detailed fashion(更为详细地); in considerable detail(相当详细地); in more detail(比较详细地); in some detail(较为详细地)等。

⑥ 范围

radically(根本上); on the whole(总的说来,大体上); in general、far and by(大体上,一般说来); essentially、basically、primarily、largely(基本上); main、chief、primary、major[主要的(定语)]; mostly、mainly、chiefly、dominantly、predominantly、in the main、in a great measure、for the most part、for the greater part(主要地,大部分); entirely、completely、utterly、wholly、to the full extent(完全地); in part、partially、partly(部分地); extremely、in the extreme、to the last degree(极其,非常)等。

⑦ 度量

on the order of(相当于……数量,大约); to within(在……精度以内); by weight(volume, area …)[按重量(体积、面积)计量]; a slope of … in(on, to)(坡度……:……); a batter of … in(斜度……:……); angle of … with(……与……成……度角); at an angle of θ to(……与……成 θ 角); at a right angle to、squared to(与……成直角); in length(width, depth …)[按长度(宽度、深度)]; 数字 + 单位 + long(wide, deep …)[长(宽、深)……]等。

例 21-79 The research of construction control in detail depends, in a considerable degree, on the data collecting from site.

例 21-80 At present, the tensional force in cables can only be measured to within ±3%.

例 21-81 The steel tube is 30 m long, and its ends are closed with a cast steel disk which is 5 cm in thickness and 30 cm in diameter.

7）比例和比率

① ……与……成比例

... (be) in the proportion of...

② 与……成正比例

be a direct measure of; be a direct dependence upon(on); vary (directly) as; vary in the direct ratio of; be (directly) proportional to; be in proportion to; be relative to 等。

③ 与……成反比例

be in relation to; be an inverse measure of; be an inverse dependence upon(on); vary inversely as; vary in the inverse ratio of; vary in the reciprocal of; be inversely proportional to; be in inverse proportion to; depend inversely as 等。

例 21-82 The acceleration of a body is directly proportional to the force acting and is inversely proportional to the mass of the body.

例 21-83 For mild steel loaded in elastic region, the stress varies directly as strain.

8）图表和公式

在科技论文中，为了更加直观、简洁和明确地表述一定的概念、理论和应用，往往采用不少图表和公式。

① 图

与图有关的词汇有：graph、diagram、drawing、chart、sketch 等，如：curve line graph（曲线图）；projection drawing（投影图）；flow chart（流程图）；diagrammatic sketch（示意图）；key diagram（概略原理图）；perspective drawing（透视图）；histogram（直方图，频率曲线）等。

在工程图纸中常用的词汇：plan（平面图）、side view（侧视图）、top view（俯视图）、elevation（立面图）、section（截面图）、detail（大样图）、scale（比例）等。

若论文较短，可将文中所有的图形按顺序依次编号，如 Fig.1、Fig.2……；对于较长的学位论文或报告，可分章节编号，如 Fig. 1-1、Fig. 1-2……或 Fig.1.1、Fig.1.2……；图名跟在其后。

例 21-84 Fig.8.4 Typical member strength simulation results for reinforced concrete beam-columns

例 21-85 As indicated in Fig.2, stress intensity is shown on the vertical axis and strain on the horizontal.

此外，若采用的图形引自其他文献，就需要在文中或图名后注明来源。注明来源的方式通常有以下几种：

source：Ellingwood, 1977（资料来源：……）

photograph, H. L. Smith（图片取自……）

Furnished by permission of ...（蒙……允许载用）

Courtesy of the ...（蒙……特许刊用）

From：... published by ...（引用……出版的……）

copyright：...（本图版权为……所有）

在论文中,有时还需要对图形中的符号及其位置等进行解释,举例如下：

at the top(在顶上)；at the top left(在左上角)；at the bottom right(在右下角)；in the middle(在中间)；upper(middle,lower) part[(一张图的)上(中、下)部]；upper(lower) half[(一张图的)上(下)一半]；top(bottom) row[上(下)一排]；top(middle,bottom,right,left) panel(plot)[指图中某一小(分)图的位置]；blackened(full,filled,solid) circle(实心圆)；open circle(空心圆)；line of circles(圆点组成的线)；(solid,open) square[(实心、空心)方块]；cross(十字符号)；dashed line (chain dash)[小线段(虚线)]；dash-dot-dash line(—— · ——线)；chain dot line (点线)；dotted-dashed line (点划线)；(solid,broken) line[(实、虚)线]；heavy(thick) solid line(粗实线)；thin(light) broken line(细虚线)；(straight,wavy) line[(直、波状)线]；(smooth,dotted) curve[(平滑、点)曲线]；(shaded,clear) area[(阴影、空白)区]；(dotted,hatched,cross-hatched) area[(布点、网状、阴影线)区]；(dark,light) shaded area[(深、线)阴影区]等。

② 表

与表有关的词汇有：table、form、list 等。表的编号、标题的位置以及对表的来源的说明等与图的要求类似。值得注意的是,英语的表格一般只列横线,尽量少列竖线,几乎没有斜线。当一页不能容纳下一张表时,则在当页下注明"to be continued"、并在下页上注明"continued"。此外,对表中项目的注释,可放在表中、也可放在表外。

在解释表格内容时,会论及 row(行)、column(列)等,例如：two rows from top(前两行)；the middle row(中间 1 行)；the third column from right(右数第 3 列)；the second row from bottom(倒数第 2 行)等。

③ 公式

公式或方程在科技论文中比比皆是。与图、表类似,公式的编号可按顺序依次进行、或按章节分开进行,其位置一般在公式的右侧靠边。下面给出一些描述公式推导的例子和相关词汇。

例 21-86　Assuming that ... , the solution takes ...

其中,可用 by setting(putting,letting) ... 替代 Assuming that ...；与 take 相近的词有：result in、yield、give、get、have、arrive at、find、obtain、produce、follow 等。

例 21-87　By analogy to Eq. (2) the equation can be rewritten in the form of ...

其中,by analogy to(by analogy with,on the analogy of)表示"根据……类推"。in the form (of the form,in…form)表示"以……形式",类似的还有：in linear form(以线性形式)；in equation form(以方程形式)；in finite-difference form(以有限差分形式)；in symbolic form(以符号形式)；in nondimensional form(以无量纲形式)；in integral form(以积分形式)；in vectorial form(以矢量形式)等。

例 21-88　Substituting Eq. (3) .into Eq. (5) it follows that ...

例 21-89　Combining Eq. (1) and (2) allows us to write the expression for stress σ = xyz.

推导方程时常用到以下短语：

(by) substituting ... into ...；（by）inserting ...；（by）eliminating ... in；（by）combining ... and ...；（by）introducing ... into；（by）multiplying（dividing）...；（by）subtracting(adding) ...；（by）solving ...；（by）neglecting（ignoring, dropping）...等。

若需要对公式中的数学符号进行解释或定义,应在公式下一行开头处,用 where 或 in which(式中)引出。

例 21-90 ..., where(in which) x refers to ..., y is ... and z = ...

例 21-91 Eqs. (5)~(6) were performed on initial conditions that ...

其中, 与 perform 类似的常用词汇有：proceed、derive、simplification、approximation、(re)arrangement、algebra（代数运算）、positive、negative、condition、assumption 等。

9) 数量和单位

① 数量的表达

A. 描述数量、次数等的常用短语如下：

a lot of、lots of、a great many、a large quantity of、a great deal of（许多）；a negligible amount of（很少一点）；an insufficient quantity of（数量不大的）；a wide variety of（各种各样的）；a mass of、a volume of、a world of（大量的）；a series of、a train of（一系列的）等。

例 21-92 Scores of experiments have been done and a lot of observations need to be processed.

B. 描述数量的大小、近似、范围等的常用短语如下：

about、around、some、roughly、approximately、in the vicinity of、a matter of、in the neighborhood of、of(in, on) the order of（大约）；order of magnitude（量级）；range from ... to ...、in the range of ...（在……范围内变化）；up to ...［最大,(高、多……)达……］；down to ...［最小,(低、少……)达……］；as high(low, many, few) as ...［高(低、多、少……)达……］；in excess of、over、above、more(greater, higher) than（超过,……以上）；below、under、less(fewer) than（低于,……以下）；increase(differ, decrease, change) by ...［增加(相差、减少、变化)……]等。

例 21-93 The live load caused by passengers on bridges is about 3.5 kN/m^2.

② 单位的表达

A. "以……为单位"表达为"in unit of ..."、"in ..."或"by ..."

例如：an angle in radians（以弧度为单位的角）；weight in tons（重量以吨计）；vehicle's speed in m/s（车速的单位为 m/s）；by weight(volume, area)［以重量(体积、面积)计］等。

B. 当以数量词为单位时,用"in + 基数词的复数"的形式

例如：in hundreds（以百个为单位）；in thousands（以千个为单位）；in dozens（以打为单位）；in millions of US dollars（以百万美元为单位）等。

10) 常用词汇

① 表达"调查和研究"的词汇

investigate、inquire、explore、examine、look into、inspect、study、consider、search、 seek、seek out、analyze 等。

②表达"设计和准备"的词汇

design、scheme、project、plan、propose、arrange、dispose、organize 等。

③表达"实验和试验"的词汇

experiment、test、trial、try out、measure、record、equipment 等。

④表达"处理和操作"的词汇

examine、deal with、handle、treat、process、sort out、operate、conduct、activate、control、manage、function 等。

⑤表达"举例和例外"的词汇

example、instance、case、illustration、exception、exclusion 等。

⑥表达"极值和均值"的词汇

maximum、upper(上限)、minimum、lower(下限)、average 等。

⑦表达"准备和精确"的词汇

accurate(准确)、precise(精确)、correct(正确)、exact 等。

21.3　技能训练

1. 从土木工程中文专业期刊或"中国知网"上寻找若干篇专业论文,对照中文,检查这些文章的英文题名、英文摘要、英文关键词等是否存在问题,并写出调研报告。

2. 认真阅读和研究一篇被 SCI、EI 或 ISTP 检索的土木工程英文论文,该文的作者应是以英语为母语的,文章长度 5~10 页为宜(如果文章很长,则重点研究其中若干章节)。研读后请从以下几方面来撰写研读报告:(1)全文提纲;(2)每一段落的展开方法;(3)重要句型;(4)重要单词和对应中文;(5)习惯用法,包括主谓搭配、动宾搭配、形容词与名词的搭配、副词与动词的搭配、介词与名词的搭配以及短语动词等;(6)重点研究若干段落中的冠词用法和时态的用法;(7)参考引用他人作品的方式;(8)全文的写作特点;(9)从英语写作的角度出发,请回答:通过认真地阅读和研究该文章,你有何收获?

Unit 22　Learning Tools of Special English

22.1　英语词典

在土木工程专业英语的阅读、写作和翻译过程中，我们不可避免地会遇到一些生词和不确定的表达，查询英语词典是解决这一问题的重要方法之一。但目前市面上的英语词典林林总总、品种繁多，本节结合英语词典的分类，介绍当前流行的著名英语词典及其在英语学习与实践中的作用。

22.1.1　英语词典按语言类别分类

1. 英英词典

英英词典是纯英语的单语词典，无论是释义还是例句，都是用英语表达的，没有中文对应词和翻译。它的最大好处是提供了一个纯英文的环境。通过查词典，阅读英文释义和例句，培养使用者的语感和用英语思维的能力。

目前市场上最受欢迎、在世界范围内享有盛誉的大型英英词典是《朗文当代英语辞典（第4版）》(图22-1)。

图 22-1　朗文当代英语辞典　　　　图 22-2　朗文当代高级英语辞典

2. 英汉双解词典

英汉双解词典是在保留英语原版词典内容的基础上，为原版词典中的词条、例句和注释等提供对应的中文翻译的词典。简单地说，它是在英语原版词典的基础上加上中文释义的词典。使用英汉双解词典，读者既能了解某单词在中文中的对应词，又能通过英文释义更清楚、更准确地理解其含义，避免单纯看中文对应词而产生的词义扩大或缩小等带来的理解偏差。

《朗文当代高级英语辞典（英英·英汉双解）》(图22-2)就是一部非常优秀的英汉双解词典，其编纂理念是"一切为了英语学习"。

3. 英汉词典

英汉词典与英汉双解词典的不同在于，英汉双解词典既有中文释义，又有英文释义，而英汉词典则只有中文释义。但这并不是说,英汉词典不如英汉双解词典。我们知道,英汉双解词典都是基于英英词典的,而英英词典是英语国家的词典编纂者为本国的读者或者全世界学习英语的读者编纂的,除特例外一般并不专门针对中国读者;而英汉词典一般是由中国国内的专家结合具体的英语教学实践编纂而成的,专门针对中国学习者学习英语中遇到的难点和问题,切合中国实际,也更有的放矢。

4. 汉英词典

汉英词典,顾名思义,就是知道了汉语,查对应的英语词典。目前市面上较好的汉英词典有《汉英词典(修订版缩印本)》(图 22-3)和《新世纪汉英大词典》。

图 22-3 汉英词典

图 22-4 英汉汉英词典

5. 英汉汉英词典

英汉汉英词典,俗称双向词典,也就是说,词典分为两部分,一部分为英汉词典,一部分为汉英词典。一部双向词典在手,可以顶两部词典用。《外研社·柯林斯英汉汉英词典》(图 22-4)和《外研社现代英汉汉英词典》都是市场上非常畅销的双向词典。

22.1.2 英语词典按收词量或者使用者英语水平分类

1. 初阶英语词典

初阶英语词典的收词量约为 1 万~2 万,主要适合小学高年级学生和初中学生使用,如《朗文初阶英汉双解词典》。

2. 中阶英语词典

中阶英语词典的收词量约为 3 万~5 万,主要适合初中学生和高中学生使用,如《朗文中阶英汉双解词典》。

3. 高阶英语词典

高阶英语词典的收词量约为 6 万~10 万,主要适合高中学生及以上英语学习者使用,如《朗文高阶英汉双解词典》(图 22-5)。对于土木工程专业的大学生或从业人员而言,最好应备一本高阶英语词典,以满足将来学习和工作的需要。

图 22-5 朗文高阶英汉汉英双解词典

22.1.3 英语词典按功能分类

1. 学习型英语词典

英语学习型词典是专为母语不是英语的学习者编纂的英语词典,其特点是:

1) 选词适当,根据语料库数据分析词频后选用英语学习者最需要的常用词汇。

2) 释义简明,释义词汇控制在一定数量之内(如 2 000～3 500 个常用词汇),英文解释简单易懂,故学习者使用方便。

3) 例句丰富,最新的英语学习型词典都根据语料库的语料,提供生活中的真实例句,地道、自然,同时时代感鲜明,可以即学即用,与一些老词典的过时内容形成对比。

4) 不仅语法详解、还提供句型,对于疑难之处有重点说明。

5) 用法详细,提供使用说明、词汇搭配、词语辨析,学习功能齐全。

目前市面上常见的《朗文当代高级英语辞典(英英·英汉双解)》《朗文高阶英汉双解词典》《麦克米伦高阶英汉双解词典》等都是优秀的学习型词典。

2. 阅读型英语词典

阅读型词典的功能在于帮助读者了解单词的意思,其特点为:

1) 收词量一般比较大。

2) 例句很少。

3) 释义严谨、简短,但释义用的单词难度较大。

图 22-6　牛津现代
英汉双解词典

目前,《外研社最新简明英汉词典》、《牛津现代英汉双解词典》(图 22-6)、《牛津袖珍英汉双解词典》等都属于优秀的阅读型词典。

值得说明的是:学习型词典和阅读型词典不能说孰优孰劣,对于需要掌握英语的学生而言,学习型词典比较适用;对于只需要了解单词意思即可的使用者而言,阅读型词典反而更为适用。

22.1.4　英语词典按地域分类

1. 英式英语词典

英式英语词典主要有朗文、牛津、柯林斯、麦克米伦和剑桥五大国际品牌。这五大品牌的词典由于知名度不同,所以在市场上的接受度也各有不同;但实际上质量都差不多,特点也各有千秋。

2. 美式英语词典

美式英语词典著名的有 Webster(《韦氏词典》)和 American Heritage(《美国传统词典》)两大品牌。Webster 词典名满天下,自 1828 年 Noah Webster 编纂出第一部韦氏词典以来,已有 180 年的历史,尤受到报考托福、GRE、GMAT 人士的追捧。American Heritage 词典的历史要短一些,但也有几十年了,也是响当当的品牌,金山词霸中就收有这个品牌的词典。

比起英式英语词典来,市面上的美语词典不多,单语词典有《麦克米伦高阶美语词典》等几部词典,英汉双解词典则更少一些,比较优秀的有《韦氏高阶美语英汉双解词典》和《美国传统英汉双解学习词典》。

22.1.5　英语词典按专业分类

根据英语词典所收录词条专业背景的不同,英语词典可分为非专业性英语词典和专业性英语词典两大类。以上所介绍的四类英语词典均为非专业性英语词典,我们在学习和实践土木工程专业英语的过程中,由于非专业性英语词典对土木工程专业词汇的收录量有限,因此除了应用这些非专业性英语词典外,对于那些专业性较强的词汇,通常还必须借助土木工程专业英语词典。

目前,市面上的土木工程专业英语词典亦不少,比较流行的有以下几本:

1.《新英汉建筑工程词典(第二版)》(夏行时,中国建筑工业出版社,2008 年版)(图 22-7)

该词典为土木工程专业英汉词典,内容包括建筑设计、结构设计、智能建筑、建筑构造、建筑施工、建筑材料以及城镇规划、给水排水、道路桥梁、工程管理和环境保护等,共收土木工程专业词汇 5 万条。

2.《汉英建筑工程词典(第二版)》(张人崎,中国建筑工业出版社,2005 年版)(图 22-8)

该词典为汉英词典,主要涉及建筑工程领域,具体收录建筑、结构、采暖、通风、空调、上下水、电气、建材以及施工等方面的词汇 33 000 多条;此外还有插图约 800 幅,约 1 500 词条附有简注。较之该词典的第一版,第二版主要在房地产开发原理、城市设计、环境设计、中外古建筑及电脑辅助设计等方面增加了 8 000 余条新词,并对第一版中与国家标准、规范不统一的词汇进行了修正。

图 22-7 新英汉建筑工程词典

图 22-8 汉英建筑工程词典

图 22-9 简明英汉汉英道路工程词汇

3.《简明英汉汉英道路工程词汇》(刘开生,人民交通出版社,2008 年版)(图 22-9)

该书是一本简明、方便实用的英汉汉英专业技术词典,主要收录土木工程专业道路与交通工程建设方面的专业词汇,可满足专业英语英译汉、汉译英等多种需要。

22.2 网络搜索引擎

电子计算机和互联网的出现,极大地改变了我们的生活、学习与工作方式,亦为我们提供了新的英语学习与实践工具。浏览器的搜索引擎对我们撰写或翻译土木工程专业英语有一定的帮助,特别是在判断一种英语表达正确与否时具有重要的作用。但值得指出的是,利用 Google、百度等搜索引擎来进行土木工程专业英语查询,所得的结果仅供参考,由于网络信息来源的不确定性,其中可能会存在错误;比较专业的做法是查询英语语料库。

22.3 计算机英语词典软件、在线词典和在线翻译

22.3.1 计算机(手机)英语词典软件

计算机英语词典软件以电子计算机为运行载体,可以载入包括《英汉综合大辞典》《现代汉语词典》《英汉朗文综合电脑词典》等在内的大量词库,以及包括机械、冶金、土木、建筑、计算机、医疗、会计、贸易、商务、经济等在内的专业词库。此类软件通常具有取词、查词、查句等经典功能,一般还收录了单词的纯正真人发音。目前流行的该类软件主要有金山公司的金山词霸、网易公司的有道词典等。使用时通常需先在网站下载并安装词典软件,然后下载本地词典包文件,安装后便可使用。

随着智能手机的普及,金山词典和有道词典等手机词典应用软件(APP)的出现,克服了传统纸质英语词典和计算机携带不方便的缺点,却和计算机词典软件的功能同样强大。

22.3.2 在线词典和在线翻译

目前,互联网上的在线词典和在线翻译工具很多,比较常用的有"金山词霸在线词典和在线翻译""有道词典和有道翻译""Google 翻译"等。

22.3.3 CNKI 翻译助手

与上述一般英汉互译工具不同的是,CNKI(中国知网)翻译助手是以 CNKI 总库所有文献数据为依据。它汇集了从 CNKI 系列数据库中挖掘整理出的 800 余万常用词汇、专业术语、成语、俚语、固定用法、词组等中英文词条以及 1 500 余万双语例句、500 余万双语文摘,形成海量中英在线词典和双语平行语料库。CNKI 翻译助手的数据库实时更新,内容涵盖自然科学和社会科学的各个领域,包括土木工程专业英语领域。

22.4 语料库

在应用土木工程专业英语的过程中,利用网络搜索引擎来辅助翻译和写作,其最大优点是应用操作方便、搜索效率高、搜索结果简单明了,但其也有很大不足之处——由于网络资源的原作者并不一定是以英语为母语的人士,故通过搜索引擎搜索到的专业英语资料中可能存在一些错误,有时甚至会有比较严重的英语表达错误。

相对于网络搜索引擎,英语语料库的出现为我们提供了非常权威的语言工具。英语语料库是以电子计算机为载体来承载语言知识的基础资源,库中存放着大量在语言实际使用中真实出现过的语言材料。因此,可以很好地辅助我们进行土木工程专业英语的语言实践。

目前世界上流行的英语语料库主要包括 British National Corpus(英国国家语料库,简称 BNC)、Corpus of Contemporary American English(美国当代英语语料库,简称 COCA)、Time Corpus(时代杂志语料库)、Collins Wordbanks Online English Corpus Sampler(柯林斯语料库)等,本节介绍其中最为实用的"美国当代英语语料库"和"柯林斯语料库"。

22.4.1　美国当代英语语料库(COCA)

Corpus of Contemporary American English(美国当代英语语料库,简称 COCA)是由美国 Brigham Young University 的 Mark Davies 教授所开发的当今世界上最大的英语平衡语料库。COCA 于 2008 年首次在互联网上正式推出,此后每年至少更新两次。目前该语料库中收集了最近 30 年(1990 年—2019 年)美国境内多个领域的语料资料,含有高达 5 亿多个美国最新当代英语词汇,并以每年约 2 000 万词汇的趋势不断更新。因此,COCA 不仅是我们学习英语的一个宝库,亦是帮助我们进行土木工程专业英语写作和翻译的一个重要工具。

值得一提的是,与当前其他著名语料库不同的是,COCA 是免费在线供大家使用的,使用者只需在 COCA 网站上利用 Email 进行注册并激活,便可在有效期内使用 COCA 的功能。

COCA 美国当代英语语料库涵盖美国这一时期的口语、小说、流行杂志、报纸和学术期刊五大类型的语料,默认情况下语料库分类区的查询为 IGNORE 状态,即忽略语料库和时段的分类,在所有语料中查询。此处应注意的技巧是可以在点选了一个语料库后按住计算机键盘上的 Ctrl 键继续选择多个语料库或时段。对于我们进行土木工程专业英语的查询而言,最主要是应用其中的学术期刊语料库。

我们在撰写和翻译土木工程专业英语文章时,可以运用网络语料库来确定句法结构和词汇搭配、辨析不同文体中近义单词的运用情况、并丰富专业英语表达。

22.4.2　柯林斯语料库

Collins Wordbanks Online English Corpus Sampler(柯林斯在线英语语料库)中收录了高达 56 万当代英语书面词汇和口语词汇。通过设定语料类型,按照一定的查询格式输入单词或单词组合、并正确使用通配符等,我们可以搜索到相关的结果。柯林斯语料库的使用原理与 COCA 语料库具有一定的相似性,但在检索格式等细节方面存在一定的差别。

22.5　技能训练

1. 在进行土木工程专业英语的翻译或写作过程中,可利用哪些英语词典? 这些词典各有何优缺点? 请就这一问题开展小组讨论。

2. 通过土木工程专业英语的翻译和写作实践,分析并体会计算机和互联网在专业英语实践中的作用。

3. 通过使用美国当代英语语料库和柯林斯语料库,体会语料库在土木工程专业英语学习和实践中的作用,并对比两种语料库使用方法的区别。

4. 以"In What Way Can a Corpus Help Us to Solve Problems in Special English Writing?"为题撰写一篇 300~500 词的英文论文,要求不重复使用教材中的例子,介绍自己使用语料库的体验。

Addenda answer

Unit 1   Civil Engineering and Civil Engineer

Lead-in:

the Grand Canal, the Pyramids, the Greek Parthenon, the Colosseum Amphitheater

Vocabulary Warm-up:

c f h e a i b d l k g j

Exercises:

1. Fill in the Blanks.

1) design, construction, maintenance, canals, dams

2) municipal, federal, individual, international

3) physical and scientific, physics and mathematics, materials, mechanics

2. Write down the words with the following prefix or suffix.

geo-   geotechnical, geology, geological, geographical, geography

sub-   sub-discipline

non-   non-military

-port   airport, seaport, transport, import

-ology   technology, geology, hydrology

Unit 2   Building Types, Components and Design

Lead-in:

the Eiffel Tower, the Leaning Tower of Pisa, Empire State Building, the Burj Khalifa Tower

Vocabulary Warm-up:

d f h j a e g b c i

Exercises:

1. Fill in the blanks.

1) load-carrying, elevators, superstructure, substructure

2) compaction, Uneven, lean, crack, inoperative, collapse

2. Write down the words with the following prefix or suffix.

super-   superstructure, supervise

multi-   multidisciplinary, multi-story, multiple

trans-   transport, transparent, translate

sur-   surpass, surface, surround

-bility   accessibility, permeability, stability, variability

Unit 3   Building Structures and Seismic Resistance

Vocabulary Warm-up:

d f a b g c e

Exercises:

1. Write down the words with the following prefix or suffix.

post-   post-elastic, post-war

mid-   mid-face, mid-span

seis-   seismic, seismicity, seismometer, seismological, seismologist

-sphere   lithosphere, atmosphere

-tude   magnitude, amplitude, altitude

Unit 4   Road Design

Vocabulary Warm-up:

d g a j i b c f h e

Exercises:

1. Write down the words with the following prefix or suffix.

uni-   unit, united, unify, uniform

out-   outstanding, outline, outlay

ex-   excerpt, exception, exclusive, exit, extensive, expansion

vis-   visual, visualize, television, provision, supervise, revise

-ography   topography, geography, biography

Unit 5   Road Subgrade and Pavement Engineering

Vocabulary Warm-up:

d f b a h c g e

Exercises:

1. Write down the words with the following prefix or suffix.

de-   deflect, descend, deformation, detriment, deterioration, devastating

con-   consist, compose, concrete, compaction, consolidation

max-   maximum, maximize, maximal

cross-   cross-sectional

-icity   simplicity, elasticity

Unit 6   History, Types and Structures of Bridges

Lead-in:

3. b a c e d

Vocabulary Warm-up:

d c a i g e b j h f

Exercises:

1. Decide whether the following statements are true(T) or false(F) according to the text.

F T F F T

3. Write down the words with the following prefix or suffix.

eletro-   electro-chemical

semi-   semicircular, semi-permanent

para-   parallel, parabola, parapet, parameter

-duct   viaduct, aqueduct, conduct

-ment   abutment, reinforcement, element, segment, equipment, attachment, movement

Unit 7   Design, Construction, Maintenance and Rehabilitation of Bridges

Vocabulary Warm-up:

c a b

Exercises:

1. Write down the words with the following prefix or suffix.

ultra-   ultra-high, ultraviolet, ultraman

poly-   polymer, polyester, polytechnic

post-   post-tensioned, postpone

mini-   minimal, minimum, minimize, minimization

-ance   assurance, performance, maintenance, clearance, acceptance, resistance

2. Decide whether the following statements are true(T) or false(F) according to the text.

F T F T

Unit 8   Rail Transit Engineering

Vocabulary Warm-up:

d f a e h b c g

Exercises

2. Choose the best answer to each question.

B C D A B

3. Write down the words with the following prefix or suffix.

over-   overlap, overhead, overcome, overspeed

auto-   automobile, automatic, automatically, automated, automation

mono-   monorail, monotonous

-sist   assist, consistent, resistance

-ify   specify, classify, justify

Unit 9   Tunnel Engineering

Vocabulary Warm-up:

e c h a f b g d

Exercises:

1. Write down the words with the following prefix or suffix.

inter-   interlocked, interface, intermediate

under-   underground, undersea, underpass, underneath

dis-   discover, displace, discharge, disadvantage, disembark

-vert   invert, divert, convert, vertical

-struct   structure, construct, construction, destruct, obstruct, substructure

Unit 10   Geotechnical Engineering and Underground Engineering

Vocabulary Warm-up:

b d f a h c j g e i

Exercises：

1. Decide whether the following statements are true（T）or false（F）according to the text.

F T F T T

2. Write down the words with the following prefix or suffix.

syn-/sym- synonymous, synthetic, symmetry, asymmetric

-medi- medium，intermediate，immediately

-ward downward，upward

-ness stiffness，usefulness，compactness

-th strength，length，depth，growth

Unit 11 Airport Engineering

Vocabulary Warm-up：

b d f a h c e g

Exercises：

1. Decide whether the following statements are true（T）or false（F）according to the text.

F T F F F

2. Write down the words with the following prefix or suffix.

air- airport, aircraft, airbase, airside, airspace, airfield

aero- aerotren, aeronautical, aerodrome

-cess access，excess，success，process

-less- less-developed，seamless，jointless

-ity security, density, uniformity, capacity, majority, ambiguity, complexity, authority

Unit 12 Civil Engineering Materials

Vocabulary Warm-up：

c f a b d e

Exercises

1. Decide whether the following statements are true（T）or false（F）according to the text.

F F

2. Write down the words with the following prefix or suffix.

man-/ manufacture manufacture, manual, manipulate

hydr- hydration, dehydration, hydroxide

visco- viscous, viscoelastic

-work brickwork, framework, formwork, earthwork

-crete concrete, shotcrete, rollcrete

Unit 13 Civil Engineering Equipments

Vocabulary Warm-up：

c f j a d i b h g e

Exercises：

1. Decide whether the following statements are true（T）or false（F）according to the text.

T T

2. Write down the words with the following prefix or suffix.

therm(o)-thermal, thermodynamic, thermostat, geothermal

sim-　　　simultaneously, similar, simulate

liqu-　　　liquid, liquefy

-press-　　pressure, compress, impress, depress

-um　　　aquarium, medium, optimum

Unit 14　Civil Engineering Construction

Vocabulary Warm-up:

e a i d h b j f c g

Exercises:

1. Decide whether the following statements are true (T) or false (F) according to the text.

T F

2. Write down the words with the following prefix or suffix.

co-　　　cooperation, coefficient, coordinate

pro-　　　project, procedure, process, progress, proportion, provision

-tract　　contract, subcontract, contractor, subtract, detract

-age　　　sewage, drainage, shrinkage, leakage, storage, percentage

-ation　　innovation, litigation, implementation, variation, excavation, installation, elevation

Unit 15　Civil Engineering Management Ⅰ: Cost Estimation

Vocabulary Warm-up:

e a g h b l c k j f d i

Exercises:

1. Write down the words with the following prefix or suffix.

pre-　　　prediction, prepare, preliminary, precast, prefabricated

re-　　　revise, remind, reinforce, rehabilitation, revolution, resurgence, reuse, rejuvenate

nov-　　　innovation, renovation

-gress　　progress, regression

-meter　　parameter, diameter, thermometer

Unit 16　Civil Engineering Management Ⅱ: Construction Planning

Vocabulary Warm-up:

h c a j g d k l f b i e

Exercises:

1. Write down the words with the following prefix or suffix.

ill-　　　ill-structured, ill-defined

-fore-　　before, fore-runner, unforeseen

-form-　　formal, formulate, conformance, performance, uniform

-cur-　　　occur, incur, current

-pose　　propose, impose, expose, suppose

Unit 17   Civil Engineering Management Ⅲ：Quality Control

Vocabulary Warm-up：

b d e a c

Exercises：

1. Write down the words with the following prefix or suffix.

micro-/macro-　　microeconomics, macro-texturing

-pact　　compact，impact

-ject-　　project，reject，subject，objective

-gram　　program，histogram，diagram

-sis　　analysis，basis，emphasis

Unit 18   Environment Engineering and Aesthetics in Civil Engineering

Vocabulary Warm-up：

g e d h a c b f

Exercises

1. Fill in the blanks with the information given in the text.

1) appropriate　　2) impressive　　3) toxieity abatement　　4) irritation　　5) vibration aviation

6) permanent　　7) rhythm　　8) grouchy

2. Write down the words with the following prefix or suffix.

anti-　　　　antifreeze, anti-drumming, antidazzle

bi- / di-　　　bilateral，dioxide，divert

bio-　　　　biological，bionical

-oxid(e)-　　dioxide, monoxide, oxidant

-proof　　　fireproof，rustproof，waterproof

Reference documentation

[1] 段兵廷.土木工程专业英语[M].武汉：武汉工业大学出版社，2000.

[2] 李嘉.专业英语（公路、桥梁工程专业用）[M].北京：人民交通出版社，2000.

[3] 赵娅丽.交通运输工程专业英语[M].上海：同济大学出版社，2007.

[4] 从丛,李咏燕.学术交流英语教程[M].南京：南京大学出版社，2003.

[5] 钱永梅,庞平.土木工程专业英语（建筑工程方向）[M].北京：化学工业出版社，2004.

[6] 郑家顺.大学英语语法全解[M].2版.南京：东南大学出版社，2017.

[7] James K. Wight，James G. MacGregor. Reinforced Concrete：Mechanics and Design. 5th Edition. London：Prentice Hall，2008.

[8] Jane F. Garvey. Flexibility in Highway Design. U. S. Department of Transportation，Federal Highway Administration，2001.

[9] David E. Newcomb，Richard Willis，David H. Timm. Perpetual Asphalt Pavements：A Synthesis. AsphaltRoads. org，2010.

[10] Eugene J. Hall. The Language of Civil Engineering in English（English for Careers）. London：Prentice Hall，1977.

[11] 过小宁,李成明.江苏省研究生英语教学的调查和思考[J].学位与研究生教育，2004(4)：10-13.

[12] 蒋琍,董祥.试析职校英语类课程教学中学生综合能力的培养[J].职业教育研究，2007(12)：92-93.

[13] 陆红,施鸣鸣.非英语专业硕士生写作元认知水平调查与分析[J].中国外语，2007(4)：38-41.

[14] 施鸣鸣.关于非英语专业硕士生博士生翻译课的性质与任务的几点思考[J].学位与研究生教育，1998(4)：41-44.

[15] 施鸣鸣.英译汉中的理解[J].江苏外语教学研究，1999(1)：74-78.

[16] Augustine Fredrich. Sons of Martha：Civil Engineering Readings in Modern Literature. Civil Engineering－ASCE，1990,60(5)：71-73.